The Moral Brain

Jan Verplaetse · Jelle De Schrijver · Sven Vanneste ·
Johan Braeckman
Editors

The Moral Brain

Essays on the Evolutionary and Neuroscientific Aspects of Morality

Springer

Editors
Jan Verplaetse
Ghent University
Universiteitstraat 4
9000 Gent
Belgium
jan.verplaetse@ugent.be

Jelle De Schrijver
Ghent University
Blandijnberg 2
9000 Gent
Belgium
jelle.deschrijver@ugent.be

Sven Vanneste
University Hospital Antwerp
Dept. Neurosurgery
Wilrijkstraat 10
2650 Edegem
Belgium
sven.vanneste@ua.ac.be

Johan Braeckman
Ghent University
Blandijnberg 2
9000 Gent
Belgium
johan.braeckman@ugent.be

ISBN 978-1-4020-6286-5 e-ISBN 978-1-4020-6287-2
DOI 10.1007/978-1-4020-6287-2
Springer Dordrecht Heidelberg London New York

Library of Congress Control Number: 2009926809

Printed on acid-free paper

Springer is part of Springer Science+Business Media (www.springer.com)

Contents

Introduction 1
Jan Verplaetse, Johan Braeckman and Jelle De Schrijver

The Immoral Brain 45
Andrea L. Glenn and Adrian Raine

"Extended Attachment" and the Human Brain: Internalized Cultural Values and Evolutionary Implications 69
Jorge Moll and Ricardo de Oliveira-Souza

Neuro-Cognitive Systems Involved in Moral Reasoning 87
James Blair

Empathy and Morality: Integrating Social and Neuroscience Approaches . 109
Jean Decety and C. Daniel Batson

Moral Judgment and the Brain: A Functional Approach to the Question of Emotion and Cognition in Moral Judgment Integrating Psychology, Neuroscience and Evolutionary Biology 129
Kristin Prehn and Hauke R. Heekeren

Moral Dysfunction: Theoretical Model and Potential Neurosurgical Treatments 155
Dirk De Ridder, Berthold Langguth, Mark Plazier, and Tomas Menovsky

Does It Pay to be Good? Competing Evolutionary Explanations of Pro-Social Behaviour 185
Matthijs van Veelen

How Can Evolution and Neuroscience Help Us Understand Moral Capacities? 201
Randolph M. Nesse

Runaway Social Selection for Displays of Partner Value and Altruism 211
Randolph M. Nesse

The Evolved Brain: Understanding Religious Ethics and Religious Violence 233
John Teehan

An Evolutionary and Cognitive Neuroscience Perspective on Moral Modularity 255
Jelle De Schrijver

Index 271

Contributors

C. Daniel Batson Department of Psychology, The University of Kansas, Lawrence, KS, USA, dbatson@ku.edu

James Blair Mood and Anxiety Program, National Institute of Mental Health, National Institutes of Health, Department of Health and Human Services, Bethesda, MD, USA, jamesblair@mail.nih.gov

Johan Braeckman Department of Philosophy and Moral Sciences, Ghent University, Belgium, johan.braeckman@ugent.be

Jean Decety Departments of Psychology and Psychiatry, and Center for Cognitive and Social Neuroscience, The University of Chicago, Chicago, IL, USA, decety@uchicago.edu

Ricardo de Oliveira-Souza Cognitive and Behavioral Neuroscience Unit, labs-D'Or Hospital Network and Gaffrée e Guinle University Hospital Rio de Janeiro, Brazil, rdeoliveira@gmail.com

Dirk De Ridder BRAI^{2}N & Department of Neurosurgery, University Hospital Antwerp, Belgium, dirk.de.ridder@uza.be

Jelle De Schrijver Department of Philosophy and Moral Sciences, Ghent University, Belgium, jelle.deschrijver@ugent.be

Andrea L. Glenn Department of Psychology, University of Southern California, Los Angeles, CA, USA, aglenn@sas.upenn.edu

Hauke R. Heekeren Neuroscience Research Center, Berlin NeuroImaging Center and Department of Neurology, Charité University Medicine Berlin, Germany; Max-Planck-Institute for Human Development, Berlin, Germany; Max-Planck-Institute for Human Cognitive and Brain Sciences, Leipzig, Germany, heekeren@mpib-berlin.mpg.de

Berthold Langguth Department of Psychiatry, University of Regensburg, Germany, berthold.langguth@medbo.de

Tomas Menovsky BRAI^{2}N & Department of Neurosurgery, University Hospital Antwerp, Belgium, tomas.menovsky@uza.be

Jorge Moll Cognitive and Behavioral Neuroscience Unit, Neuroscience Center LABS-D'Or Hospital Network Rio de Janeiro, Brazil, mollj@neuroscience-rio.org

Randolph M. Nesse The University of Michigan, Ann Arbor, MI, USA, nesse@umich.edu

Mark Plazier BRAI^{2}N & Department of Neurosurgery, University Hospital Antwerp, Belgium, mark.plazier@uza.be

Kristin Prehn Neuroscience Research Center, Berlin NeuroImaging Center and Department of Neurology, Charité University Medicine, Berlin, Germany; Department of Psychology, Humboldt University, Berlin, Germany; Max-Planck-Institute for Human Development, Berlin, Germany, kristin.prehn@fu-berlin.de

Adrian Raine Department of Psychology, University of Southern California, Los Angeles, CA, USA, araine@sas.upenn.edu

John Teehan Department of Religion, Hofstra University, Hempstead, New York, USA, john.p.teehan@hofstra.edu

Matthijs van Veelen Department of Economics (AE), Amsterdam University, Amsterdam, Netherlands, c.m.vanveelen@uva.nl

Jan Verplaetse Department of Law, Ghent University, Belgium, jan.verplaetse@ugent.be

Introduction

Jan Verplaetse, Johan Braeckman and Jelle De Schrijver

Science at the Edge of Science Fiction

The moral brain teases the imagination and triggers the fantasy of many people, both laymen and scientists. Engineering human morality is a recurring and popular theme in science fiction, from Robert Louis Stevenson's *Dr. Jekyll and Mr. Hyde* and the French fin-de-siècle author Albert Robida's "moral bacilli" to Anthony Burgess' *A Clockwork Orange*. Even the renowned Spanish neurologist Santiago Ramon y Cajal dedicated one of his fiction tales to this fantasy. All these stories are based on similar scripts: callous doctors convert harsh criminals into docile individuals who, once operated upon, become dull and lose their critical capacities, or, conversely, neurosurgeons remodel exemplary citizens in remorseless warriors who are unstoppable from committing cruelties afterwards.

At short distance from these dreamy novelists, down-to-earth scientists explored the human brain in search of a moral center. Historically, this quest to localize morality went through favorable and unfavorable times. Although limited successes were attained in the past, most statements turned out to be scientifically untenable. While some scientists proposed localizations of morality in the human brain, more skeptical brain researchers urged for reservation and patience. Scientists opposing this project even typified these bold and unsupported hypotheses as omnipotent fantasies aired by overambitious colleagues. Nowadays, the climate is once again encouraging and attractive. Brain scientists show optimism, and the belief prevails that a crucial breakthrough is near. Soon, the basic architecture of the moral brain will be disentangled. Unequalled in the history of behavioral science, expectations run so high. Disappointment might be unequalled too.

This volume presents an overview of the current research in the field. It will distinguish scientific fact from science fiction by bringing together contributions of leading experts continuing this long-lasting scientific project aimed at explaining how our brain processes moral emotions, judgments and behavior.

J. Verplaetse (✉)
Department of Law, Ghent University, Belgium
e-mail: jan.verplaetse@ugent.be

J. Verplaetse et al. (eds.), *The Moral Brain*, DOI 10.1007/978-1-4020-6287-2_1,

This project started at the beginning of the 19th century. Around that time, medical science altered our view on mankind entirely. Human capacities, even our most advanced as morality, ceased to be seen as phenomena of an immaterial soul as religious and philosophical doctrines dictated since ancient times. Morality was transferred from the soul to the body, in particular the brain, a switch that transformed moral processes into bodily or organic phenomena. Terminology mirrored this psychological revolution. Scientists and philosophers introduced modern concepts that stressed the empirical and naturalistic quality of morality, such as 'moral sense' or 'moral instinct', thus restyling the old-fashioned concept of 'conscience'.

Besides medical science, Charles Darwin's theory of evolution gave a powerful impetus to this change. Evolutionists no longer considered morality a human privilege. Processes resembling moral behavior in humans governed the social life of non-human species as well. No clear-cut demarcation separated the social life of animals from moral phenomena in humans. Anthropologists who studied the development of morality among human ancestors, sought to understand the natural mechanisms that hold groups of non-humans together. Evolutionary theory's impact on the moral brain project was substantial indeed. In spite of the theoretical problems this theory encountered – and to a certain degree still encounters – to appreciate altruistic behavior from the perspective of evolution by natural selection, evolutionary theory offered the most convincing arguments against a dualist approach. If man evolved from a soulless species by way of natural selection, why should morality be exempt from this organic evolution? Theory of evolution shifted the burden of proof from the materialist side to the dualist one. Since millions of years of survival in ancestral environments have molded our mind, psychology could not disregard evolution's impact. This equally applies to one of our most impressive capacities: human morality.

Evolutionary theory's contribution to the moral brain project is not restricted to philosophical debates of the past. Nowadays, evolutionary psychologists and behavioral economists carry on the ambition to understand human morality from a Darwinian angle. Today, evolutionary-inspired researchers continue the naturalistic shift started in the 19th century, resulting in a wealth of encouraging publications and new books on this issue. Remarkably, until now, a profound confrontation between evolutionary and neurological approaches toward human morality has been lacking. Despite significant progress in both disciplines, researchers hesitate to find fruitful inspiration in each other's work. A more inclusive crossover between evolutionary and neurological insights is advisable from a scientific point of view. This book has the ambition to fill this omission in actual research. In this introduction, we will explore the value of a more integrated approach between the theory of evolution and neuropsychology. Prior to this exploration, we present an overview of the main findings of both disciplines in the study of human morality. This overview will enable readers who are not acquainted with these fields of research to appreciate why an interdisciplinary approach is mandatory.

The Moral Brain: A New Climate – A New Technique

Localizing the moral brain is not without precedents (Verplaetse, 2009). Since the start of the 19th century, medical experts speculated about the possible existence of a moral or ethical brain center. Encouraged by remarkable clinical cases like the unfortunate Phineas Gage, medical discoveries and innovative technologies, prominent neuroscientists like Paul Flechsig, Constantin von Monakow or Oskar Vogt were not reluctant to propose localizations of human morality in the brain. In their writings, we encounter various proposals, even the most bizarre.

The Viennese neurologist Moritz Benedikt (1835–1920), for instance, situated morality in the occipital lobes (Verplaetse, 2004). Two arguments grounded this extravagant hypothesis. Inspired by a modern philosophical tradition that restyled conscience as a moral sense, Benedikt conceptualized morality as a sense organ. Just as our olfactory organ immediately distinguishes pleasant odors from repulsive ones, our moral sense universally disapproves of murder or rape and approves of empathic concern toward people in need. Already acquainted with the fact that our occipital lobes accommodated our visual perception, Benedikt related our moral sense to the same brain region. According to him, the occipital lobes constituted the seat of all sense organs. His second argument sounded more evidence-based. While measuring brains of executed criminals, Benedikt observed that their occipital lobes did not cover the cerebellum. This so-called non-coverage of the cerebellum was borrowed from zoology and played a prominent role in 19th century discussions of the hierarchical classification of apes and humans. This feature was considered a sign of inferiority. By comparing the brains of guillotined criminals with those of gorillas and Laplanders who showed a similar non-coverage, Benedikt inferred that voluminous occipital lobes facilitated normal people to keep their animalistic urges in check. Individuals who lacked this neuroanatomical feature possessed no conscience and were incapable of inhibiting criminal passions. Giants in neurology like Theodor Meynert and Theodor von Bischoff ridiculed Benedikt's original yet pseudoscientific hypothesis, which still combined oversimplification with the radicalism of a phrenological doctrine. In the first decades of the 19th century, phrenologists like Joseph Gall and Johann Caspar Spurzheim proclaimed that mankind was bestowed with various mental faculties which could be localized in a square-like brain area. Despite some controversy about the precise location of morality, most phrenologists were committed to a localizable moral sense (Renneville, 1999; Cooter, 1984).

The localization defended by Vogt (1870–1959), whose scientific prestige equaled that of Meynert, exhaled a modern spirit not found in Benedikt's speculations (Verplaetse, 2009; Hagner, 2004). Vogt played an eminent role in the development and progress of neurohistology, a novel discipline studying the microscopic architecture of the human cortex. By comparing cortical layers from various species and distinct areas spread over the entire brain, neurohistologists designed anatomical maps of the brain. Histologists designated areas with identical cellular structures with an equal number (Brodmann's method) or an equal name (Vogt's method). The ambition to reduce mental functions to anatomical structures constituted the most prominent drive behind this laborious project. Vogt's work certainly encapsulated this ambition, yet by striving to relate individual differences in personality to cortical patterns, he went even further. Firmly convinced that anatomical peculiarities mirrored talents or mental deficiencies, a particular brain area struck his attention: the *lamina pyramidalis* of the frontal lobe. When Soviet authorities assigned him to scrutinize Lenin's cortex for abnormal patterns that might explain 'the political and moral genius' of the deceased Soviet leader, he noticed an enlarged *lamina pyramidalis* abundantly punctuated with so-called association cells (Richter, 2007). Around the same time, he made opposite observations among the brains of deceased criminals who showed a narrowed lamina pyramidalis with few association cells. Unlike Lenin or normal people, these ill-fated people lacked self-discipline and moral control. Deficient brain structures made them incompetent to pair moral rules to feelings of reward or punishment, which may have explained their incapability to learn from past failures. The thinning of the frontal cortex explained why antisocial individuals made undesirable social decisions.

Vogt's localization of morality deserves the label 'modern' for several reasons. He dismissed the relevance of cranial or cerebral volumes for psychological explanations and descended into deeper cellular structures where brain processes actually take place. Here, we were far removed from 19th century phrenology that projected morality on the skull surface. Moreover, Vogt rejected the oversimplified view of morality as a singular mental phenomenon, a readymade and inborn capacity invariably present in all human individuals. Vogt disqualified the idea of a moral sense as a misleading metaphor. Morality, he argued, bundles a complex of mental representations that are linked by learning processes to accompany emotions (reward and punishment). While humans are certainly born with the organic conditions that support moral education, culture teaches us the particular moral codes and prescribes our specific morals. We encounter Vogt's view in current opinions. His view precedes the recent localizing suggestion of psychopathic personality disorder in the prefrontal region put forward by Adrian Raine, who stresses the cortical thinning in (incarcerated) psychopaths (Yang et al., 2005). Vogt's view also preceded Antonio Damasio's somatic marker hypothesis in which the coupling of mental representation to bodily sensations is considered an indispensable prerequisite for moral development (Damasio, 1994). Moreover, Vogt's ideas return in the current hype that surrounds the so-called Von Economo neurons (VENs). These spindle neurons might be unique to the prefrontal cortex of primates and less abundant in the brains of patients suffering from frontotemporal dementia (FTD) who show a diminished moral sensitivity (Seeley et al., 2006).

Benedikt and Vogt illustrate the ancestry of the project we actually call 'the moral brain'. They were no exceptions. The rise of medical science in the 19th and early 20th century stimulated a multitude of proposals (Verplaetse, 2009). Nearly all hypotheses elicited harsh replies, which demonstrated that suggesting a localization of morality was intellectually charged, even taboo. Critics characterized these adventurous localizations as impossible (Lacassagne), ridiculous (Tarde), narrow (Koch), old-fashioned (Kurella) or a waste of time (Meynert). To the psychologist

Wilhelm Wundt, who, by the way, knew exactly where to localize the brain correlates of attention and volition, these ideas had as little to do with neurology as Jules Verne's explorative expeditions had with astronomy or geology. Skeptics scoffed at this phrenology-like ambition and commended methodological accuracy and modesty, even to the extent that ignorance was preferred to uncertain knowledge. Despite this criticism, the multitude of proposals, speculations and hypotheses testifies to an ongoing ambition to connect human morality to brain matter. This project constituted no marginal endeavor. Even the most sarcastic comments could not stop this ambition. In fact, the neural mapping of morality inherently belongs to the program of the neurosciences. As long as the brain circuits underlying higher mental capacities are not satisfactorily disentangled, this program will continue. Neuroscience does not restrict itself to the neural correlates of motion and perception; it expands its curiosity to achieve a better understanding of mental phenomena like character, willpower, intelligence and morality. Scientists who do not accept that conscience is stocked in an immaterial soul will always be troubled by the question of how morality is situated in our body until a satisfying answer has been attained.

Despite the fact that Greek philosophers with a materialistic worldview already tackled this question, elaborated answers did not dawn until the intellectual climate was sufficiently favorable. Throughout Western history, this condition was rarely met. Before the 19th century, dualist dogmas dominated so vigorously that silence was the best answer for a materialist neuroscientist. During the 20th century, non-biological paradigms framed the interests of psychologists, leaving no room for neurological avenues. Post-war social sciences showed sufficient interest in moral issues but discarded neurological and biological explanations. Intrigued by the flexibility of moral behavior and the ease of influencing ethical attitudes (Milgram, Zimbardo), morality was depicted as the outcome of social interactions in a well-defined context. Moral behavior appeared to be too volatile for leaving a palpable and observable imprint in brain functioning. For a better understanding of the human mind and behavior, neuroscience was of no use. Social sciences and neurosciences belonged to the strictly separated university departments. The prefixes 'neuro' and 'socio' demarcated their isolated scientific worlds.

Since the 1990s, much has changed. A new generation of psychologists has recognized that human morality corresponds as well to more universal patterns of conduct, such as sharing and distributing food, taking risks in favor of the group or maintaining public goods. Often framed in an evolutionary context, contemporary psychology rediscovered the moral dimension in elementary social phenomena as cooperation, helping or punishment. Innovative disciplines like evolutionary psychology or behavioral economics asked for attention to similarities in moral behavior across cultures and populations. Whereas post-war social science had overemphasized the complexity and diversity of proximate mechanisms, adherents of these novel disciplines invite us to study morality from a broader anthropological perspective and to reopen the search for the ultimate origins of altruism and cooperation. This perspective created an intellectual atmosphere which is most attractive to brain researchers. A climate wherein moral behavior is viewed from a wide evolutionary angle offers a fertile ground for neurological contributions to the study of morality. The more elementary and the more general human capacities are perceived, the more researchers with a biological or medical background show an interest. The moral brain project profits from this paradigm change.

Perceiving morality as an evolutionary product certainly helped to stir the enthusiasm of experts who, until recently, had no entrance to a domain in which social scientists exercised exclusive research rights. However, one should not forget that the moral brain project would not have gotten off the ground if new technologies were not created. Without this innovative radiological equipment, the moral brain project was simply nonexistent. These new technologies make it possible to observe brain activity in vivo and in response to experimental stimuli in a way no previous technologies were able to capture. Again, this is not a new story. History of science provides us with parallel episodes where the invention of a new technique promptly boosted the self-confidence of localizers of morality. In the past, neuroscientists began to formulate their localization once they mastered a promising research method. Vogt did not start to fantasize about the moral brain until he discovered the value of staining methods that could be used to observe Lenin's *lamina pyramidalis* through the ocular of his microscope.

Brain Imaging Technologies

Since these technologies are of crucial importance for the moral brain project, a brief introduction may be helpful. *Positron emission tomography* (PET) is a nuclear imaging technique that enables us to measure metabolic brain activities using the delay period of radioactive molecules. PET is a semi-invasive scanning technique. Approximately 1 hour prior to the experiment, a radioactive, short-lived isotope is injected into the living subject. This isotope has been chemically incorporated into a metabolically active molecule that will become concentrated in the tissues of interest. If certain brain regions require more activity-depending substances, one will observe more radioactive isotopes in those very regions. The emitted positrons of the isotopes can be detected, allowing for visualizing the metabolically active brain regions. PET produces brain images with an excellent resolution, but the high costs needed to produce the short-lived isotopes limit the use of PET scanning. In the neurology of morality, the number of PET studies is rather limited.

Almost all neuroimaging studies in moral brain research use a *functional magnetic resonance imaging* technique (fMRI). Non-functional MRI is used to image every part of the body – not only cerebral mass – and to visualize contrasts between distinct kinds of tissues. In MRI, an oscillating electromagnetic field pushes hydrogen nuclei, found in abundance in the water molecules throughout the body, out of alignment with the main magnetic field. When the protons snap back into alignment, they produce a detectable radiofrequency signal. Since protons in different areas of the body realign at different speeds, the different structures of the body can be revealed. Functional MRI does not visualize body tissue but cerebral areas where the demand for oxygen is higher due to various tasks the subjects in the scanner are administered. This scanning technique takes advantage of the magnetic properties of hemoglobin in red blood cells that carry oxygen throughout the brain. Functional MRI measures the effects of stimuli, tasks and experiments on the brains of participants. If you contrast the hemodynamic response over different moments of the experiment or with a well-chosen control condition, more active or less involved brain parts can be isolated, reflecting the mental activities stimulated or inhibited experimentally. Ideally, fMRI points at the brain regions that are central to whatever function or capacity you wish to research.

Contrary to PET or MRI, *transcranial magnetic stimulation* (TMS) is not an imaging technique but an engineering device that allows neuroscientists to excite or disrupt neural activity. An eight-shaped electromagnetic coil is placed on the scalp and induces weak electric currents in the brain tissue underneath producing rapidly changing magnetic fields that influence or disrupt the neural circuitry. TMS is important in neuropsychology since it can demonstrate causality. Functional MRI allows researchers to see which regions of the brain are activated when a subject performs a certain task, but this does not prove that those regions are actually used for the task; it merely shows that a region is associated with a task. If activity in the associated region is suppressed (knocked out) with TMS stimulation, and a subject then performs worse on a task, this is much stronger evidence that the region is actually involved in performing the task. This does not mean that the use of TMS is unlimited. Since TMS might stimulate nerves or muscles on the overlying skin as well, this technique might cause discomfort or pain. Moreover, not all brain parts are eligible for this stimulation device. The electric currents do not penetrate deeper than approximately 2 cm below the scalp. Deep brain structures are inaccessible by TMS.

In spite of their limits, these techniques are undoubtedly revolutionary. They allow us to visualize and to engineer the moral brain without any historical precedence. Past neuroscientists interested in the body-mind problem could only have dreamt of these instruments. Nevertheless, they create particular problems as well. Besides being expensive, this research is time-consuming and labor-intensive. From experiment to publication, several years pass. Scientists are required to work in teams with specialized experts and are financially dependent on public budgets or private sponsoring. This massive mobilization of labor, time and money increases the pressure to obtain positive returns (say, spectacular results). It is true that peer reviews critically evaluate incoming papers, but due to a lack of time and means, editorial boards cannot ask for a replication of the original study. Since researchers, in general, try to harmonize their results with previous findings, it can take a while before one starts to doubt claims made in pioneering studies. Neuroimaging studies face a multitude of caveats. The circumstances in which the experiments are carried out are quite alienating. Helping someone on the street undoubtedly differs from pushing a joystick in the direction of a picture of a bleeding person while you are immobilized in a magnet tube wearing earplugs and headphones. High costs and limited resources restrict the number of participants to no more than a few dozen, hence augmenting the risk of inadmissible extrapolations. Even when you obtain powerful results on a small subject group, one cannot exclude that a different selection might lead to dissimilar findings.

Notwithstanding these limits, neuroimaging research is immensely popular. A stream of papers queues in the mailboxes of the editorial boards of high-ranking journals like *Science, Nature, Neuron* and *NeuroImage*. Between 2000 and 2008, around a hundred papers on the moral brain appeared. If you take the basic interests of the researchers into account, you can subdivide these studies into four groups. A first group mainly explores how our brains process well-known and well-documented distinctions in *moral psychology* or *moral philosophy*. This group examines whether and how our brain differentiates fact and norm, guilt and shame, intention and consequence, justice morality and care ethics, core disgust and sociomoral disgust, etc. How does our brain process these distinct categories which are central to our daily moral reasoning and moral intuitions? A second group of *behavioral economists* shows an interest in the neural underpinnings of altruism versus self-interested behavior. Which brain activation underlies cooperative behavior? Which brain parts are involved in the suppression of egoism? What is happening in our brain if we punish cheaters or free-riders? With the help of computerized social dilemmas mimicking common moral situations in daily life, these researchers expect to find out how our brains execute tasks like sharing, cooperating, punishing, contributing to a common good and how our brains deal with conflicts between self-interest and socially-oriented choices. A third group focuses on morally relevant *social behavior* or *social skills* such as empathy and trustworthiness and investigates how we are able to sympathize with someone in grief or pain, what it means to take someone's perspective, how we decide to help persons in need or abstain from doing this and how we detect untrustworthy people. Of all moral phenomena, empathy is certainly one of the best-studied fields of research. Up to now, nearly all components central to this fundamental social skill were subject to brain imaging studies which advanced our comprehension of the empathy matrix immensely (Decety & Lamm, 2006). A fourth and final group of investigators consists of neuropsychiatrists dealing with antisocial individuals and criminal behavior. Those familiar with the historical antecedents of the moral brain project will certainly not be surprised to read that recidivist criminals, aggressive delinquents, mentally disturbed patients and adult psychopaths are nowadays researched in a magnet. In the past, researchers have always been fascinated by these 'antiheroes of the human conscience' and assumed that something must be wrong with or different in the brain functioning of these people. A similar fascination prompts current psychiatrists attempting to disentangle the *immoral brain* of psychopaths and dangerous criminals. More quietly – with the earlier failures in the back of their mind – they share the hope that more knowledge on the immoral brain might result in innovative therapies, which could help antisocial individuals who resist treatment programs.

The Breakthroughs

What have been the major breakthroughs since the first brain imaging studies on human morality appeared in high-ranking journals? Let's summarize these in four main points.

There is No Moral Center

Unsurprisingly, current research amply demonstrates that there is no such thing as a moral center localizable in a definite cerebral locus. Notwithstanding sensational messages broadcast by news agencies that 'researchers identified the brain's moral

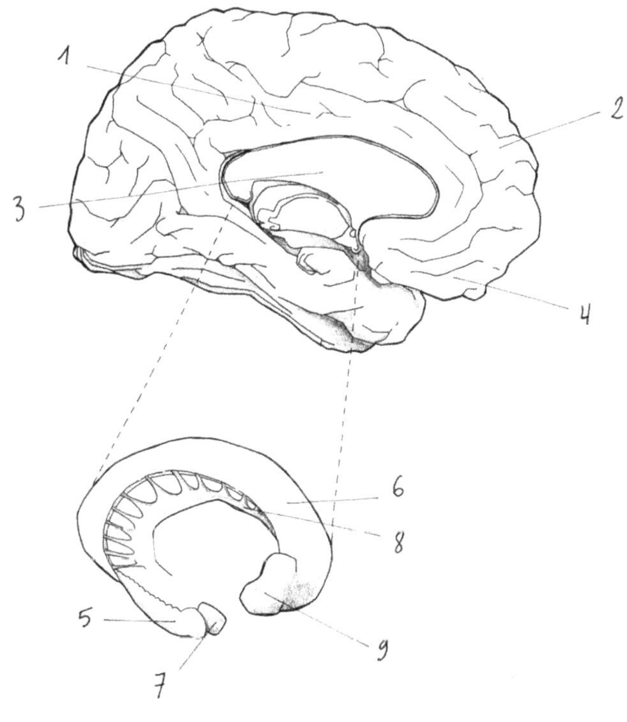

1 Cingulate cortex (CC)
2 Medial prefrontal cortex (MPFC)
3 Corpus callosum
4 Ventromedial prefrontal cortex (VMPFC)
5 Hippocampus
6 Caudate nucleus (CN)
7 Amygdala
8 Putamen
9 Nucleus accumbens (NA)

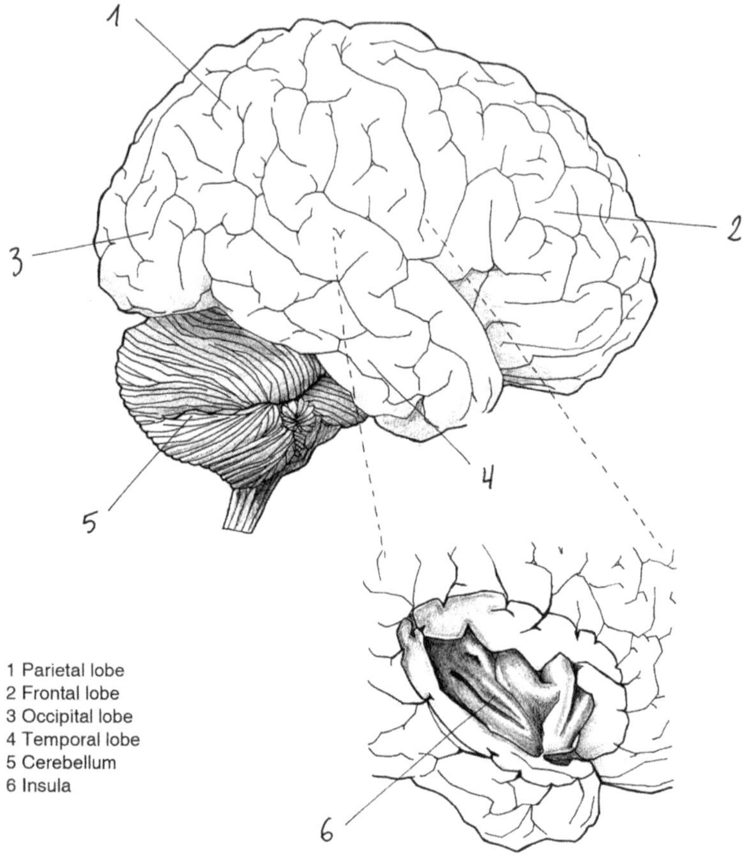

center' (Reuters Health), all researchers plainly reject the idea that a single cubic inch of brain tissue is reserved for moral tasks. Brain areas identified as central to moral tasks are involved in a mass of different activities as well. Functional research into the anterior insular cortex (AIC) illustrates this point very well. Previous studies already showed that feelings of core disgust activate this brain area (Calder, Lawrence, & Young, 2001). When this area was electrically stimulated through implanted depth electrodes, nasty and almost unbearable sensations in the throat and mouth were produced (Penfield & Faulk, 1955; Krolak-Salmon et al., 2003). Neuroimaging research confirmed the prominent role of the AIC in evoking unpleasant odors and tastes (Small et al., 2003; Zald, Lee, Fluegel, & Pardo, 1998, Royet, Plailly, Delon-Martin, Kareken, & Segebarth, 2003; Zald & Pardo, 2000; Heining et al., 2003). Now, some moral emotions have much in common with experiences of disgust. Individuals might be revolted by unfair treatment or sudden exclusion from a group. Sociomoral disgust is a moral emotion that evokes repulsion in a most direct way (Rozin, Haidt, & McCauley, 1993; Haidt, 2003). Liberal people might gag on racist comments, while child adoption by homosexual parents might turn the stomach of conservative people.

Researchers accept that the AIC is central to a whole range of strong negative moral feelings. Meanwhile, studies have confirmed the crucial role of this region in revulsion to incestuous sex (Schaich Borg, Lieberman, & Kiehl, 2008), in rejection of unfair offers (Sanfey, Rilling, Aronson, Nystrom, & Cohen, 2003), in empathic distress in observing pain experienced by loved relatives (Singer et al., 2004) and in indignation of being excluded from a group (Eisenberger, Lieberman, & Williams, 2003). These findings validate a general principle underlying most research: our brain recycles areas originally developed for 'lower' mental tasks for the realization of 'higher' mental tasks. The moral brain can be broken up into several modules whose original functions have nothing in common with morality. To produce an emotion, judgment or behavior that we call 'moral', our brain integrates these basic modules or circuits. This recycling principle guides other findings as well. Pain empathy activates the anterior cingulate cortex (ACC), a region that turns negative sensations into quick motivations (Singer et al., 2004). *Schadenfreude,* or malicious pleasure, evokes increased activity in the nucleus accumbens, a region which plays a major role in mediating satisfaction (Singer et al., 2006). The decision to cooperate with unfamiliar individuals during a prisoner dilemma game augments the activity in the caudatum, an area involved in processing reward (Rilling et al., 2002). The decision to punish an unfair partner stimulates neural activity in the (right) dorsolateral prefrontal cortex (DLPFC), a region that has been studied for its integration of emotion and cognition (Sanfey et al., 2003). No brain area has a privileged or exclusive role in the moral brain. All much-cited regions are part of different brain circuits that perform moral tasks in addition to a number of other tasks.

Are we still entitled to talk about 'the moral brain'? If no brain area is exclusively moral, why not just drop the concept of the 'moral brain' altogether? Does the moral brain not disappear among this multitude of shared-brain activities? The answer is no. When contrasting brain activity during a moral task and a non-moral task, both activities do not overlap completely. In one of his first studies, the Brazilian neuroscientist Jorge Moll asked subjects to view two categories of images: horrible images without moral content (a portrait of a man who suffered from a tumor in his face), on the one hand, and, on the other, equally disturbing pictures that evoke moral emotions (a man who attacked a woman with a knife) (Moll et al., 2002b). Moll simply asked his participants to look at the pictures while scanning their brains. If you consequently subtract all brain activity produced by looking at non-moral images from the brain images taken during confrontation with moral pictures, you will obtain 'moral' brain activity. Moll indeed found that the contrast did not erase all 'moral' brain activity. In contrast with the non-moral images, more activation was registered during moral pictures in the orbitofrontal cortex (OFC) and the superior temporal sulcus (STS). The former region is functionally associated with the representations of reward and punishment, while the latter area has been linked to the detection of intentional actions. Both functions tended to be more active in response to moral horror pictures. From these relatively simple studies, one is allowed to deduce that the moral brain – or its various circuits – can be differentiated from the neural architecture that bears comparable, but non-moral mental capacities. Studies carried out with diverse stimuli, meanwhile, confirmed the relative independence of circuits that underlie moral capacities, such as understanding factual statements versus normative expressions (Moll et al., 2002a; Heekeren, Wartenburger, Schmidt, Schwintowski, & Villringer, 2003) or noticing tactical versus moral problems (Robertson et al., 2007).

It is no longer permitted to conceive the moral brain as an isolated center specializing in processing morality. This rejection is far from original. Brain imaging studies are even superfluous to realize this. One could easily have drawn this conclusion decades ago. Piled up in medical archives are thousands of clinical case studies borrowed from patients suffering from brain tumors or cerebral traumas (Verplaetse, 2009). These abundantly demonstrated that our moral sense will not disappear if one particular brain part is pathologically affected. Today, the moral brain is perceived as an extensive network that encloses remote areas. Researchers are starting to become familiar with the basic architecture of this circuitry and are attempting to clarify the functions of its most important junctions in relation to morality. However, they also realize that this functioning is hugely complex and task-dependent. Given that current research confirms that moral tasks are spatially dissociable, we remain entitled to refer to the moral brain.

Each Moral Task Has Its Own Neural Network

The moral brain project profits from successes in translating traditional moral psychological or moral philosophical distinctions into spatially distinct brain patterns or dissociable levels of intensity. A first success was published in *Science* in 2001, notably in the dramatic week of 9/11, by the American neurophilosopher Joshua Greene. In his pioneering study, Greene showed that the human brain processes personal moral dilemmas differently from impersonal ones (Greene, Sommerville, Nystrom, Darley, & Cohen, 2001). In an impersonal moral dilemma, subjects are asked to sacrifice the life of one person in order to save several others. However, in contrast to a personal moral dilemma, the degree of personal involvement in the act of killing is rather minimal. Pushing a button or pulling over a switch suffices. Since personal moral dilemmas require more direct action – you have to eliminate unlucky individuals with your bare hands – people feel more reluctant, experience more discomfort or take more time to decide. From a rational, utilitarian perspective that solely focuses on the consequences of a deed, both dilemmas are nevertheless identical: more people are saved at the expense of one individual. But when taking our emotions into account, both scripts differ vastly. Greene demonstrated that our different moral intuitions mirror distinct brain activity. Personal moral dilemmas evoke increased activity in the medial prefrontal cortex (MPFC), the posterior cingulum cortex (PCC) and the STS. To our brains, emotions count as much as rational deliberation. In a follow-up study Greene went a step further and started to explore the neural circuitry that differentiates Kantian and utilitarian moral judgments (Greene, Engell, Darley, & Cohen, 2004). To moral philosophers, this theoretical distinction is well known. While a Kantian actor believes that he must respect certain moral principles in all circumstances (e.g. the right to live), a utilitarian believes that some situations justify the violation of these principles (killing is allowed). In this study, Greene focused on a class of difficult moral dilemmas that roused both philosophical options, such as the crying baby dilemma. In this dilemma, enemy soldiers have orders to kill all civilians. To escape being killed, you and some of your townspeople have sought refuge in the cellar of a large house. When the soldiers search the house, your baby begins to cry loudly. If the child does not stop, you will attract the attention of the soldiers who will kill all the people hiding out in the cellar. Will you smother your baby? A Kantian actor will never kill his baby since killing is never permissible, not even in this tragic situation. A utilitarian prefers to serve the greater good, however terrible the decision might be. Greene confronted participants with such dilemmas in a magnet and recorded their brain activity during the moment of decision. He found that utilitarian choices produced increased activity in the rational prefrontal cortex, more particularly in the DLPFC and the emotional PCC, the same region that showed enhanced activity during personal moral dilemmas. Intuitively, these neuroimaging findings are consistent with psychological interpretations. Since a utilitarian chooses to smother the child, he experiences more uneasiness, which can be connected to the increased activity in the PCC. At the same time, he must prevent this emotional hyperactivation from blocking his rational choice. How gruesome the decision might be, a utilitarian decides that the killing of one child is morally correct to save the other townspeople. This might explain the enhanced activity in the DLPFC.

Translating moral distinctions into specific brain patterns is the main topic of dozens of studies. Diane Robertson (2007) compared care-based and rule-based moral decisions. Care ethics apply to moral conflicts solved by a focus on the situational variable needs and relationships of the people involved and guided by social emotions such as empathy and altruism. For instance, you decide to skip work next Saturday – although much needs to be finished – because otherwise your son will have no transportation to his football match. Individual needs and empathic consideration will determine your moral decision. Rule-based judgments, or justice ethics, solve conflicts by a focus on moral principles that tend to be violated. In this case, you decide not to work next Saturday because you have already promised to accompany your girlfriend on a shopping trip. As a matter of principle, you disapprove of breaking your promises. Using real life moral issues, Robertson found that these ethical viewpoints are associated with dissociable neural processing events. Rule-based moral decisions enhanced activity in the inferior parietal lobule (IPL), a region involved in third person perspective-taking, while care-based ethical choices activated brain areas associated with emotions (ventromedial prefrontal cortex, VMPFC and PCC) and self-referential processing (MPFC) which is crucial in generating empathy. Once more, neural localization, mental function and moral task interrelated well.

Some researchers prefer to explore the neural correlates of less common distinctions. Jason Mitchell found evidence that the difference between cold and warm empathy is associated with dissociable neural networks (Mitchell, Macrae, & Banaji, 2006). He asked participants in a magnet to judge the preferences of targets that either held opposing political opinions or who concurred on ideological viewpoints. Mitchell expected that political (dis)similarity affected the kind of empathy that is necessary to answer questions such as ‘Do you think that European films are generally better than the ones made in Hollywood?’ When confronted with similar targets, subjects will refer to their own opinions, feelings and experiences in order to answer these questions. This self-referential mode is of no use, however, in responding to dissimilar individuals. To judge their preferences correctly, a different system based on deduction and inferential reasoning is applied. Mitchell discovered that the self-referential ‘warm’ empathy and the inferential ‘cold’ empathy rely on dissociable networks within the MPFC. Warm empathy is associated with a ventrally located region in MPFC, whereas cold empathy engages an area more dorsally situated in MPFC.

One could easily add dozens of similar attempts to localize psychological distinctions in dissociable neural areas. We will, however, shed light on some problems inherent to this kind of research. Successful results are not always achieved. Up to now, no study has documented functional brain differences between shame and guilt (Berthoz, Armony, Blair, & Dolan, 2002; Shin et al., 2000; Takahashi et al., 2004). Even though the distinction between these moral sentiments is evident to all of us, both feelings seem to activate similar brain patterns. In guilt and shame scenarios alike, an increased neural activity in overlapping prefrontal and temporal areas was noticed. Recent studies even cast doubt on the crucial role of the anterior insula (AIC) in sociomoral disgust. Contrary to expectations, repulsive moral transgressions that are not sex-related, such as violent attacks on innocent people, do not augment activity in that region (Schaich Borg et al., 2008; Moll et al., 2005). The interpretation of contrasts causes methodological difficulties as well. Since functional interpretations and neuroimaging findings are strongly interdependent, circular explanations are difficult to avoid. New findings are explained with the help of earlier interpretations based on previous studies that sometimes lack methodological accuracy. Furthermore, since our functional knowledge regarding brain areas is far from complete, current neurological literature attributes the most diverging functions to certain regions, so that even the most frustrating results can easily be harmonized with the existing body of evidence. After shopping through the abundant supply of functional correlates, one picks out an explanation that nicely fits the theory. Conflicting interpretations are kept silent. Double interpretations, frequently occurring in the discussion session of research papers, constitute a particular concern. Double interpretations cast methodological soundness into doubt because they always fit the data. Take the following example: Suppose that psychopaths display increased activity in the dorsolateral prefrontal cortex (DLPFC) during a task. Unfortunately, according to your theory you should expect less activation in the DLPFC in response to the stimuli. Using sound reasoning, one should conclude that the data falsifies the theory. Double interpretations sidestep this conclusion. One might argue that psychopaths expend more effort to reach an equal level of performance. Compared with healthy participants more activity in their DLPFC is required. Most readers will not notice this implicit shift in the interpretation from 'normal excitation' to 'additional effort'. Moreover, given the current state of the art, we cannot refute this ad hoc explanation. Scanners do not distinguish the brain's natural inclination from its compensatory efforts. Nevertheless, it remains a trick to circumvent disappointing results.

Engineering the Moral Brain

Undoubtedly, the study of Daria Knoch, published in *Science* in 2006, was the most spectacular moral brain study to date. This study established that the moral brain is subject to engineering (Knoch, Pascual-Leone, Meyer, Treyer, & Fehr, 2006). Inspiration was borrowed from an earlier fMRI experiment published by James Rilling and Alan Sanfey (2003) in which subjects played an ultimatum game. In this simple game, a subject proposes a division of a certain amount of money (say 10 €) in the presence of another subject. For instance, I decide to take 6 €, so that you will receive 4 €. The responder has the right to oppose. If he exercises his right, he prevents the deal, and both players get nothing. Confronted with unfair offers, a responder tends to oppose more, though this is not so wise from a rational point of view. Even if a proposer keeps 9 € for himself and offers 1 € to the responder, by using his veto this second player ultimately receives nothing. Rilling and Sanfey scanned the brains of 19 players who responded to offers made by Kelly, an accomplice. In each trial, Kelly communicated the division she proposed to the subjects in the magnet. Subsequently, the subjects were free to accept or to oppose. Following an unfair trial, increased brain activity was registered in the AIC, the ACC, and most importantly, in the rDLPFC, a region that is accessible to brain stimulation. Knoch wondered what would happen if that region was blocked using TMS. Would subjects then accept more unfair offers, or would they, in contrast, become infuriated and punish the smallest deviation from an equal distribution? The obtained results are remarkable. A third of the participants accepted all offers, even extremely unfair ones. No subject of the control group who got a placebo or sham treatment was willing to accept these unfair offers. If you take a look at a prototypical unfair allocation of 8 € to the proposer and 2 € to the responder, 45% of the TMS-treated subjects accepted. In contrast, no more than 10% of controls showed an equal benevolence. Although the manipulation certainly did not affect the behavior of all participants (55% of TMS-treated individuals reacted as controls did), it evoked a significant effect. Remarkably, the TMS-treatment had no impact on the moral opinions of the treated participants. Questionnaires revealed that all participants, treated or not, equally disapproved unfair offers. Moreover, the TMS group was not numbed or apathetic. Their mood did not alter over the course of the experiment. They probably forfeited the ability to convert moral indignation into the appropriate veto response.

It has been abundantly documented that the rDLPFC plays an eminent role in conflict management and harmonizing emotion, cognition and action. Disrupting this region prevents a sound integration of moral opinion, ethical sentiment and moral behavior.

More recent studies further documented the prospects of TMS for the moral brain project. Lucina Uddin and her team disrupted performance on a self-other discrimination task (Uddin, Molnar-Szakacs, Zaidel, & Iacobini, 2006). She asked subjects to look at pictures in which the subject's face was gradually morphed with a stranger's face. The subjects answered whether or not they could still recognize their own face in the pictures. She found that TMS over the right inferior parietal lobule (IPL) significantly impaired the subjects' performance to discriminate self faces from other faces. This is the first evidence that TMS treatment might knock out a part of the circuitry associated with distinguishing ourselves from others. In the field of moral judgment, this ability is of vital importance. Without this mental power, the well known distinction between justice ethics and care ethics would simply be nonexistent.

TMS is a time-consuming and rather ponderous manipulation device that disturbs the normal activity in a certain brain region. In participants, this instrument may cause uneasiness or even pain. In 2005, Michael Kosfeld, published a study in *Nature* using a less invasive technique to influence our social behavior (Kosfeld, Heinrichs, Zak, Fischbacher, & Fehr, 2005). His subjects received an intranasal dose of oxytocin, while a control group got a placebo substance. After substance administration, he asked both groups to play a trust game. In a trust game, two subjects are invited to participate in a potentially lucrative transaction. They interact anonymously and play either the role of an investor or a trustee. The investor has the option of choosing a risky trusting action by giving money to the trustee. If the investor decides to transfer money to the trustee, this amount is automatically tripled by an experiment leader, but the trustee ultimately chooses how to divide the investment. He might violate the investor's trust by keeping the invested amount for himself. In an anonymous, one-shot trust game, rational investors are inclined to transfer none or only small amounts of the money, in view of the fact that egocentric trustees might dishonor the investors' trust. Kosfeld found that an administrated dose of oxytocin, a neuropeptide that plays a key role in social attachment and affiliation in non-human mammals, augments the level of money transfer. Oxytocin affected the investors' willingness to accept the risk of dealing with egocentric trustees. Compared with the control group, they assigned more money to the trustees. Remarkably, oxytocin did not influence the sharing decisions made by the trustees.

Currently, oxytocin nasal sprays can be found on the shelves of all laboratories involved in the moral brain project. Parallel research demonstrated that oxytocin increases mind reading in humans. Volunteers who received this active substance showed improved performance on a test that required inferring emotions from facial expressions (Domes, Heinrichs, Michel, Berger, & Herpertz, 2007). In spite of these successes in the field of neurochemistry of morality, a lot of issues are unaddressed. Why does oxytocin influence some individuals more than others? Why does it not affect the trustees' cooperative decision? As Damasio put forward in a comment, one certainly does not need to worry that political operators will generously spray the crowd with oxytocin at candidate rallies (Damasio, 2005).

Nevertheless, these pioneering studies illustrate that the moral brain is subject to engineering. Morality is a matter of flesh and blood. With the help of innovative technologies and chemical substances, researchers are able to manipulate the biochemical processes underlying our moral behavior. Obviously, TMS treatment does not switch off our moral sense but only disconnects one brain region. A complete knockout would definitely contradict the network structure of moral processes that we emphasized before. The moral brain cannot be localized in a definite brain region but connects remote neural areas during specific moral activities. However, if technologies are developed that simultaneously disrupt brain activity at different spots, new breakthroughs might be expected. Then, it would be feasible to influence the detailed brain pattern that has been associated with a particular moral task. Such a revolutionary technology is, at this moment in time, only depicted in science fiction.

The Immoral Brain

A group of people that instantly crosses our minds if we daydream about engineering the moral brain are criminals and therapy-resistant psychopaths. Historically, brain scientists have always been fascinated with this group of antisocial individuals who violate the basic rules of conduct in civilized society (Renneville, 1999). As soon as imaging techniques entered the scientific domain, researchers commenced investigating the brains of criminals and psychopaths in search of deficiencies (Sakuta & Fukushima, 1998). In contrast to moments of enthusiasm in the past, current researchers have obtained rewarding results offering interesting perspectives. If we restrict our focus to psychopathy research, we might discern two distinct approaches that are nevertheless not mutually exclusive.

According to Raine, who had already started his neurobiological research program in the 1980s, these antisocial individuals lack so-called *somatic markers*, which are presumably localizable in the VMPFC. Somatic markers, a term coined by Damasio, pair moral or social rules to unconscious bodily sensations. For instance, if I am in a hospital where a highly attractive nurse looks after me, then my somatic markers warn me that pawing her is not an option. At a subconscious level, my body obstructs sexual harassment. My somatic markers inform me about the negative consequences of such action. A loss of respect or punishment will be my penalty. In normal circumstances, these emotional-labeled mental representations guide our social conduct. Individuals who lack these markers do not learn to avoid social conflicts or take too many risks. Using simple tests, such as the Iowa Gambling Task that assesses risk aversion, Damasio demonstrated that rare patients who suffered from brain damage in the VMPFC caused by tumors or traumas during childhood misjudged dangers and risks. These unfortunate patients did not learn from misfortune, disadvantage or punishment. They persisted in risky and antisocial actions, meanwhile scoring low on various morality testing devices.

What available evidence may demonstrate that developmental psychopaths are deficient of somatic markers as well? In the spring of 2005, Yang and Raine published results showing a thinning of the prefrontal cortex among incarcerated psychopaths. This group of psychopaths showed a 22.3% reduction in prefrontal gray matter volume compared with control subjects (Yang et al., 2005). However, non-incarcerated psychopaths, who Yang and Raine found in job centers, did not reveal prefrontal structural deficiencies. In spite of its only partial success, this finding is nevertheless remarkable. Let's not forget that no structural abnormalities could be observed in the psychopaths' brains throughout the entire medical history of the concept. The psychiatric history of the concept goes back to the 1830s (James Prichard) and went through several periods of time that favored a neurobiological orientation. Unfortunately except for Vogt's microscope, no research instrument revealed abnormalities in the past.

From these studies, it would be scientifically unsound to conclude that psychopaths have an abnormal prefrontal cortex. The technology that Yang and Raine applied is incapable of displaying any spatial details of the prefrontal cortical thinning. Damasio's somatic marker hypothesis predicts reduced volumes of VMPFC and unaffected DLPFC, but, however, no study confirmed this prediction until now. A trickier problem concerns the negative result among non-incarcerated psychopaths. These calculating and callous antisocial persons seem smart enough to avoid sentencing and imprisonment. How do we understand their personality disorder from a neurological point of view? What deviating brain processes give rise to the psychopathic features of this group? If no neural defects or brain abnormalities can be found, this is most unsuitable. This would imply that the psychopath *par excellence*, who besides being affectively disturbed and extremely manipulating is also gifted with sufficient intelligence to escape detection of his crimes, has a perfectly normal brain. This is a conclusion that neuropsychiatrists deplore.

In the research of James Blair and Niels Birbaumer, one encounters a distinct approach. According to these researchers, psychopathic features are not so much associated with a lack of somatic markers, which couple moral rules to elementary bodily sensations, but are linked to a diminished sensitivity of certain emotions that are crucial for inhibiting socially unacceptable conduct. In particular, psychopaths seem insensitive to the suffering their crimes cause. They seem unimpressed by the strong negative feelings normal people generally experience if they commit violent or damaging acts. Psychopaths lack the deep social instincts inhibiting human aggression in normal circumstances. If ordinary men notice people in trouble or perceive signs of submission during a violent encounter, their aggression ceases at once. One does not continue to kick a person who begs for mercy. In normal people, cues of fear and sadness block aggression and violence. It has been hypothesized that psychopaths detect these emotional cues less accurately and, consequently, experience less distress. Psychological tests confirmed this presumption (Blair, 1995; Hastings, Tangney, & Stuewig, 2008; Marsh & Blair, 2008). Psychopaths showed a deficit in discriminating fearful and sad faces. In response to different emotions (for instance, disgust) their sensitivity seems intact. In an original study, Christopher Patrick, Bradley, and Lang (1993) found that psychopaths show a less responsive eye-blink to images depicting mutilated or distressed people when a hard noise randomly interrupted these pictures. Normal controls subconsciously expected the hard noise during this set of pictures and seemed better prepared to blink their eyes once the noise effectively occurred.

Previous neurological research consistently linked the processing of fear and the recognition of fearful faces to the amygdala (Adolphs et al., 1994, 2005). Therefore, it has been alleged that this basal brain region, rather than the prefrontal cortex, is dysfunctional in psychopathy. Experiments undertaken by Birbaumer, Veit, and Lotze (2005) provide us with direct evidence that the functioning of psychopaths' amygdala indeed differs from that of normal subjects. Birbaumer analyzed the neural activity in the amygdala of psychopaths during an aversive learning task in which he confronted the volunteers with two kinds of faces. If the subject observed a mustached face, painful pressure in theirs arms caused by a pneumatic device followed within a few seconds. No pain was applied if they observed a picture of a mustacheless face. Compared to healthy controls, psychopaths showed equal arousal to the administrated pain. However, reduced levels of skin conductance were recorded if the psychopaths looked at mustached faces, thus indicating that they responded with less sensitivity to stimuli that announced pain or punishment. Brain imaging analysis led to a similar conclusion. In the control group, enhanced activity in the (left) amygdala accompanied confrontation with the mustached person, whereas the amygdala of psychopaths failed to show increased activity. These findings support the theory that psychopaths lack the neural basis for anticipating aversive events.

After 170 years of investigating the immoral brain, we have finally obtained observable deficiencies. Current brain imaging techniques indeed reveal structural and functional abnormalities in the brains of psychopaths and criminals which cannot be ignored any longer. Despite these promising results, final conclusions ought to be warded off. Researchers are reluctant to speculate about the causal relation between antisocial conduct and brain anomalies. To what extent do impairments in the amygdala or VMPFC attribute to the development of the personality disorder? Most researchers do not believe that structural and functional impairments inescapably cause psychopathic behavior. They find it more likely that complex disruption of neural circuitry predisposes one to psychopathic traits. Besides prefrontal and amygdalar deficiencies, neural abnormalities have been recorded elsewhere in the brains of psychopaths. Reviews of neural dysfunctions in psychopaths document a dozen brain regions showing deviating activity patterns (see Glenn in this volume). As hypoactivity in the amygdala and VMPFC straightforwardly fits theoretical assumptions, these deficits are predominantly mentioned. However, psychopathy remains an extremely complex disorder whose neuropsychological components are not fully understood yet. As tempting as it might be to describe antisocial individuals by referring to classical dichotomies in moral psychology – no feelings of guilt, no empathy – more profound investigations seems to reject this oversimplified view. As Joshua Greene elegantly stated: "Psychopathy is not Nature's controlled experiment with amorality." Future research into the immoral brain ought to illuminate whether psychopaths and recidivist criminals are indeed the archetypes of human immorality, our genuine 'antiheros of the conscience'.

Prospects and Limits

Investigating the moral brain is hot. Novel technologies and recent successes currently generate a stimulating climate for ambitious research programs. Scientific dreams that for decades were filed away in laboratory logbooks revive these days. Our increased knowledge of the moral brain inspires confidence, to the extent that current researchers reanimate old ambitions. Let us list three of these prospects.

Neuromorality tests. To date, psychologists have not succeeded in developing a satisfactory device for measuring moral progress in individual children or adolescents. All existing morality tests (Defining Issue Test, Moral Judgment Test) apply solely to groups. If we can achieve a detailed picture of the neural circuitry underlying moral processes, measurement of an individual's moral development comes within reach, probably by combining psychological and neural parameters. Such a neuropsychological assessment would prove useful for diagnoses in medico-forensic and pedagogical settings that require more insight into the moral maturity of delinquent individuals.

Neurotherapy. Currently, treatments intended to treat criminals, psychopaths or sexual deviants by means of medication, brain electrodes or deep brain stimulation remain science fiction (but see De Ridder in this volume). However, if sufficient knowledge of the moral brain is attained, it might be possible within a few years to assess the effectiveness of psychotherapies. Many therapies in forensic settings aim to increase the moral sense in incarcerated criminals or to heighten awareness about the detrimental consequences of their crimes. It might be reasonable to expect that if we accumulate insight into the empathy matrix, the neural underpinning of excuses and justifications or even feelings of guilt, neuroimaging might help to evaluate the advantages of the therapy.

Controversies in moral philosophy. In spite of the recurrent criticism moral philosophers spout on the moral brain project (i.e. neuroscience should reduce complex phenomena to basic dichotomies such as altruism or egoism, etc.), they might nevertheless benefit from it. While neurosciences will probably not resolve ancient disputes in philosophy, they might facilitate the reformulation of metaphysical enigmas in scientific questions that are empirically answerable. For instance, since the 1970s, a fierce debate has divided moral philosophers over whether women preferentially endorse care ethics while men favor justice ethics. Psychological investigations offered contradictory results. Why not try to resolve this dispute via scanner research?

Undoubtedly, current limitations are a very good reason for not doing this. Many ambitious applications demand individual diagnosis. We like to gauge the individual moral development, not the mean percentage in a group of participants. Unfortunately, current brain imaging research is mainly group-based. Research results are presented in a virtual universe full of probabilities, mean percentages and statistical significance. Statements like 'psychopaths show a thinning of the prefrontal cortex' are deceptive if researchers refrain from warning that average results differ from universal features. In group analyses, high-scoring individuals will compensate for low-scoring subjects. Accordingly, not all incarcerated psychopaths will suffer from a smaller prefrontal cortex. The disentanglement of individual brain variation is still in its infancy, though some very recent studies are promising. In one study, it was found that neural activity in nucleus accumbens, nucleus caudatum and insula during the course of an obligatory moral task accurately predicted performance in a voluntary moral task (Harbaugh, Mayr, & Burghart, 2007). Neural activity at an earlier moment of time was instructive for predicting the social response of the individual subject at a later moment. A recent study carried out by Kristin Prehn and Hauke Heekeren (see this volume) documented the neural impact of differences in individual moral maturity on a simple moral task. These studies unquestionably open perspectives for the development of a 'neuromorality test'. However, we immediately call attention to the rarity of such studies.

If imaging techniques improve in both resolution and quality, we will encounter more individual-based studies in the near future. Current technical limitations prevent us from reconstructing the moral brain in a more detailed manner. Moreover, a phrenology-like approach, in which mental functions are spatially correlated to definite brain areas, still dominates current neuropsychological research. Although this approach has been rejected as erroneous and even misleading, limitations such as low resolution imaging and a lack of connectivity studies thwart the dissemination of more complex and realistic models. If future research incorporates connectivity, chemical and temporal variations, we will certainly move towards mature models that mirror the theoretical notion of a neural network or neural circuitry much better than the phrenology-like activity centers. In spite of these limitations, we should not be dismissive of the great progress that has been achieved. The continuing stream of studies that started at the end of the 1990s increased our knowledge enormously. As a matter of fact, this progress weakens the position of critics of the moral brain project. Apart from some exceptions, current researchers no longer disqualify this adventurous project as impossible or ridiculous or even as a waste of time. Certainly, the philosophical spirit has changed since the 19th century. Old-fashioned dogmas disappeared. The belief in an immaterial soul, or its materialist opposite, no longer shapes the agenda of neuroscience. The amphitheater's marble, to paraphrase the 19th century French novelist Barbey D'Aurevilly, no longer constitutes the bedstead of materialistic sciences but rather has become the bed of all sciences. Those who at present proclaim that the human brain carries the highest and most distinctive human faculties state the obvious: 'Ohne Phospor, keine Gedanke' (without phosphor, no thoughts) – this 19th century provocation of the German–Dutch physiologist Jacob Moleschott turned out to become commonplace, though the phosphor is presently replaced by synapses, neurotransmitters and action potentials. Advancements over the last decade ultimately substantiate that science is capable of capturing our innermost moral experiences in neurobiological terms, now already in gross lines, later on in more detail.

Morality and Evolution

In his second masterwork *The Descent of Man* (1871) Charles Darwin wrote that he was fully in agreement with those who believed that the possession of a moral sense and a conscience were the main features which distinguished man from all other animals. Nevertheless, he was equally convinced that these features – like all other adaptive features – were the product of evolution by a process of natural selection. For this reason he thought it likely that some traces of morality could also be found elsewhere in non-human nature. Whilst it is true that animals do not possess the typically human ability to understand and assess the difference between good and evil in a reasoned manner, many other species clearly have social instincts and moral sentiments. According to Darwin, this is all that is necessary to make the evolutionary development of a moral sense possible – and probable. In particular, it can be expected that animals which live in groups will more quickly develop a kind of basic 'moral code', since this is necessary to maintain their relatively complex social structure. Even in the 19th century Darwin was able to give dozens of anecdotal examples.

In modern times, scientific research has taken the study of such matters many stages further. This research has shown that the following characteristics are present in (amongst other species) anthropoid apes, dolphins and whales: bonding between mothers and children; co-operation and mutual help between members of the group; certain forms of sympathy and empathy; various forms of altruism; feelings of justice; attempts to solve problems or to maintain peaceful relations; deception and the discovery of deception; a sense of being part of a group; the existence of implicit social rules. Be that as it may, humans still possess all these characteristics to a far higher degree than any other animal. Moreover, not only do men and women have a fundamental sense of morality, but we also have the ability to reflect upon the nature, origin, validity and usefulness of moral rules. This allows us to bring such rules into question, or to change them, or even to reject them. As far as we know, animals are not yet able to do these things. However, the presence of social instincts and moral sentiments in some species makes an evolutionary leap towards a more human type of morality quite feasible. Present-day evolutionary biology, supported by disciplines such as primatology, behavioural ecology and neurobiology, is carrying out a wide range of investigations into the nature, origin and evolution of moral behaviour. Many of the insights on which this new work is based date back to the pioneering research carried out by several eminent 20th century Darwinians, such as David Lack, George Williams, George Price, William Hamilton and Robert Trivers. Between them, these scientists largely solved the problems relating to the origin of altruism in animals: a crucial element in the development of morality.

Altruistic acts – which are best described as acts which reduce the evolutionary fitness of the doer and increase the fitness of the receiver – are central to the question of morality. It has been evident for many years that man is not the only species capable of altruistic acts. Some types of animal – in particular the so-called eusocial insects, such as *Hymenoptera* (bees, wasps and ants) and *Isoptera* (termites) – are altruistic to an extreme degree. We are all familiar with the image of the bee, stinging any predator who threatens its hive, thereby giving up its own life for the safety of the group. The sting is painful for the predator, but it is fatal for the bee. By leaving its sting stuck in the flesh of its adversary, the bee effectively secures its own fate. Minutes later it dies. Even before making this ultimate sacrifice, the bee will have done everything possible throughout its life to secure the well-being of the queen bee and the other members of the colony, and this in a number of different ways: helping to bring up the young, guarding the entrance to the hive, gathering nectar for food, etc. In short, the bee is continually concerned with the welfare of its wider community, to such an extent that it no longer even has time to procreate. As a result – and perhaps even more amazingly – the worker bee (like other categories of worker insects) has actually become infertile. This leads us to an obvious question: how did this extreme type of altruism first come into existence?

In Pre-Darwinian times, this type of behaviour (together with all other types of adaptive collaboration between animals) was ascribed to the intervention of an Intelligent Creator. It was more difficult to justify in terms of the theory of natural selection. In fact, it was almost impossible. According to Darwin, the selection process operates at the level of the individual. The characteristics which the individual organism possesses will therefore be functional and beneficial for the individual itself. If an animal were to be discovered whose characteristics are functional for another species – for example, a rabbit whose calls are designed to attract foxes – this would destroy the theory of natural selection: or so Darwin believed. In other words, he was well aware of the vulnerability of his own arguments. In this respect, the existence of sterile workers in colonies of social insects was a major stumbling block. How was it possible for such self-sacrificial behaviour to develop and persist? Surely the aim of every creature was to secure the future of its type through reproduction? To answer this conundrum, Darwin was required to develop a theory of group selection (see also van Veelen in this volume). This theory argued that while selection normally operates at the level of the individual, in certain special cases characteristics are selected which are useful for the species – and therefore not necessarily for the individual organism. Thanks to the selfless behaviour of the sterile workers, the group *as a whole* has a greater chance of survival. This is a line of reasoning which, at first sight, seems plausible and it still has many supporters, even today. In addition to social insects, several types of social mammals and birds also seem to act in the wider interests of the group, or even the species. The British Biologist V.C. Wynne-Edwards developed this idea in his book *Animal Dispersion in Relation to Social Behaviour* (1962). He cited examples of various organisms which, he argued, 'consciously' choose not to reproduce, even if they are not sterile. The 'decision' of these organisms is based on an 'assessment' of the natural resources available to the group or species. If there is room enough and food enough, the organism will reproduce. If there is not, the organism will sacrifice its reproductive chances for the benefit of the group or species. Most biologists accepted the basis of this argument, without realising that it is problematic when looked at from a Darwinian point of view.

David Lack and George Williams were amongst the first to suggest that the theories of Wynne-Edwards and his supporters did not really hold water – no matter how plausible they might sound. One of their most important counter-arguments was the fact that group selection – which develops characteristics which are beneficial for the group but not necessarily for the individual – is extremely vulnerable to exploitation *from the inside*. An organism that is prepared to profit from the altruistic behaviour of its fellow organisms, without making any altruistic contribution of its own, is almost guaranteed to be successful in terms of survival and reproduction. This type of 'free rider' would quickly undermine the social cohesion of any group based on altruism. But if this is true, how do we explain the existence and behaviour of sterile insects? This mystery was unravelled by George Price and William Hamilton, who were able to chart the genetic basis of such self-sacrificial conduct. Their hypothesis argued that natural selection occurs at a level much lower than Darwin had believed. They gradually came to see – and to convince others – that organisms can be understood as 'vehicles' for genetic material. The transmission of genes from generation to generation is central to the theory of evolution and the individual organisms are the method by which this transmission takes place. This does not undermine the belief in a possible moral sense in animals, but it does offer new insights into the origin and evolutionary development of that sense.

We do not find it surprising that parents – whether they be animal or human – try to help their offspring. Nowadays, it is generally accepted that genetics play a crucial role in explaining this phenomenon. In humans, a biological father 'invests' 50% of his genetic material in his child. The same is true of the biological mother. Viewed from a genetic standpoint, it is self-evident that parents will be far more likely to display self-sacrificial behaviour for a child that they themselves have created, rather than a child created by someone else. This premise has inspired many recent studies – both with regard to human and non-human species – into a wide range of subjects, including food sharing, infanticide, pregnancy, etc. These studies have yielded results which would have been unthinkable in the days prior to the genetic approach to evolutionary theory. The case of the altruistic, sterile insects illustrates this point well. William Hamilton realised that there was a link between the altruistic behaviour of the Hymenoptera and the manner in which their sex is determined. The sex of an insect from the Hymenoptera family is dependent upon its procreative origin: it either originates from an unfertilised egg or from an egg fertilised with sperm. The unfertilised egg – which therefore only contains the chromosomes of the mother – develops into a haploid male, i.e. a male with a single set of chromosomes. This means that when such a male becomes an adult, it automatically produces genetically identical haploid sperm. Fertilised eggs develop into diploid females, which means that each cell contains two sets of chromosomes: one set from the mother and one set from the father. If a female mates with a single male, her resulting daughters will all have the same set of fatherly chromosomes. This is not the case with many others species, including humans. In humans a child will obtain 50% of its genetic material from its diploid mother and the other 50% from its diploid father, resulting in an average genetic agreement of 50% between different children of the same parents. But in Hymenoptera the average agreement between the genetic material of the various offspring is 75%. In other words, the brothers and sisters of the Hymenoptera are more closely related to each other than the brothers and sisters of many other species (again, including humans). This means that, viewed from a genetic standpoint, it is 'wiser' for a wasp, bee or ant worker to sacrifice itself for the safety of its sisters, since it has more genetic material in common with its sisters than with any offspring that it might ever have been able to produce itself.

The social insects are not the only species which can appeal to our human imagination by virtue of their self-sacrificial instincts. There are also species of mammals which have developed unusual methods of reproduction and collaboration, similar to those practised by the Hymenoptera. The most outstanding example is the naked mole rat, in particular the *Heterocephalus glaber*. This rodent lives an underground life in the drier regions of Kenya, Ethiopia and Somalia. It survives by eating roots, is blind and lives in colonies of ten to three hundred individuals, all of whom are closely related to each other in terms of genetics. The mole rats display all the characteristics of the eusocial insects: the colonies house overlapping generations, there is a division of labour based on reproductive capability, individuals who are unable to reproduce still help with the rearing of the young, etc. In each colony, there is a single female who gives birth to all the offspring – sometimes as many as 28 babies per litter or more than 100 per year. The members of the colony all help the 'queen' to raise her young, either by gathering food or by protecting the burrow. They carry out these tasks in rotation, so that there is a system of 24-h care. The harsh conditions in which these animals live make social collaboration an essential pre-requisite for survival. Mole rats who try to live alone, or who refuse to co-operate with others, quickly weaken and die.

In addition to the altruistic behaviour of the eusocial insects and of certain mammals, such as the naked mole rat (whose behaviour we can now understand, thanks to the genetic insights of Hamilton and others), forms of altruism also exist in some other species, where there is no close genetic link. This is perhaps particularly true of humans, but we are by no means unique in this respect. Similar phenomena can be observed in several other types of animal – sometimes where we might least expect it. A good example is the so-called vampire bat (*Desmodus rotundus*). If a bat is unable to feed on blood for a period longer than 60 h, it will lose up to a quarter of its body weight. As a result, it will be unable to maintain its body temperature and will soon die. This means that a bat needs to drink 50–100% of its body weight each night, in order to remain healthy. This is not always possible, but in the course of their evolutionary development vampire bats have discovered a way of helping each other. Once an individual bat has consumed enough blood for its own requirements, it will 'donate' any surplus to a less fortunate colleague. The next night the roles might be reversed. This type of co-operation is known as 'reciprocal altruism' and its logic was first deciphered by the American biologist, Robert Trivers. The advantages of the system are, by any standards, spectacular. Without this form of collaboration, a bat would live to an average age of about three years. Thanks to the operation of the blood donor system, many individuals can reach ages of up to 15 years. Trivers was able to demonstrate that reciprocal altruism is theoretically possible in a range of species which have the capacity to recognise each other and which have a sufficiently developed memory to remember the fellow-animals they have helped – and from whom they might therefore expect help in return. This ability to recognise is essential, since the mutual help system might otherwise be undermined by a 'free rider'. Consequently, this means that the species which practise 'reciprocal altruism' are also capable of detecting deceit and perhaps even of interpreting the intended behaviour of their fellow-creatures. As a result, there is a much-increased chance in animals of this type that the conditions exist for the development of an elementary morality by means of natural selection.

As previously mentioned, in addition to man it is also possible to detect similar capabilities in anthropoid apes, dolphins and whales. How these capabilities have developed and how they can best be interpreted are matters which are still open to considerable discussion. Apart from Darwin in the 19th century and Hamilton and his colleagues in the 20th century, it is only in recent times that scientific research has begun to concentrate seriously on the evolutionary origins and development of moral qualities in both man and animals.

Indirect Reciprocity and Strong Reciprocity

Hamilton's and Trivers' theories of kin selection and reciprocal altruism may falsely elicit the idea that altruism can be reduced to 'gene-centred egoism', for it seems as if the altruistic element of altruism is removed by emphasizing the (genetic) benefits. Indeed, in the end 'reciprocal altruism' relies on the return of favours to the altruist. 'But are you truly altruistic if you help, while expecting help in return later on?' one may object. This type of questions evokes a cloud of misunderstanding due to a mix-up of proximate and ultimate causation. Whereas the former explains the physiological or motivational causes of behaviour, ultimate causation pertains to the long-term evolutionary explanations. Though in the process of evolution adaptations and characteristics are selected that increase success of the 'selfish' genes (ultimate causation), this does not imply that the motivations of the individuals are selfish (proximate causation). Similarly, the very noisy process of welding iron plates during the construction of a ship does not cause boats to be noisy as there is a difference between the process and the product. Thus, on an individual level, a person doesn't necessarily help because he wants a favour, but because his emotions and motivations which have been selected on an ultimate level, steer him to do so.

Still, a debate ensued concerning 'strong reciprocity', which is, according to its proponents, a predisposition to cooperate with others at a personal cost, the occurrence of selfless acts that do not provide any benefit to the individual (Gintis, Bowles, Boyd, & Fehr, 2003). In experimental settings next to a willingness to punish at a net cost to oneself, cooperation was observed in several one-shot, anonymous games. For instance in an anonymous dictator game where a subject is free to distribute a given amount of money between himself and an anonymous partner, subjects are willing to share though they could easily profit by keeping the money. Although the participants were fully informed that they couldn't expect a favour in return later in life, they still didn't behave selfishly (Camerer, 2003). Also, in an ultimatum game, which is a dictator game where the receiving partner has the right to veto the deal, offers were rejected if they seemed too low. Yet, in a rejected deal both parties were worse of, so self-interest seems to be sacrificed. This occurred in cultures across the planet (Henrich et al., 2005). Was this the ultimate proof that humans are 'truly' altruistic beings?

To some it is. They reinstate the old notion of group selection to explain the occurrence of 'strong reciprocity'. Yet, as was pointed out by Hamilton, as soon as non-altruists infiltrate a group of altruists, the occurrence of strong reciprocity is undermined. As migration during human evolutionary history may have disturbed any process where a group of altruists develops, genetic group selection is a rather weak evolutionary force. However cultural group-selection (or gene-culture co-evolution) may have provided solutions to this dilemma (Fehr, Fischbacher, & Gachter, 2002). When groups with differing culturally acquired norms compete, the groups endorsing altruistic norms will prevail. Particularly norms which are beneficial to the group but costly to the individual (strong reciprocity) could spread (Boyd & Richerson, 2002) . Once group-beneficial norms are culturally acquired, it is assumed these may be secured in the genetic make-up of the individuals. Thus, the generous offers and costly punishment observed in the experimental games is argued to be the result of the genetic incorporation of group-beneficial norms that originally prevailed thanks to cultural selection (Fehr & Henrich, 2003; Hagen & Hammerstein, 2006).

Yet both the occurrence of strong reciprocity and the role of group selection are disputed. It may as well be explained with theories based on the notion that selection is foremost taking place at the level of the genes and individuals. First, one may object that humans didn't evolve in a social context where anonymous one-shot encounters took place. Therefore, the cooperation seen in experimental games can be regarded as the byproduct of a mental mechanism that evolved to deal with cooperation in a social context with repeated encounters. There is a mismatch between the current experimental setting and social environment and the evolutionary background humans evolved in (Trivers, 2004). Though the subjects in the experiments are informed they won't be able to interact later on with their partners, they are unable or unwilling to switch of their intuitions to cooperate anyway.

Second, altruism among unrelated individuals may not only have evolved through reciprocal altruism, but as well by indirect reciprocity. Instead of 'I help you and you help me' this entails that 'I help you and somebody else helps me.' The evolution of cooperation by indirect reciprocity leads to reputation building (Nowak & Sigmund, 2005). With a good reputation you will have more chance of being helped later on. Therefore one is not enclined to defect or cheat as this may undermine one's status. Subtle cues can affect the generosity, for as soon as one has the impression someone's watching, it is advantageous to be seen as generous. In experimental game settings it was demonstrated that cooperation increases when visual cues of someone are present such as stylized eyespots (Haley & Fessler, 2005). With this in mind it is argued that the instances of strong reciprocity observed in experimental settings thus far may be the result of the experimenter's inability to convince the subjects they are totally anonymous as reputation effects were impossible to ignore (Hagen & Hammerstein, 2006).

Can the study of the brain help us out to settle the feud between advocates and opponents of strong reciprocity? Probably not. However, brain studies may point out which brain processes are taking place in the subject's mind and to what extent emotions are involved in the 'calculations' during experimental games, enlightening us about the proximate mechanisms involved.

Towards a Deep History of Human Morality

As fervent supporters for more cross-pollination between the neurosciences and evolutionary-inspired research, we sympathize with the popular metaphor in which the human brain is compared with a rich archaeological site full of ancient objects or an old mountain range with clearly distinguishable strata and interesting fossils. This metaphor symbolizes the belief that neurological research might be of central importance to a *deep history* of the human mind. Just as geologists uncover the history of mountains and archaeologists excavate ancient settlements by delving into ancient layers, neuroscientists are able to uncover our evolutionary history by looking into the human brain (Smail, 2008). The possibility of using the brain as a device for rebuilding our deep history had already been conceived in 1912 by the American historian James Harvey Robinson who thought that "we are now tolerably well assured that could the human mind be followed back, it would be found to merge into the animal mind, and that consequently the recently developing study of animal or comparative psychology is likely to cast a great deal of light upon certain modes of thought" (Robinson, 1912). Whereas Robinson proposed a history that paid attention to comparative psychology, contemporary neurosciences has to be credited for the recent upheaval of deep history. Due to its progress, an evolutionary narrative of the human mind is now possible.

The logic behind this comparison sounds reasonable at first glance. Just like our physical capacities evolved from evolutionary pressures, our mental powers result from adaptations to recurrent problems in our ancestral environment. Neural structures that are fossilized throughout our brains generated these mental abilities. Our brain architecture mirrors the cognitive challenges of prehistoric man and the adaptive solutions which human evolution offered to these. Inversely, we will not encounter these brain structures among species that have not faced these challenges or found divergent solutions. So, neurosciences promise us a time machine that allows us to scroll back to our earliest mental development. With the help of scanners, brain stimulations devices, single cell monitoring and sophisticated microscopy, we should be able to reconstruct the origin of uniquely human behavior, emotions and cognitive abilities.

For those interested in the evolutionary roots of morality, a territory that embodies the quintessence of human nature, this perspective is a promising one. Final answers to longstanding questions suddenly come within reach. A clear picture will form of how and when our brains developed the necessary structures underlying altruism. Neurosciences enable us to trace back the diversity of moral systems (kin selection, reciprocal altruism, indirect and strong reciprocity) to different neural patterns, and their respective origins can be dated on the evolutionary timescale. The differences between human and animal morality will become perfectly clear since neurology will provide us with detailed atlases of the moral brain modules that we share with other animals or that are exclusively human. Finally, neurology will solve the enigma of strong reciprocity visible in altruistic punishment, third-party punishment or initial benevolence during singular encounters with strangers. To date, this enduring debate was mainly fueled by theoretical and speculative arguments. Neurology might possess the key for deciphering these mysteries. Once the evolutionary history of the moral brain is reconstructed, we will know when and how humans started to act altruistically towards strangers, began to punish free-riders and cheaters without personal profit or started to show benevolence during cooperative deals. We refer to our analogy again. Just like a sought-after fossil that documents crucial steps in animal evolution, it might be assumed that our brain contains the sediments of the transition from egoism to altruism and from mutual or indirect altruism to strong reciprocity. Once these sediments are excavated, new empirical data will settle the remaining controversies surrounding the evolution of morality.

A Misleading Analogy?

Unfortunately, our analogy is somewhat misleading. In contrast to a site or cliff face, the human brain happens to be a biological or organic object, which entails the incompleteness of any strictly anatomical description, however exhaustive this picture nevertheless might be. Morphology and histology are just not enough to comprehend the developmental history of the brain. More profound knowledge of the genetic basis of these structures, patterns and mechanisms is required to help answer fundamental questions. After all, of the approximately 30,000 genes of the human genome, a substantial amount is involved in brain processes. Although our knowledge regarding the relationship between genome, brain and behavior continually increases, we are still utterly ignorant about the neurogenetic basis of human morality. Studies that shed a first light on the connection between genes and moral behavior are exceptionally rare (Knafo et al., 2008). In spite of this scarcity, the importance of a genetic perspective on the moral brain has been fully recognized. So, even if the human brain is an open book that directly shows which brain structures are essential to perform moral tasks, we need to identify the genes and proteins encoding these structures. This necessary work will take decades or longer. Impatient readers with high-running expectations are warned that this volume does not contain information approaching this gap.

In addition, the geological analogy gives the wrong impression that brains are transparent, easily accessible and well-conserved organs, just like an archaeological site barely eroded by time and decline or a cliff rising above the seashore. However, we all know that the human brain is an extremely stubborn object hampering research with its extraordinary complexity, its microscopic scale and perishable nature. These misfortunes discouraged optimistic and impatient scientists in the past and will continue to conceal essential data in the future. Our knowledge about the brains of extinct humanoid ancestors, for instance, is inherently limited. Fossilized human skulls have been measured and endocranial casts gave us a clue of the hemispheric brain surface of our ancestors, but this data provides us with only a tremendously rough picture of the brain of *Homo erectus* (1.7 million years ago) or *archaic Homo sapiens* (500,000 years ago). Our understanding of the brain of living man shows serious gaps as well. We still know very little about the detailed connectivity and functional interdependencies inside the human brain. Today, neuroscience is simply not mature enough to settle issues about how specific changes in the neuronal connections affect overall brain function (Striedter, 2005). We do not know in what respect these connections vary from those in other species and how these variations relate to functional differences. Little is known about the genesis, development and plasticity of these connections and circuits. It is widely accepted that neurons and synapses develop during periods of excessive profusion, after which they stabilize or disappear following the 'use it or lose it' principle (Oppenheim, 1985). Beginning, duration and scope of this maturation process seem to vary from brain region to brain region. Our visual cortex matures earlier than our prefrontal cortex (Huttenlocher, 2002). But how this brain maturation generates particular functions or develops in different species remains largely unknown. Even less is known about restorative anatomical processes following brain damage or spontaneous morphological modifications during learning and memory processes.

Finally, the cliff analogy falls short if one realizes that the human brain barely mirrors our mental history, at least not its most interesting epochs. Although our brain is certainly stratified just like the proverbial rock face, neuroscientists fail to discern unique human 'layers' offering valuable information about our mental abilities. All vertebrate brains are divisible into telencephalon, diencephalon, mesencephalon and rhombencephalon (Ariëns Kappers, Huber, & Crosby, 1936; Nieuwenhuys, Ten Donkelaar, & Nicholson, 1998). Each of these archetypical regions is, in turn, composed of two or more highly conserved major divisions. The vertebrates' telencephalon encloses the limbic system (amygdala, cingulate gyrus) and the basal ganglia (nucleus caudatum, putamen, nucleus accumbens, globus pallidus). We share these brain structures with all vertebrates, such as reptiles, fishes, birds and amphibians. In mammals, the dorsal part of the pallium, a compartment of the telencephalon that is visible in amphibians as well, gave rise to the development of the neocortex that typifies mammalian brains. However, in mammals, obvious structural brain differences cannot easily be discerned. Rats and cats have prefrontal cortices as well. If we compare our human brain with that of other primates, differences become extremely hard to find. All Brodmann areas, even the most sophisticated located in our prefrontal cortex, are also present in the brains of macaques (Petrides & Pandya, 1994). Our DLPFC is not exclusively human, even though this celebrated brain area performs our highest cognitive tasks.

If any differences do exist, they are far outweighed by the overwhelming similarities between brains of primates (Lynch & Granger, 2008). This observation, which echoes Richard Owen, Alfred Russel Wallace and Alfred Vulpian in the 19th century, thwarts the prospect for building a historical narrative of the moral brain. It even questions the fertility of a cross-pollination between evolutionary theory and neurosciences, which is the core theme of this volume. Since our morality evidently differs from that of rats, macaques or chimpanzees, but no neuroanatomical variations can be detected to explain these differences, neurosciences might teach us nothing about the origin and evolution of human morality. Although brain imaging might be interesting in mapping active brain regions among humans, it provides no information for answering evolutionary issues. All brain regions so frequently mentioned in neuropsychological papers are present in the brains of primates, mammals or even vertebrates that fail to perform cognitive tasks administered to human participants. This observation might force us to conclude that human morality is driven by neural structures that we share with a multitude of other animals. All mammals make use of the same hardware, but humans utilize this universal machinery apparently for distinct behavioral goals. Internal differences are merely functional, not structural. Consequently, we better drop our geological metaphor and return to the classic computer image. To some, human morality does not seem more than a piece of software or a computer game run by our neural PlayStation.

The Importance of Brain Imaging

In our view, this drastic conclusion is unwarranted. In the first place, human uniqueness in moral issues is overemphasized. Morality is certainly not privileged to humans. Mammals and humans share a strikingly large amount of behavior that can be considered 'moral', although major differences cannot be ignored. If we broaden the definition of morality in conformity with current approaches in moral psychology, the fixed boundaries between animal and human morality promptly cease. Leaving rational ethics aside, which is obviously human, the following moral abilities are documented among other mammals: emotional contagion, helping behavior, attachment, care, violence inhibition, incest avoidance, reciprocal altruism, concern for reputation, inequality aversion, altruistic punishment, etc. Just like humans, animals possess a multitude of adapted mechanisms to help relatives or that are dedicated to start and maintain cooperation. Even 'immoral behavior' is not uncommon among animals. Examples of plain warfare, mutilation and rape are well reported. If we wish to obtain a better understanding of how morality evolved, we must not remain stuck on the most recent strata of the mammalian or primate brain, in which separated structures are indeed exceedingly thorny to uncover. Ancient strata common to all mammals might be equally informative to the evolution of morality. Given that homologous brain structures largely process equal functions across species, comparative brain research might reopen the possibility of a deep history of shared moral powers. By pinpointing the brain structures that are essential to moral tasks in man and animals alike, or, inversely, are absent in species that lack these moral competencies, we regain the prospect to reconstruct the roots of moral propensities.

Brain imaging research (fMRI, PET, DTI) has to play a key role in this project. Humans are, by large, the most prominent species to participate in such in vivo research. During scanning experiments, we exemplify the neural circuits that we share with other animals. For once, we seem to be the guinea pigs that provide knowledge about the neural correlates of the animal psyche. Human participants grant the possibility to learn more about the neural underpinning of reciprocal altruism in chimpanzees or inequality aversion in capuchin monkeys. Brain mapping across humans will put forward candidates for interspecies neural networks underlying shared moral functioning. Once these networks are accurately mapped, we might successfully start reconstructing the evolutionary roots of these moral abilities.

Of course, not all expressions of human morality are already present in other mammals or primates. Whether theory of mind (ToM), which is an indispensable prerequisite for genuine empathic concern and complex social interaction, is present in humanoid primates is still a matter of controversy (Call & Tomasello, 2008; Focquaert & Platek, 2007). Less controversial is the apparent absence of sociomoral disgust or third-party punishment in non-human animals, which can be considered uniquely human. Furthermore, frequently occurring moral phenomena among humans, such as altruistic punishment or even reciprocal altruism, are rarely observed in non-human primates, presumably because their occurrence requires a number of conditions that are difficult to fulfill. Although a definitive demarcation of human versus animal morality has not been settled yet, major differences in moral behavior, emotion and judgment are beyond dispute.

So, how can we explain these differences if brain variation between species is minimal or even nonexistent? Our simple answer is that we cannot. Since essential deviations in behavior must parallel neural differences at some level, the similarity thesis that holds a complete likeness between human and non-human brains must be flawed somewhere. Admittedly, this thesis historically aided arguing the evolutionary origins of man, as Darwin already advanced in his *Descent of Man* (1871), but boomeranged as soon as this line of reasoning ruled out biological explanations of typical human capacities such as morality or intelligence. From old, neurologists attempted to escape this unfavorable consequence and continued to propose unique neurological features that might account for human characteristics. As already noted, Benedikt regarded the voluminous occipital lobes as uniquely human, and Vogt stressed the specific histological structure of the *lamina pyramidalis*. Both unsuccessful proposals were intended to accommodate morality in a brain part that deviated from that in primates.

Currently, better candidates are being proposed. We already referred to the spindle cells, or Von Economo neurons, in the frontal insular and anterior cingulated cortex of primates and large cetaceans. In humans these cells are more abundant and their number continues to increase throughout early life. Other experts rediscovered the potential significance of our large brain. Human brains deviate from the relative or expected brain size more than any other primate. Our brains are about 2.3 times larger than would be expected in other primates. This observation continues feeding speculations about an increased cortical connectivity and our astonishing pallet of behavioral options (Lynch & Granger, 2008). Still, other neurologists stress the large size of the lateral prefrontal cortex in humans that enfolds our DLPFC, a much-cited region in moral brain studies. Expansion of this brain area might explain why humans exert more control over other brain regions than other primates do (Striedter, 2005).

This is not the place to discuss all candidates for unique human brain features which are associable with higher mental powers. We have mentioned some of them merely to show that the similarity thesis is currently under attack. Innovative microscopic and brain imaging technologies have multiplied the number of plausible candidates that might discern human from non-human brains. Modern genetics reveal distinctive genes and proteins that could be linked to dissimilarities in brain structure and mental functioning. The recovered belief in the existence of neurological dissimilarities between humans and non-humans offers an opportunity for a more integrative collaboration between neurosciences and evolutionary psychology.

Again, brain scanning studies might be of great interest in unraveling possibly dissimilar features. Although their resolution is certainly too limited to reveal microscopic or chemical variations, scanners yield valuable information about neural circuits and brain parts involved in typically human activities. This information might stimulate researchers with diverse backgrounds to look for deeper deviating structures or patterns in the neural circuits exposed by fMRI or DTI research. Conversely, evolutionary-inspired studies dealing with morality might encourage future neurological research. Certainly, these studies will further illuminate which moral capacities are exclusively human and which are shared with other animals. They will go on discussing our uniqueness from the psychological side. Yet equally important, by developing clever experiments, evolutionary psychology strives to discern moral intuitions that are molded by adaptive selection from patterns of moral behavior that are merely the outcome of cultural processes. The first group of moral expressions is likely to be developed into firmly fixed circuits in our brains. Evolutionary psychology might be of great help to define these deeply rooted moral intuitions which should be excellent candidates for successful neural localization.

This volume mirrors the belief that brain research is useful for a deep history of our mental lives and sets the stage for a cross-pollination between evolutionary and neurological insights toward morality. For centuries, the moral sense was regarded as a chalice of divine beauty awarded to the human species as a token of god's special consideration. Morality was the playground for philosophers, impenetrable to the natural sciences. Today, both evolutionary sciences and cognitive neuroscience are closing in on human morality. Evidently, both disciplines take different perspectives. Cognitive neuroscience focuses on the neurological causes of behavior, the mental machinery that enables us to behave as we do. Niko Tinbergen calls this type of explanation the proximate causes of behavior. Yet, to give a full explanation one has to expand the explanation by investigating the question of why this machinery is as it is. Why has evolution favored certain types of mental mechanisms? This type of ultimate question is answered in an evolutionary approach.

We consider both types of explanations to be complimentary. They are inseparable companions in exposing the chalice that was guarded so long by some philosopher-guards. Just as geologists and archaeologists go hand in hand when they are uncovering the roots of vanished societies, evolutionary-inspired researchers and neurological experts will profit from more collaboration as well. However, not all contributors in this volume share our optimism. As the reader will notice, some authors, certainly overwhelmed by the premature nature of our current knowledge, hesitate to speculate about this integration and have restricted their contributions to empirical data that does not go beyond their particular discipline. Other contributors take this volume as an opportunity to express some of their ideas touching the integrative purpose of this book. Both data and ideas are intended to encourage new scientific research at the crossroad of both disciplines.

Plan of the Book

This book is a greenhouse. As the authors explore the evolution and neurology of the moral sense from different operating bases, this confrontation may allow for cross-pollination. By stitching together the different paradigms, the patchwork of morality can be revealed. In this overview of the book chapters, the new research approaches that this cross-pollination has already yielded are pointed out. The first six chapters are written from a cognitive neuroscience perspective, whereas the latter three take an evolutionary approach. A final chapter compares both schools of thought.

Studying the blind may help us to understand vision. Similarly, when **Andrea Glenn and Adrian Raine** study subjects diagnosed with psychopathy or antisocial personality disorder showing serious disruption of their moral sense, they focus on the core of the moral brain. For indeed, the subjects show malfunctions in several structures that are commonly implicated in moral decision-making, including the VMPFC, MPFC, the amygdala and the angular gyrus. This confirms that these areas might be key regions involved in morality. In their contribution, Glenn and Raine elaborate on the neurological and evolutionary aspects of antisocial behavior. In accordance with Linda Mealy (1995), they hypothesize that immoral behavior is not necessarily a dysfunction from an evolutionary perspective but may rather be adaptive. This is demonstrated in a prisoner's dilemma wherein cooperation and non-cooperation are modeled. Cheating flourishes as long as there are enough cooperators. But the prisoner's dilemma helps to uncover neurological aspects of cooperation. Hereby, the authors demonstrate that a game theoretic device that was used originally in evolutionary studies proves to be a powerful experimental tool that even helps to uncover the neurological aspects of cooperation. The use of cooperation games in fMRI research is a fine illustration of the crossover between cognitive and evolutionary sciences that has been taking place over these last years.

What sets us apart from other animals? This question surfaces in many of the contributions to the book. **Jorge Moll and Ricardo de Oliveira-Sousa** remark that it is human's ability to attach themselves to symbols. Human society, they note, is impregnated by abstract objects, such as norms and ideologies. But how is this capacity to die for a country, fight for symbols or cooperate with group members fastened in our brains? In their essay, the authors propose that an extended form of attachment is an important ingredient. The ancient mechanism supporting basic forms of attachment in other species may have evolved to enable the uniquely human ability to attach to cultural objects. To prove their thesis, the authors survey evidence from comparative and human neuroanatomical data, social psychology and cognitive neuroscience. For instance, they show that the brain regions involved in attachment are activated when one is motivated to help or cooperate. Thus, the authors explain how an ancient mechanism evolved to settle child-mother attachment may have gathered new functions to play a crucial role in our moral minds.

However, 'in some cases an ideology or sense of duty may be at odds with a feeling of attachment, in some instances even suppressing empathy (e.g., a soldier that inflicts pain to other humans on behalf of his commitment to his country). But a mechanism of extended attachment (and guilt avoidance) may well be at work for someone to feel committed to a given moral duty (e.g., the soldier feels attached to his country and to the values associated with it),' they observe. Perhaps extended attachment plays a role in the endorsement of moral rules focused on care, whereas it is absent in situations where one appeals to duty (principled moralities)? Indeed, morality seems to be a flag that covers a mixed cargo, ranging over care, reciprocity or disgust-related moral rules. The authors' hunch is that extended attachment may be involved in the endorsement of many different values and ideologies, not only those that are focused on care.

There may be something wrong with Moll's frameworks, Blair remarks in an addendum (this volume). Blair charges his colleague with holding a unitary position towards the moral brain. This means that we would have only one system that processes all social rule-related information. 'His criticisms simply do not do apply to our research,' Moll objects. The seemingly unitary definition of morality Moll and Oliveira-Souza started with was only an operational definition when they were the first to explore the neural correlates of morality in 2001, he explains. Further he does not object that our moral sense may be the result of different interacting subsystems. Just as visual perception relying on a well specified brain region does not imply that all visual experience (such as motion, color, space) relies on 'exactly' the same neural substrates, neither does pointing out a stable neural architecture in diverse aspects of moral cognition lead one to assert a unitary view on the phenomenology or neural architecture of morality, Moll observes.

However, this metaphor about the visual system leads Blair to assert that there really is a difference in opinion between them. Moll remarks that the visual system is *one* system. The visual stimuli are processed by its subcomponents. In Blair's multiple moralities view, however, 'there are different classes of stimuli that are processed by largely independent systems (to continue the metaphor, a visual and an auditory system).' Moll's suggestion that 'extended attachment' may be involved in the attachment to a wide variety of different symbols, therefore fits easier in the view that one common element could underlie many different aspects of morality, whereas Blair asserts the contrary. We should conversely note that Moll never speaks of extended attachment as a single mechanism underlying morality but as one among many others.

James Blair further elaborates on the 'multiple moralities' approach, the view that there are multiple systems which mediate largely separable forms of social rule-related processing that are lumped together within a culture's concept of morality. Blair actively seeks the confrontation between a neuroscience and evolutionary perspective as he confronts his opinions with the anthropologist Jonathan Haidt's, who studies human morality from an evolutionary perspective. Blair suggests the existence of at least four partially separable neurocognitive systems that mediate different aspects of social rule-related processing which can be considered moral: care-based, reciprocity, social convention and disgust-based. A central role in these semi-independent systems is played by stimulus-reinforcement learning. The observer's sensibility for the pain of others, for instance, is what allows him to learn the care-based moral rules. In comparison to Haidt, Blair puts more emphasis on these emotional learning mechanisms and further points at the role of non-affect-based reasoning in moral judgment.

Jean Decety and **Daniel Batson** investigate the relation between empathy and morality. Rudimentary forms of empathy are present in other mammals and apes. However, human empathy is unique as it is a much more layered construct that not only relies on mirroring the emotions of others but is further characterized by the realization that the feelings experienced are not one's own but rather someone else's. The authors argue that this may induce altruistic motivation. Evidence is presented from neuroimaging studies that the awareness of agency – a third person perspective – may lead to empathic concern – a caring reaction towards a person in need – whereas a first person perspective is associated much more so with stress and fear and selfish reactions. Furthermore, they argue that, in order to respond to people in need, the sensation of empathy can be regulated and modulated in differing contexts. However, being empathic does not necessarily lead to being moral, the authors conclude. For instance, helping a sad infant because one feels empathy for him, while one does not help a child that is worse off at the other end of the planet may not be the most moral thing to do. Though human empathy may lead to a reaction of concern, this does not mean that the reaction is morally correct. Thus, the authors point out that empathy forms a building block of our moral sense provided by evolution, that contributes yet does not suffice in itself to make a 'moral person'.

In the last decades, a so-called affective revolution swept through behavioral and cognitive sciences: the role played by emotions in the guidance of human behavior became widely acknowledged. In the study of moral psychology, a similar shift from cognition to affect took place. However, the pendulum may have swung too far; cognitive processes at the reasoning end of the spectrum may be undervalued. This is a point that **Kristin Prehn and Hauke Heekeren** make in their chapter. While it is true that several experiments have provided strong evidence for the impact of emotions on moral judgment, others, however, have shown that emotions are less important, at least during simple and unambiguous moral decision-making tasks. The authors point at the fact that further studies even indicated that emotional responses should be suppressed or regulated when utilitarian moral judgments are required. To integrate the competing models and conflicting empirical results, Prehn and Heekeren present a functional approach, a working model of moral judgment that takes into account many different factors like situational context, emotional and cognitive processes as well as individual differences in information processing.

So, just as Blair, who points at the contribution of non-affect-based reasoning in moral judgment processes, and Decety and Batson, who stress that altruistic motivation is more than emotional contagion alone, Prehn and Heekeren tap into a similar vein when they point at the role of reasoning processes in moral decision-making. They present morality as consisting of two inseparable, yet distinguishable aspects: a person's moral orientations and principles, and a person's competence to act accordingly. In contrast to Blair, who emphasizes the different independent emotion-based learning mechanisms, Prehn and Heekeren focus on the difference between on one hand, the lower affective responses and, on the other, the higher cognitive processes canalizing the affect.

With the drive of a practicing neurosurgeon, **Dirk De Ridder** and colleagues take the discussion a level further and investigate potential treatments of persons diagnosed with psychopathy or pedophilia. As our moral sense is based on the functioning of our brain, implantation of electrodes may modulate these capacities and help those who have difficulty incorporating moral and social rules. Though it sounds like science fiction, it may not be as fanciful as we imagine. To demonstrate this, the authors give an overview of the brain processes involved in moral judgment and point out the different stages where a neurosurgeon could interfere. A crucial element is the observation that 'moral decision making could be considered as the result of positive or negative reinforcement.' Therefore, 'morally "good" or "bad" could reflect the hedonic response of the opiate part of the mesolimbic reward system to internally imitated social stimuli.' Influencing this reward system may, in the end, influence behavior and help people with psychopathy or pedophilia, for 'neuromodulation can change the rewards the individuals receive when performing antisocial behavior.' Different techniques such as TMS or implanted electrodes may induce (temporary) changes in brain functioning and, thus, symptoms. The authors demonstrate how and where we could interfere to change behavior for the better. Do scenes of *A Clockwork Orange* loom ahead? Perhaps, but knowledge that similar interventions are within the bounds of possibility can only help open the discussion.

In contrast to that of the cognitive neuroscientists introduced above, the prime intention of evolutionary scientists is to give answers to the ultimate question of why morality evolved as it did. What are the evolutionary rationales, invisible to the individual, that explain why and how cooperation and reciprocity could evolve?

For long, it seemed from an evolutionary perspective contradictory that the process of evolution might yield self-sacrificing virtues and motivations. The aim of the contributions of evolutionary-inspired scientists is, in the first place, to demonstrate how this apparent barrier was overcome. The roots of morality are, at least in part, seen as solutions to a problem our ancestors faced: cooperation, and by extension, social interaction and all the problems involved. **Matthijs van Veelen** discusses a wide range of explanations that have been offered thus far, for not only kin selection, reciprocal altruism or sexual selection, but also different forms of group selection have been called upon to explain altruism. Mathematical models, Van Veelen observes, may allow us to test different hypotheses. However, it seems that *different* evolutionary processes may have contributed to the creation of our moral senses. Just as one of the mouth's original functions was the intake of food and only later it developed as a speech organ, adaptations that originally promoted kin altruism may have evolved over time. An adaptation to help kin may count as a trait signaling one's mating value. This may have led to the exceptional forms of self-sacrifice we observe in modern humans, he states.

Randolph Nesse makes a similar investigation and suggests that a runaway social selection process is essential in shaping the building blocks of our moral sense, such as empathy, self-esteem, guilt, anger and tendencies to display moral traits and to judge others. 'Social selection,' he argues, 'is the subtype of natural selection that results from the social behaviors of other individuals. Competition to be chosen as a social partner can, like competition to be chosen as a mate, result in runaway selection that shapes extreme traits.' This means that displaying moral traits is critical for survival, as it is crucial to be selected as a social partner. Nesse suggests that the capacity for commitment and altruism may be the result of a runaway process. Just as the peacock would not find any mates if its tail were absent, people will not find social partners if they lack these moral building blocks, as only those who possess a higher level of these traits are chosen. In other words: 'If others view you as moral, you will thrive in the bosom of a human group. If, however, others view you as immoral, you are in deep trouble.'

Though evolution may have provided us with a toolbox of instruments that guide us along the dangers and temptations of social life and deliver mutual benefits to cooperators, these instruments may fail once the group size grows too big. 'The larger and more complex a society becomes, the greater the temptation to defect from social cooperation, and the greater the chance of doing so successfully', the philosopher **John Teehan** observes. Religion, he analyzes, may form the vehicle extending the reach of our inborn moral intuitions in large societies. Of course, religion provides 'a sense of community and a code of in-group behavior' that creates an atmosphere wherein altruism and commitment to group members is stimulated. For instance, 'showing oneself to be a member of a religion by having mastered the rituals and practices of that religion signals that one has already made a significant contribution of time and energy to the group and a willingness to follow the code that governs the group. That is, it signals that one is a reliable partner in social interactions and can be trusted to reciprocate.' Thus, Teehan offers a perspective on religion that makes sense in an evolutionary framework.

This book is a composite of contributions from scientists with either a cognitive neuroscientific or an evolutionary psychological background. It is never easy to bring those who are raised separately under the same roof, for prejudices may undermine cooperation. Besides, the danger exists that differences in opinion may be coated with a layer of varnish that suggests unanimity but actually hides the state of affairs. It is central, then, to uncover the cracks that lurk behind the stucco, for only knowledge of these differences will help us to blend them into a new approach. Therefore, in a final chapter, **Jelle De Schrijver** investigates if and how both scientific schools can be reconciled with regard to the notion of moral modularity. Moral modularity entails that different more or less independent brain systems contribute to our moral sense. Whereas one module may focus on care-related phenomena such as the suffering of others, the disgust of, for instance, incest may be the outcome of another module. The author observes that in cognitive neuroscience a stronger emphasis is placed on the importance of learning processes and the role of higher cognition, whereas in evolutionary psychology the focus lies primarily on the innateness of the modules. 'Whereas some evolutionary psychologists suggest that most of the preferences, motivations and values are determined by evolutionary forces, the cognitive neuroscientist's view entails the occurrence of several more or less independent systems that "learn" a quite diverse set of motivations,' De Schrijver states. He concludes that at least part of the discrepancies disappear when one takes into account the different perspectives – either zoomed in or zoomed out.

Conclusion

The philosopher Joshua Greene observed that the clinical case of 'abasketballia' does not occur (Greene, 2005), there are no patients who are selectively bereft of their ability to play basketball. After all, a large range of independent brain processes are involved in catching and throwing the basketball. A stroke that wipes out some or all of these capacities to play the game will necessarily affect the capacities to walk, dance or juggle. In contrast, the neural mechanisms of visual cognition may be more tightly constrained, as Bill Casebeer observes (Casebeer, 2003). Certain brain hemorrhages cause specific deactivations of processes involved in visual perception; some patients fail to see movement, others lose their ability to see color. Vision is, therefore, an easier target for neurobiological research. Studying the neural correlates of basketball would be harder as it would mix up a whole range of independent processes that are hard to fixate. Besides, playing basketball is culturally contingent. There is ample evidence of cultures where basketball has not been 'invented', whereas there are no cultures where people are not able to see, as vision is part of the general makeup of human nature. This is not surprising, since vision is the result of a long process of Darwinian selection, whereas this singling out process may have been absent for playing basketball. In this sense, visual perception is a more robust, *natural kind* than the game of basketball. We may actually regard vision, on one hand, and playing basketball, on the other, as two poles in a spectrum, dividing natural kind from cultural invention.

The case of (acquired) psychopathy seems to demonstrate that a stroke may deprive one of the moral sense, the capacity to respond appropriately in the social environment. However, this interpretation is incorrect, Greene observes. Unlike those with disorders in visual perception, persons with psychopathy display a disturbance that is not restricted to only a specific domain. For instance, psychopathy leads to failing decision-making in risky environments. There is not a single central brain system involved in moral cognition. Moreover, similar to basketball, studying morality unveils a whole range of processes that are at least partially independent, such as the emotional learning mechanisms introduced by Blair. However, despite variations in social conventions and their particular emphasis, different human cultures display common moral rules, such as those governing parental care, hierarchical relations or justice. In the dipole of vision and basketball, the human moral sense seems to float in between.

What we understand as morality is not a *natural kind,* but it is instead the result of the interaction between different independent neural processes and social contexts. It is a patchwork of semi-independent processes, the union of independent domains, which as a result of different adaptive pressures - or even as different solutions to a similar problem: cooperation in group as is pointed out by the evolutionary inspired scientists in this book – resulted in different emotional learning mechanisms.

This patchwork is uncovered from different perspectives, either focusing on proximate or ultimate origins of the process. New theories, traditions or techniques allow for the unraveling of the characteristics of the moral sense. As a greenhouse, this book confronts the different approaches, and we hope for cross-pollination. Do we see archaeologists or geologists unveiling the patchwork on the dipole between vision and basketball? Both or perhaps neither. Metaphors may help to enlighten an impenetrable topic, but they may blur the narrative as well. As a start, this book is a meeting point between both cognitive neuroscientists and evolutionary scientists. It is time to for them to take stage.

References

Adolphs, R., Tranel, D., Damasio, H., & Damasio, A. R. (1994). Impaired recognition of emotion in facial expressions following bilateral damage to the human amygdala. *Nature, 372*, 669–672.

Adolphs, R., Gosselin, F., Buchanan, T. W., Tranel, D., Schyns, P., & Damasio A. R. (2005). A mechanism for impaired fear recognition after amygdala damage. *Nature, 433*, 68–72.

Ariëns Kappers, C. U., Huber, C. G., & Crosby, E. C. (1936). *Comparative anatomy of the nervous system of vertebrates, including man*. New York: Macmillan.

Berthoz, S., Armony, J. L., Blair, R. J. R., & Dolan, R. J. (2002). An fMRI study of intentional and unintentional (embarrassing) violations of social norms. *Brain, 125*(8), 1696–1708.

Birbaumer, N., Veit, R., & Lotze, M. (2005). Deficient fear conditioning in psychopathy: A functional magnetic resonance imaging study. *Archives of General Psychiatry, 62*, 799–805.

Blair, R. J. R. (1995). A cognitive developmental approach to morality – investigating the psychopath. *Cognition, 57*, 1–29.

Boyd, R., & Richerson, P. J. (2002). Group beneficial norms can spread rapidly in a structured population. *Journal of Theoretical Biology, 215*(3), 287–296.

Calder, A. J., Lawrence, A. D., & Young, A. W. (2001). Neuropsychology of fear and loathing. *Nature Reviews Neuroscience, 2*(5), 352–363.

Call, J., & Tomasello, M. (2008). Does the chimpanzee have a theory of mind. 30 years later. *Trends in Cognitive Science, 12*(5), 187–192.

Camerer, C. (2003). *Behavioral game theory: Experiments in strategic interaction*. Princeton, NJ; Woodstock, UK: Princeton University Press.

Casebeer, W. D. (2003). Moral cognition and its neural constituents. *Nature Reviews Neuroscience, 4*(10), 840–846.
Cooter, R. (1984). *The cultural meaning of popular science: Phrenology and the organization of consent in nineteenth-century Britain*. Cambridge: Cambridge University Press.
Damasio, A. R. (1994). *Descartes' error. Emotion, reason and the human brain*. New York: Putnam.
Damasio, A. R. (2005). Brain trust. *Nature, 435*, 571–572.
Darwin, C. (1859). *The origin of species*. London: John Murray.
Darwin, C. (1871). *The descent of man and selection in relation to sex*. London: John Murray.
Decety, J., & Lamm, C. (2006). Human empathy through the lens of social neuroscience. *The Scientific World Journal, 6*, 1146–1163.
Domes, G., Heinrichs, M., Michel, A., Berger, C., & Herpertz, S. C. (2007). Oxytocin improves mind-reading in humans. *Biological Psychiatry, 61*(6), 731–733.
Eisenberger, N., Lieberman, M. D., & Williams, K. (2003). Does rejection hurt? An fMRI study of social exclusion. *Science, 302*, 290–292.
Fehr, E., Fischbacher, U., & Gachter, S. (2002). Strong reciprocity, human cooperation, and the enforcement of social norms. *Human Nature-an Interdisciplinary Biosocial Perspective, 13*(1), 1–25.
Fehr, E., & Henrich, J. (2003). Is strong reciprocity a maladaptation? On the evolutionary foundations of human altruism. In P. Hammerstein (Ed.), *Genetic and cultural evolution of cooperation* (pp. 55–82). Cambridge, MA: MIT Press in Cooperation with Dahlem University Press.
Focquaert, F., & Platek, S. M. (2007). Social cognition and the evolution of self-awareness. In S. M. Platek, J. P. Keenan, & T. K. Shackelford (Eds.), *Evolutionary cognitive neuroscience* (pp. 457–497). Cambridge, MA: MIT Press.
Gintis, H., Bowles, S., Boyd, R., & Fehr, E. (2003). Explaining altruistic behavior in humans. *Evolution and Human Behavior, 24*(3), 153–172.
Greene, J. D. (2005). Cognitive neuroscience and the structure of the moral mind. In P. Carruthers, S. Laurence, & S. Stich (Eds.), The *innate mind: Structure and contents* (pp. 338–353). Oxford: Oxford University Press.
Greene, J. D., Sommerville, R. B., Nystrom, L. E., Darley, J. M., & Cohen, J. D. (2001). An fMRI investigation of emotional engagement in moral judgment. *Science, 293*, 2105–2108.
Greene, J. D., Nystrom, L. E., Engell, A. D., Darley, J. M., & Cohen, J. D. (2004). The neural base of cognitive conflict and control in moral judgment. *Neuron, 44*, 389–400.
Hagen, E. H., & Hammerstein, P. (2006). Game theory and human evolution: A critique of some recent interpretations of experimental games. *Theoretical Population Biology, 69*(3), 339–348.
Hagner, M. (2004). *Geniale Gehirne. Zur Geschichte der Elitegehirneforschung*. Göttingen: Wallstein.
Haidt, J. (2003). The moral emotions. In R. J. Davidson, K. R. Scherer, & H. H. Goldsmith (Eds.), *Handbook of affective sciences* (pp. 852–870). Oxford: Oxford University Press.
Haley, K. J., & Fessler, D. M. T. (2005). Nobody's watching? Subtle cues affect generosity in an anonymous economic game. *Evolution and Human Behavior, 26*(3), 245–256.
Harbaugh, W. T., Mayr, U., & Burghart, D. R. (2007). Neural responses to taxation and voluntary giving reveal motives for charitable donations. *Science, 316*, 1622–1625.
Hastings, M. E., Tangney, J. P., & Stuewig, J. (2008). Psychopathy and identification of facial expressions of emotion. *Personality and Individual Differences, 44*, 1474–1483.
Heekeren, H. R., Wartenburger, I., Schmidt, H., Schwintowski, H. P., & Villringer, A. (2003). An fMRI study of simple ethical decision-making. *Neuroreport, 14*(9), 1215–1219.
Heining, M.,Young, A. W., Ioannou, G., Andrew, C. M., Brammer, M. J., Gray, J. A., et al. (2003). Disgusting smells activate human anterior insula and ventral striatum. *Annals of the New York Academy of Sciences, 1000*, 380–384.

Henrich, J., Boyd, R., Bowles, S., Camerer, C., Fehr, E., Gintis, H., et al. (2005). "Economic man" in cross-cultural perspective: Behavioral experiments in 15 small-scale societies. *Behavioral and Brain Sciences, 28*(6), 795–855.

Huttenlocher, P. R. (2002). *Neural plasticity: The effects of environment on the development of the cerebral cortex*. Harvard: Harvard University Press.

Knafo, A., Israel, S., Darvasi, A., Bachner-Melman, R., Uzefovsky, F., Cohen, L., et al. (2008). Individual differences in allocation of funds in the dictator game associated with length of the arginine vasopressin 1a receptor RS3 promoter region and correlation between RS3 length and hippocampal mRNA. *Genes, Brain and Behavior, 7*, 266–275.

Knoch, D., Pascual-Leone, A., Meyer, K., Treyer, V., & Fehr, E. (2006). Diminishing reciprocal fairness by disrupting the right prefrontal cortex. *Science, 314*, 829–832.

Kosfeld, M., Heinrichs, M., Zak, P. J., Fischbacher, U., & Fehr, E.. (2005). Oxytocin increases trust in humans. *Nature, 435*, 673–676.

Krolak-Salmon, P., Henaff, M. A., Isnard, J., Tallon-Baudry, C., Guenot, M., Vighetto, A., et al. (2003). An attention modulated response to disgust in human ventral anterior insula. *Annals of Neurology, 53*, 446–453.

Lynch, G., & Granger, R. (2008). *Big brains. The origins and future of human intelligence*. New York: Palgrave Macmillan.

Marsh, A. A., & Blair, R. J. R. (2008). Deficits in facial affect recognition among antisocial populations: A meta-analysis. *Neuroscience & Biobehavioral Reviews, 32*, 454–465.

Mealy, L. (1995). The sociobiology of sociopathy: an integrated evolutionary model. *Behavioral and Brain Sciences, 18*, 523–599.

Mitchell, J. P., Macrae, C. N., & Banaji, M. R. (2006). Dissociable medial prefrontal contributions to judgments of similar and dissimilar others. *Neuron, 50*, 655–663.

Moll, J., Oliveira-Souza, R., Bramati, I. E., & Grafman, J. (2002a). Functional networks in emotional moral and nonmoral social judgments. *NeuroImage, 16*, 696–703.

Moll, J., Oliveira-Souza, R., Eslinger, P. J., Bramati, I. E., Mourao-Miranda, J., Andreiuolo, P. A., et al. (2002b). The neural correlates of moral sensitivity: A functional magnetic resonance imaging investigation of basic and moral emotions. *Journal of Neuroscience, 22*, 2730–2736.

Moll, J., Oliveira-Souza, R., Moll, F. T., Ignacio, F. A., Bramati, I. E., Caparelli Daquer, E. M., et al. (2005). The moral affiliations of disgust: A functional MRI study. *Cognitive and Behavioral Neurology, 18*, 68–78.

Nieuwenhuys, R., Ten Donkelaar, H. J., & Nicholson, C. (1998). *The central nervous system of vertebrates*. Berlin: Springer Verlag.

Nowak, M. A., & Sigmund, K. (2005). Evolution of indirect reciprocity. *Nature, 437*(7063), 1291–1298.

Oppenheim, R. W. (1985). Naturally occurring cell death during neural development. *Trends in neuroscience, 8*, 487–493.

Patrick, C. J., Bradley, M. M., & Lang, P. J. (1993). Emotion in the criminal psychopath: Startle reflex modulation. *Journal of Abnormal Psychology, 102*, 82–92.

Penfield, W., & Faulk, M. E. (1955). The insula: Further observations on its function. *Brain, 78*, 445–470.

Petrides, M., & Pandya, D. N. (1994). Comparative architectonic analysis of the human and macaque frontal cortex. In F. Boller & B. Grafman (Eds.), *Handbook of neuropsychology* (Vol. 9, pp. 17–58). Elsevier: Amsterdam.

Renneville, M. (1999). *La médecine du crime: essai sur l'émergence d'un regard médical sur la criminalité en France (1785-1885)*. Villeneuf d'Ascq: Septentrion.

Richter, J. (2007). *Pantheon of brains*: The Moscow Brain Research Institute 1925-1936. *Journal of the History of Neurosciences, 16*, 138–149.

Rilling, J. K., Gutman, D. A., Zeh, T. R., Pagnoni, G., Berns, G. S., & Kilts, C. D. (2002). A neural basis for social cooperation, *Neuron, 35*(2), 395–405.

Robertson, D., Snarey, J., Ousley, O., Harenski, K., DuBois Bowman F., Gilkey, R., et al. (2007). The neural processing of moral sensitivity to issues of justice and care. *Neuropsychologia, 45*, 755–766.

Robinson, J. H. (1912). *The new history: Essays illustrating the modern historical outlook*. New York: Macmillan.

Royet, J. P., Plailly, J., Delon-Martin, C., Kareken, D. A., & Segebarth, C. (2003). fMRI of emotional responses to odors: Influence of hedonic valence and judgment, handedness, and gender. *NeuroImage, 20*, 713–728.

Rozin, P., Haidt, J., & McCauley, C. R. (1993). Disgust. In M. Lewis & J. Haviland (Eds.), *Handbook of emotions* (pp. 575–594). New York: Guilford Press.

Sakuta, A., & Fukushima A. (1998). A study of abnormal findings pertaining to brain in criminals. *International Medical Journal, 5*, 283–292.

Sanfey, A. G., Rilling, J. K., Aronson, J. A., Nystrom, L. E., & Cohen, J. D. (2003). The neural basis of economic decision-making in the ultimatum game. *Science, 300*(5626), 1755–1758.

Schaich Borg, J., Lieberman, D., & Kiehl, K. A. (2008). Infection, incest, and iniquity: investigating the neural correlates of disgust and morality. *Journal of Cognitive Neuroscience, 20*(9), 1529–1546.

Seeley, W. W., Carlin, D. A., Allman, J. M., Macedo, M. N., Bush, C., Miller, B. L., et al. (2006). Early frontotemporal dementia targets neurons unique to apes and humans. *Annals of Neurology, 60*, 660–667.

Shin, L. M., Dougherty, D. D., Orr, S. P., Pitman, R. K., Lasko, M., Macklin, M. L., et al. (2000). Activation of anterior Paraimbic structures during guilt-related script-driven imagery. *Biological Psychiatry, 48*(1), 43–50.

Singer, T., Seymour, B., O'Doherty, J., Kaube, H., Dolan, R. J., & Frith, C. D. (2004). Empathy for pain involves the affective but not sensory components of pain. *Science, 303*, 1157–1162.

Singer, T., Seymour, B., O'Doherty, J., Stephan, K. E., Dolan, R. J., & Frith, C. D. (2006). Empathic neural responses are modulated by the perceived fairness of others. *Nature, 439*, 466–469.

Smail, D. L. (2008). *On deep history and the brain*. Berkeley: University of California Press.

Small, D. M., Gregory, M. D., Mak, Y. E., Gitelman, D., Mesulam, M. M., & Parrish, T. (2003). Dissociation of neural representation of intensity and affective valuation in human gustation. *Neuron, 39*, 701–711.

Striedter, G. F. (2005). *Principles of brain evolution*. Sunderland: Sinauer.

Takahashi, H., Yahata, N., Koeda, M., Matsuda, T., Asai, K., & Okubo, Y. (2004). Brain activation associated with evaluative processes of guilt and embarrassment: an fMRI study. *NeuroImage, 23*(3), 967–974.

Trivers, R. (2004). Genetic and cultural evolution of cooperation. *Science, 304*(5673), 964–965.

Uddin, L. Q., Molnar-Szakacs, I., Zaidel, E., & Iacoboni, M. (2006). rTMS to the right inferior parietal lobule disrupts self-other discrimination. *Social Cognitive and Affective Neuroscience, 1*, 65–71.

Verplaetse, J. (2004). Moritz Benedikt's (1835–1920) localization of morality in the occipital lobes: origin and background of a controversial hypothesis. *History of Psychiatry, 15*, 305–328.

Verplaetse, J. (2006). *Het morele brein. Een geschiedenis over de plaats van de moraal in onze hersenen*. Antwerpen/Apeldoorn: Garant.

Verplaetse, J. (2009). *Localizing the moral sense. Neuroscience and the search for the cerebral seat of morality, 1800–1930*. New York & Berlin: Springer.

Wynne-Edwards, V. C. (1962). *Animal dispersion in relation to social behaviour*. London: Oliver & Boyd.

Yang, Y., Raine, A., Lencz, T., Bihrle, S., LaCasse, L., & Colletti, P. (2005). Volume reduction in prefrontal gray matter in unsuccesful criminal psychopaths. *Biological Psychiatry, 57*, 1103–1108.

Zald, D. H., & Pardo, J. V. (2000). Functional neuroimaging of the olfactory system in humans. *International Journal of Psychophysiology, 36*(2), 165–181.

Zald, D. H., Lee, J. T., Fluegel, K. W., & Pardo, J. V. (1998). Aversive gustatory stimulation activates limbic circuits in humans. *Brain, 121*, 1143–1154.

The Immoral Brain

Andrea L. Glenn and Adrian Raine

It has been argued that both moral and immoral behavior is the product of evolution, and that there is a neurobiological basis to both. The concept of morality may have evolved to facilitate the adaptive, mutually beneficial strategy of cooperation between individuals. Brain imaging studies are beginning to uncover the brain regions involved in moral decision making. The understanding of morality has also been enhanced by the study of individuals who primarily engage in immoral or antisocial behavior. It has been argued that psychopathy, the personification of immoral behavior, is actually an alternative evolutionary strategy; when few in number, psychopaths may successfully use deception, violence, and "cheating" to obtain resources and maximize reproductive fitness (Barr & Quinsey, 2004; Crawford & Salmon, 2002; Mealey, 1995; Raine, 1993). Imaging studies of antisocial individuals have revealed differences in brain structure and functioning (Birbaumer et al., 2005; Kiehl, 2006; Yang et al., 2005), suggesting there may be neural underpinnings to this alternative strategy. Several of the brain regions found to function differently in antisocial individuals correspond to those implicated in moral decision-making; these regions may be key to understanding both moral and immoral behavior.

In this chapter, an overview of the evolution and neurobiology of morality is first presented, followed by an overview of the evolution and neurobiology of immoral behavior, with specific reference to psychopathy. A cognitive neuroscience approach is applied to both, emphasizing the functional significance of the regions implicated in moral and immoral behavior. The Prisoner's Dilemma is used as a model of both the evolutionary and neurobiological explanations. Neurobiological mechanisms associated with the evolution of deception and, conversely, the evolution of mechanisms associated with detecting and punishing cheating behavior are also discussed. Finally, the implications of this research for the newly emerging field of neuroethics are presented.

A.L. Glenn (✉)
Department of Psychology, University of Southern California, Los Angeles, CA
e-mail: aglenn@sas.upenn.edu

J. Verplaetse et al. (eds.), *The Moral Brain*, DOI 10.1007/978-1-4020-6287-2_2,

The Evolution of Moral Behavior

Morality has been described as the product of evolutionary pressures on social animals to promote reciprocal altruism (Moll, Oliveira-Souza, & Eslinger, 2003). Reciprocal altruism is the act of spending time, energy, and resources to help others who are genetically unrelated to us. Altruistic acts of kindness and generosity towards friends, acquaintances, and even strangers occur frequently in our society. We often make sacrifices and put ourselves at risk to help others, inhibiting our selfish tendencies to focus primarily on the survival and success of ourselves and our families. We learn to set aside our short-term impulses for the greater benefits of cooperation in the long term. By performing altruistic acts, we expect that others will cooperate and return the favor at some point in the future. In this sense, we have evolved into a highly interdependent and cooperative society. In an environment of limited resources in which we interact repeatedly with others, the strategy of reciprocal altruism appears to be the most beneficial in the long run. Through development, humans transition from a selfish focus on short-term rewards, to engaging in generous, cooperative, and unselfish behavior, thus becoming sophisticated, civilized, and what we refer to as "moral" individuals.

It is likely that many of our moral values are shrouded in a deep evolutionary history in which emotions constituted the driving force of moral behavior. Social emotions have evolved as a mechanism for maintaining the cooperative, altruistic behavior which is an adaptive strategy for maximizing one's reproductive fitness (Trivers, 1971). When individuals cooperate, they are rewarded by the feelings of friendship, liking, and camaraderie. These positive emotions are rewarding and serve to reinforce cooperative, altruistic behavior, as well as to motivate us to continue such behavior in the future. Negative social emotions have evolved to deter us from engaging in selfish, non-cooperative behavior. Feelings of guilt, shame, and remorse serve to punish violations of social-norms and to steer us away from such acts. Thus, both positive and negative social emotions help to guide our behavior in what we consider to be a morally appropriate way.

The Prisoner's Dilemma Model of Social Interactions

Game theory models attempt to provide a simplified example of human social interactions, both cooperative and non-cooperative. The Prisoner's Dilemma (PD) is a case in which two men are held on suspicion of robbing a bank and murdering a bank employee. They are separated for questioning and cannot communicate with each other. The authorities need a confession from at least one of the men in order to convict them. If neither confesses, they will both be charged of a lesser crime and get 5 years in prison. If one testifies against the other, while the other remains silent, the testifier walks out a free man, while the one who remained silent gets 30 years in prison. If both testify against the other, each gets 10 years in prison because they both testified. Ideally, both would remain silent. However, they are

		PLAYER A	
		Coop	Defect
PLAYER B	Coop	$2(2)	$3(0)
	Defect	$0(3)	$1(1)

Fig. 1 Payoff matrix for the four outcomes in the Prisoner's Dilemma Game. Player A's choices (Coop or Defect) are listed atop columns and Player B's choices (Coop or Defect) are listed aside rows. Dollar amounts in bold are awarded to player A. Amounts in parentheses are awarded to player B (Rilling et al., 2007)

separated and both are worried that the other will testify against them. The dilemma is whether to remain silent and get either 5 years or 30 years, or to testify against the partner and get 0 or 10 years. The best overall outcome of the dilemma is if both stay silent; however, the rational choice for each man would be to cheat and testify against the partner. The Prisoner's Dilemma models a one-time social interaction and demonstrates that cooperation involves risking one's own well-being to trust another individual, despite the fact that cheating is the most rational strategy. In the real world, however, interactions between individuals are typically not one-time occurrences; individuals interact repeatedly and are aware that present decisions to cooperate or cheat will be remembered by others in the future. Thus, the most beneficial strategy in the long run is to establish a reputation as one who cooperates by repeatedly seeking out and engaging in mutually cooperative interactions.

An iterated version of the Prisoner's Dilemma has been implemented in several studies to model the repeated social interactions of individuals in the real world. Researchers have adjusted the PD scenario in various ways, including replacing prison sentences with monetary values (see Fig. 1) for implementation in laboratory settings.

The Prisoner's Dilemma also provides a model of the previously discussed social emotions which guide cooperative behavior. Nesse (1990) describes the reported emotions associated with different outcomes of the Prisoner's Dilemma: when both partners cooperate (and both make $2), they experience friendship, love, trust, or obligation; when both defect (both only make $1), they feel rejection and hatred; when one cooperates and the other defects, the cooperator (who receives $0) feels anger and the defector (who receives $3) feels anxiety and guilt. In both the Prisoner's Dilemma and the real world, cooperation is generally viewed as the morally appropriate choice, while defecting and taking advantage of another individual for your own benefit is considered selfish and immoral. When individuals interact repeatedly, it appears that mutual cooperation is most beneficial to the individual in the long run. Social emotions may be advantageous because they facilitate such cooperative behavior. As will be outlined in the following section, the experience of social emotions is deeply rooted in the functioning of certain brain regions.

The Neurobiology of Morality and Social Emotions

The evolution of moral behavior has shaped the functioning of different neural structures (Moll et al., 2003). Brain imaging studies have used the Prisoner's Dilemma and other tasks to elicit social emotions such as guilt and camaraderie, which serve to guide moral behavior. Mutual cooperation during the Prisoner's Dilemma has been associated with activation in the orbitofrontal cortex and the ventral striatum (Rilling et al., 2002), both of which are thought to be part of the brain's reward system (Schulz, 1998). Guilt has been associated with activation in the medial prefrontal cortex and the angular (posterior superior temporal) gyrus (Takahashi et al., 2004). The ventrolateral and dorsomedial prefrontal cortices have been shown to play a role in response reversal (Rolls, Hornak, Wade, & McGrath, 1994) that in a particular social context may facilitate a change in behavior in response to aversive social cues such as the observation of a moral transgression (Finger, Marsh, Kamel, Mitchell, & Blair, 2006). For example, in the Prisoner's Dilemma, these areas may be involved in the decision to discontinue cooperation in the event that a partner defects.

Moral decision-making is a complex process that involves not only the experience of social emotions, but many other aspects such as theory of mind (in both a cognitive and affective sense), self-reflection, and the cognitive appraisal of emotion. Thus, a wide variety of regions have been implicated in studies of moral judgment. The most commonly identified regions appear to be the angular gyrus, posterior cingulate, amygdala, and medial and ventral regions of the prefrontal cortex (for reviews see Greene & Haidt, 2002; Moll et al., 2003; Raine & Yang, 2006).

The medial prefrontal / frontal polar region has been implicated in numerous moral judgment studies. The involvement of the medial prefrontal cortex is not too surprising, as it appears to be involved in the processing of social and emotional stimuli, as well as information about the self and others (Ochsner et al., 2004). The medial prefrontal cortex has been found to play a role in self-reflection (Larden, Melin, Holst, & Langstrom, 2006), guilt and embarrassment (Takahashi et al., 2004), and the cognitive appraisal of emotion (Ochsner, Bunge, Gross, & Gabrieli, 2002), all of which are important in moral decision-making. The medial prefrontal / frontopolar cortex has been implicated in moral tasks including the passive viewing of pictures depicting moral versus nonmoral violations (Harenski & Hamann, 2006; Moll, Oliveira-Sousa, Bramati, & Grafman, 2002b), passive viewing of morally disgusting versus nonmorally disgusting statements (Moll et al., 2005), making judgments about auditory moral versus nonmoral sentences (Oliveira-Sousa & Moll, 2000), moral decision-making versus semantic decision-making (Heekeren, Wartenburger, Schmidt, Schwintowski, & Villringer, 2003), judgment on moral versus nonmoral actions (Borg, Hynes, Van Horn, Grafton, & Sinnott-Armstrong, 2006) sensitivity to moral versus nonmoral issues (Robertson et al., 2007), difficult versus easy moral dilemmas, personal versus impersonal moral dilemmas, and utilitarian moral decision-making (e.g. sacrificing life for the greater good) versus "nonutilitarian" decision-making (e.g. prohibiting a loss of life even though more lives could be saved (Greene, Nystrom, Engell, Darley, & Cohen, 2004). It has been hypothesized that the medial prefrontal cortex is important in moral judgment because it may be involved in processing the emotional and social component of moral stimuli, and assessing the perspectives of the self and others.

The ventral prefrontal cortex, which includes the orbitofrontal cortex, ventrolateral prefrontal cortex, and gyrus rectus, has been found during passive viewing of pictures of moral versus nonmoral violations (Moll, Oliveira-Sousa, et al., 2002b), responding to unpleasant moral versus unpleasant nonmoral statements (Moll, Oliveira-Sousa, Bramati, & Grafman, 2002a), passive viewing of morally disgusting versus nonmorally disgusting statements (Moll et al., 2005), judgments on moral versus nonmoral actions (Borg et al., 2006), automatic moral judgment to high versus low immoral stimuli (Luo et al., 2006), moral versus semantic decision-making (Heekeren et al., 2005, 2003), reading about moral transgressions (Finger et al., 2006), and social situational versus rule-based moral issues (Robertson et al., 2007).The ventral prefrontal cortex is highly involved in decision-making (Bechara, 2004), emotion regulation (Ochsner et al., 2005), and the affective component of theory of mind (Shamay-Tsoory, Tomer, Berger, Goldsher, & Aharon-Peretz, 2005). In moral decision-making, the ventral prefrontal cortex may be important in integrating moral knowledge with emotional cues, understanding the emotional states of others, and inhibiting antisocial impulses.

The angular gyrus (also referred to as the posterior superior temporal gyrus) is also commonly activated in moral judgment tasks, including the passive viewing of pictures depicting moral versus nonmoral violations (Harenski & Hamann, 2006; Moll, Oliveira-Sousa, et al., 2002), making judgments on auditory moral versus nonmoral sentences (Oliveira-Sousa & Moll, 2000), moral versus semantic decision-making (Heekeren et al., 2005, 2003), personal versus impersonal moral dilemmas (Greene et al., 2004; Greene, Sommerville, Nystrom, Darley, & Cohen, 2001), judgments on moral versus nonmoral actions (Borg et al., 2006), difficult versus easy moral judgments (Greene et al., 2004), and sensitivity to moral versus nonmoral issues (Robertson et al., 2007). Activation in an adjacent region of the superior temporal sulcus was observed by Moll et al. (2002a) when responding to unpleasant moral versus unpleasant non-moral statements. It has been suggested that the angular gyrus is important in complex social cognition and linking emotional experiences to moral appraisals (Moll, Oliveira-Sousa, et al., 2002b).

The posterior cingulate has been activated during personal versus impersonal moral dilemmas (Greene et al., 2001), moral versus semantic decision-making (Heekeren et al., 2005), passive viewing of pictures depicting moral versus nonmoral violations (Harenski & Hamann, 2006; Moll, Oliveira-Sousa, et al., 2002b), difficult versus easy moral decisions, utilitarian versus non-utilitarian moral decisions (Greene et al., 2004), and sensitivity to moral versus nonmoral issues (Robertson et al., 2007). The posterior cingulate may be important in the recall of emotional memories (Maratos, Dolan, Morris, Henson, & Rugg, 2001), the experience of emotion (Mayberg et al., 1999), and self-referencing (Johnson et al., 2006).

The amygdala is thought to be important in aversive conditioning, and in associating the pain and distress in others to one's own behavior. It has been found to be active in several tasks involving moral judgments (Berthoz, Grezes, Armony, Passingham, & Dolan, 2006; Greene et al., 2004; Harenski & Hamann, 2006; Luo et al., 2006; Moll, Oliveira-Sousa, et al., 2002b).

Some areas have been activated in only a few studies, so replication is necessary before firmer conclusions can be reached regarding the involvement of these areas in moral judgment. These areas include the temporal pole (Heekeren et al., 2005, 2003; Moll, Oliveira-Sousa, et al., 2002a; Oliveira-Sousa & Moll, 2000), the anterior cingulate (Berthoz et al., 2006; Greene et al., 2004), and the insula (Greene et al., 2004; Moll, Oliveira-Sousa, et al., 2002b).

Overall, moral decision-making is a complex process that involves decoding of information, integration of emotion, attachment of moral relevance, and cognitive reappraisal. Each process may involve multiple structures, and abnormalities in different brain regions may lead to different types of impairments in moral judgments and behavior.

The Evolution of Immoral Behavior

The rate at which individuals engage in morally appropriate, cooperative behavior versus cheating varies from person to person and is also influenced by situational factors. Most individuals engage in occasional "cheating" behavior, but for the most part, our society operates under the assumption that individuals will cooperate and abide by moral standards. We have developed a highly interconnected societal structure with countless exchanges of goods and services. Each exchange is a social interaction, as represented by one round of the Prisoner's Dilemma. We place a great amount of trust in other individuals, assuming that exchanges are fair and that others will not try to take advantage of us. However, there are a small number of individuals who engage in especially high rates of cheating behavior. Psychopathy has been associated with high rates of recurrent and severe antisocial behavior (Hare, 2003). As early as the 1800s, the nature of psychopaths was described as "moral insanity," (Prichard, 1835) as they seemed to lack concern for moral issues, despite knowing the difference between right and wrong.

Some have argued that psychopaths are individuals who pursue an evolutionarily stable strategy consisting primarily of "cheating" behaviors (Barr & Quinsey, 2004; Crawford & Salmon, 2002; Mealey, 1995; Raine, 1993). In this view, the behavioral, emotional, cognitive, and neuropsychological features of psychopaths are seen as organized, specified mechanisms which facilitated a viable reproductive social strategy during human evolutionary history (Crawford & Salmon, 2002). Psychopathic traits such as risk taking, aggressiveness, social manipulativeness, and early and high mating effort involving short-term, uncommitted relationships with multiple partners all may serve to maximize reproductive fitness. Under certain conditions, these individuals can pursue a life history strategy of manipulative and deceptive social interactions to gain advantage. By using charm and charisma to persuade others, mimicking social emotions, developing elaborate schemes, or moving frequently from place to place and person to person, they can cheat their way through life, taking advantage of the generally trusting nature of others and avoiding retribution. In most human societies (see Harpending & Draper, 1988 for exception), cheaters tend to exist at relatively low frequencies because, as we see in the Prisoner's Dilemma, cheaters only benefit from interactions with cooperators. Too many cheaters in a society will lead to "cheater-cheater" interactions (i.e. $1 / $1 interactions versus $3 / $0 interactions), which are less beneficial to the cheater. At low frequencies, psychopaths may be able to successfully maintain cheating as an evolutionarily adaptive strategy. Harpending and Sobus (1987) used game theory research to show that cheaters can achieve reproductive success when they are difficult to detect, are highly mobile, are verbally facile, and are skilled at persuading females to mate.

Antisocial behaviors such as lying, theft, burglary, and rape are all means by which psychopaths may cheat, taking advantage of others in order to gain status, resources, and to pass on genes with minimal effort (Raine, 1993). As evidence that psychopaths may pursue a life history strategy of high mating effort and a lack of parental investment, psychopathy has been associated with an increased number of sexual partners (Halpern, Campbell, Agnew, Thompson, & Udry, 2002; Lalumiere & Quinsey, 1996), engaging in sexual behavior at an earlier age (Harris, Rice, Hilton, Lalumiere, & Quinsey, 2007), an uncommitted approach to mating, increased mating effort and sexual coersion (Lalumiere & Quinsey, 1996), many short marital relationships, sexual promiscuity (Hare, 2003), and poor performance as parents (Cleckley, 1976). In modern societies, to gain access to resources and gain social status, psychopaths may engage in white collar crimes such as fraud, embezzlement, and insider trading, taking advantage of the trusting nature of others in business settings which rely heavily on the cooperation and integrity of employees.

An essential component to successfully pursuing a strategy that primarily involves cheating behavior would be the lack of a core moral sense (Raine & Yang, 2006). Reduced moral emotional responsivity may allow psychopaths to easily engage in antisocial behaviors with little fear or concern for potential consequences. They may be less hindered by emotions and thus able to employ more rational strategies that maximize their short-term benefits, as they are unconcerned with the pain or distress that their actions cause in others. They may not be deterred by the discomforting social emotions of guilt, shame, and remorse. Without the experience of the social emotions that guide morally appropriate decisions, and the lack of emotion necessary to experience empathy, psychopaths may easily engage in immoral, antisocial behavior (Mealey, 1995).

An essential aspect to the theory that cheating may be an alternative evolutionary strategy is that there must be a genetic basis. Individuals must genetically pass on antisocial traits, including lack of emotional responsiveness thought to inhibit immoral behavior, which are adaptive in pursuing a life history strategy of cheating behavior. Research suggests that genetic components may account for as much as half of the population variability in antisocial behavior (Coccaro, Kavoussi, &

McNamee, 2000). This may indicate that some or all of the alleles underlying antisocial behavior evolved because they conferred a fitness advantage, although the advantage may only exist in situations where it occurs at low frequencies in the population (Barr & Quinsey, 2004). The genotype for antisocial, cheating behavior may provide a fitness advantage to individuals living in a population primarily consisting of cooperators.

It is important to note that the theory of psychopathy as an evolutionary life strategy does not rule out the possibility that some individuals may develop psychopathic traits or persistent antisocial behavior by way of immediate environmental factors, including brain damage or childhood abuse. Another point, one that has been noted by Crawford and Salmon (2002), is that it is not necessary that a behavior be currently adaptive in order to implement the theory of evolution by natural selection to help in understanding it. Certain genes may be retained by natural selection because their effects were beneficial in ancestral societies. In modern society, psychopathic individuals often end up incarcerated, in severe financial debt, and may be more likely to incur personal injuries or death due to extreme risk-taking or due to aggressive interactions with others. Conversely, it has been suggested that the psychopathic life history strategy may actually be becoming increasingly adaptive; in our highly mobile, technologically advanced societies, psychopaths may gain advantages that they may not have had in smaller face-to-face populations (Crawford & Salmon, 2002).

In summary, persistent immoral behavior can be thought of as an alternative evolutionary strategy that can be beneficial at low rates in society. By lacking the emotional experiences that serve to deter immoral behavior, and by using deception and manipulation, individuals may be able to successfully cheat their way through life.

The Neurobiology of Immoral / Antisocial Behavior

Brain imaging research and lesion studies have provided a wealth of information regarding the functional neural correlates of antisocial behavior. Immoral behavior is a prominent feature of psychopathy and antisocial personality disorder. Brain imaging studies in these individuals have revealed differential functioning in several brain regions, some of which may be associated with immoral behavior (Table 1). Psychopathy is a complex construct comprised of 20 core features (Hare, 1991); in addition to immoral behavior, psychopaths also demonstrate other characteristics such as impulsivity, poor planning, superficial charm, and grandiosity, which are constructs that may or may not overlap with immoral behavior. Thus, we would not expect for the brain regions implicated in psychopathy to exactly match those implicated in moral judgment. Similarly, there are aspects of moral judgment, such as self reflection or information processing that may not be impaired in psychopaths. It appears, however, that some overlap does exist between brain regions implicated in moral judgment and regions impaired in psychopaths. We would hypothesize that

Table 1 Overview of major brain regions implicated in moral judgments and in psychopathy / antisocial personality

Brain region	Suggested functions	Moral judgment studies	Findings of reduced structure / functioning in APD / psychopathy
Medial / ventral prefrontal cortex	• Processing social and emotional stimuli • Self reflection • Guilt and embarrassment • Cognitive appraisal of emotion • Emotion regulation • Theory of mind (affective component) • Shifting behavior when rewards change	Harenski and Hamann (2006) Moll et al. (2002b) Heekeren et al. (2003) Borg et al. (2006) Robertson et al. (2007) Greene et al. (2004) Moll et al. (2005) Luo et al. (2006) Heekeren et al. (2005) Finger et al. (2006)	Laakso et al. (2002) Birbaumer et al. (2005) Viet et al. (2002) Völlm et al. (2004) Rilling et al. (2007)
Angular gyrus	• Complex social emotion • Linking emotional experiences to moral appraisals	Harenski and Hamann (2006) Moll et al. (2002a) Oliveira-Sousa and Moll (2000) Heekeren et al. (2005, 2003) Greene et al. (2004, 2001) Borg et al. (2006) Robertson et al. (2007)	Kiehl et al. (2004) Raine et al. (1997) Soderstrom et al. (2000)
Posterior cingulate	• Recalling emotional memories • Experiencing emotion • Self-referencing	Greene et al. (2001) Heekeren et al. (2005) Harenski and Hamann (2006) Moll et al. (2002b) Greene et al. (2004) Robertson et al. (2007)	Kiehl et al. (2001)

Table 1 (continued)

Brain region	Suggested functions	Moral judgment studies	Findings of reduced structure / functioning in APD / psychopathy
Amygdala	• Aversive conditioning • Associating pain of others to one's own actions • Enhancing attention to emotional stimuli	Berthoz et al. (2006) Greene et al. (2004) Harenski and Hamann (2006) Luo et al. (2006) Moll et al. (2002b)	Tiihonen et al. (2000) Wong et al. (1997) Raine et al. (1997) Kiehl et al. (2001) Birbaumer et al. (2005) Viet et al. (2002) Sterzer et al. (2005) Rilling et al. (2007)

these regions might be mostly involved in the emotional aspect of moral judgment, but future research is necessary to elucidate this.

In examining the brain regions found to function differently in psychopathy and antisocial behavior it is important to consider both the regions implicated, and their functional significance in order to understand how each region contributes to characteristics and behaviors of psychopathy. Numerous brain regions have been implicated in antisocial behavior. Each region may contribute in a unique way to increase the probability of antisocial behavior, or may be a part of a larger circuit that contributes more generally. Researchers have proposed that key regions such as the amygdala and the orbitofrontal cortex may be more significant in the development of psychopathy, but individual lesions to either of these areas do not replicate the disorder completely (Blair, 2005). While the amygdala and orbitofrontal cortex likely contribute significantly to the development of antisocial behavior, research in antisocial and psychopathic individuals demonstrates differential functioning in other regions as well; thus, it is important to consider and explore the potential contributions of these additional regions through examination of their functions and connectivity with other regions.

Brain Imaging Studies. Some of the brain regions associated with moral judgment have been found to function differently in psychopaths and antisocial individuals, and thus may contribute to the development of immoral behavior. These regions include the medial and ventral prefrontal cortex, the amygdala, the angular gyrus, and the anterior and posterior cingulate. In the prefrontal cortex, reduced prefrontal glucose metabolism has been observed in murderers compared with normal controls (Raine et al., 1994), and reduced blood flow has been correlated with increased antisocial and aggressive behaviors (Gerra et al., 1998; Kuruoglu, Arikan, Vural, & Karatas, 1996; Oder et al., 1992; Soderstrom et al., 2002; Soderstrom, Tullberg, Wikkelso, Ekholm, & Forsman, 2000). A reduction in prefrontal gray matter volume has been observed in antisocial and psychopathic adults (Raine, Lencz, Bihrle, LaCasse, & Colletti, 2000; Yang et al., 2005). Two studies have shown reduced gray matter volumes specifically in the orbitofrontal cortex in antisocial individuals (Laakso et al., 2002; Yang, Raine, & Narr, 2006). Functional imaging studies have observed reduced activity in psychopaths in the orbitofrontal cortex during fear conditioning (Birbaumer et al., 2005; Viet et al., 2002) and in antisocial individuals during inhibitory control (Völlm et al., 2004). The orbitofrontal cortex is thought to play a role in shifting behavior when reward contingencies change.

The amygdala is a region thought to be essential for moral socialization (Blair, 2006). The amygdala is necessary for the formation of stimulus-reinforcement associations which allow an individual to learn to associate their harmful actions with the pain and distress of others. The amygdala is also necessary for aversive conditioning and for enhancing attention to emotional stimuli, which facilitates empathy for victims (Blair, 2006). Reduced volume of the amygdala has been reported in two studies of violent offenders (Tiihonen, Hodgins, & Vaurio, 2000; Wong, Lumsden, Fenton, & Fenwick, 1997) and one study of psychopathic individuals (Yang, Raine, Narr, Lencz, & Toga, 2006). Functional asymmetries of the amygdala have been observed in murderers (Raine, Buchsbaum, & Lacasse, 1997), showing reduced left and increased right amygdala activity. In fMRI studies, reduced activity in the amygdala during the processing of emotional stimuli has been observed in criminal psychopaths (Kiehl et al., 2001), psychopathic individuals (Birbaumer et al., 2005; Viet et al., 2002) and in adolescents with conduct disorder (Sterzer, Stadler, Krebs, Kleinschmidt, & Poustka, 2005). However, two studies have reported *increased* amygdala activation in antisocial individuals while viewing negative visual content (Müller et al., 2003) and during aversive conditioning (Schneider et al., 2000).

In the adjacent region, reduced volume of the temporal lobe has been found in several antisocial groups, including children with conduct disorder (Kruesi, Casanova, Mannheim, & Johnson-Bilder, 2004), incarcerated psychopaths (Dolan, Deakin, Roberts, & Anderson, 2002), and antisocial personality-disordered individuals (Barkataki, Kumari, Das, Taylor, & Sharma, 2006). Functional impairments in the temporal lobe have been shown in aggressive patients (Amen, Stubblefield, Carmichael, & Thisted, 1996; Volkow & Tancredi, 1987) in aggressive children with epilepsy (Juhasz, Behen, Muzik, Chugani, & Chugani, 2001), in antisocial personality disordered patients (Goethals et al., 2005) and violent offenders (Raine et al., 2001). Reduced blood flow in the temporal cortex has also been correlated with psychopathy (Soderstrom et al., 2002). Reduced metabolism (Wong et al., 1997) and reduced blood flow (Hirono, Mega, Dinov, Mishkin, & Cummings, 2000) in the anterior inferior temporal cortex have been observed in violent patients. Reduced glucose metabolism has also been found in the medial temporal cortex in psychiatric patients with a history of violent behavior (Volkow et al., 1995) and in violent offenders (Seidenwurm, Pounds, Globus, & Valk, 1997). In impulsive violent criminals (Soderstrom et al., 2000) found reduced blood flow in the medial temporal cortex. However, one study (Raine et al., 1997) failed to detect differences in temporal lobe glucose metabolism in murderers.

Deficits in the angular gyrus (posterior superior temporal gyrus) have been found in psychopathic and antisocial individuals during a semantic processing task (Kiehl et al., 2004). Raine et al. (1997) also found reduced glucose metabolism in the left angular gyrus in murderers, while (Soderstrom et al., 2000) found reduced blood flow in the right angular gyrus in impulsive, violent criminals. Reduced functioning of the posterior cingulate, which may be involved in self-referencing and experiencing emotion, has been observed in an fMRI study of psychopaths (Kiehl et al., 2001).

There are additional areas that have been implicated in antisocial behavior that have not frequently been associated with moral judgment. For example, reduced gray matter volumes in the dorsolateral prefrontal cortex has been observed in alcoholics with antisocial personalities (Laakso et al., 2002), and reduced metabolism has been found in aggressive patients (Hirono et al., 2000) and aggressive children with epilepsy (Juhasz et al., 2001). Abnormal dorsolateral prefrontal cortex functioning has been observed in two fMRI studies (Schneider et al., 2000; Völlm et al., 2004). The functional integrity of the hippocampus has been found to be abnormal in murderers (Raine et al., 1998), in criminal psychopaths (Kiehl et al., 2001), and in violent offenders (Soderstrom et al., 2000).

Research has shown that some of the individual differences observed in brain structure in humans have a genetic basis (Thompson et al., 2001). We hypothesize that some of the individual differences in the structure and functioning of the moral neural circuit may be due to genetic influences, thus representing the genetic basis to psychopathy as an evolutionary stable strategy. Future research will be needed to elucidate whether this is the case.

*Lesion Studies.*While we argue that psychopathy may be a product of evolution, certain psychopathic features, including immoral behavior, can also be observed in individuals who have incurred brain damage. By studying the effects of impaired functioning in particular brain regions, we can gain a clearer understanding of the role of those regions in both moral and immoral behavior. Lesion studies have shown that early damage to the polar and ventromedial cortex produces personality traits and behaviors strikingly similar to those observed in psychopathy (Anderson, Bechara, Damasio, Tranel, & Damasio, 1999). Two patients with early childhood damage to these areas demonstrated deficient moral reasoning, as measured by the Standard Issue Moral Judgment (SIMJ) task. Both individuals displayed significant rule breaking, lying, impulsivity, failure to hold jobs, failure to plan for the future or form goals, and financial irresponsibly. They were described as lacking empathy, guilt, remorse, and fear, and were unconcerned with their behavioral transgressions. Both demonstrated early and irresponsible sexual behavior, and both had children which they subsequently failed to care for properly (Anderson et al., 1999). Such a demonstration of pervasive disregard for social and moral standards closely resembles that of psychopaths, and suggests that the polar and ventromedial prefrontal cortex likely play a significant role in the development and maintenance of morally appropriate behavior.

Individuals who incur brain damage as adults also show disturbances in moral behavior and decision-making (Damasio, 1994). In moral decisions involving highly conflicting considerations of aggregate welfare versus emotionally aversive behaviors (i.e. smothering one's baby to save a group of people), patients with ventromedial prefrontal cortex damage demonstrate an abnormally utilitarian pattern of judgments compared to controls (Koenigs et al., 2007). It is suggested that the ventromedial prefrontal cortex is crucial for the generation of emotional responsiveness to the aversive acts. While the patients show intact explicit knowledge of social and moral norms, their moral decisions are not affected by the emotional component to the same degree as normal controls.

Immoral behavior has also been observed in individuals with frontotemporal dementia (FTD) (Mendez, 2006), a neurodegenerative disorder that affects the frontal lobes, temporal lobes, or both. The transgression of social norms is a core feature of FTD; patients have engaged in such behaviors as stealing, shoplifting, inappropriate sexual behavior, physical violence, and financial irresponsibility (Mendez, Chen, Shapira, & Miller, 2005).

By studying individuals who develop psychopathic characteristics due to brain damage, we begin to see that the functioning of certain brain structures, particularly prefrontal and temporal regions, is essential for morally appropriate behavior.

Neurobiology of Psychopathy in the Prisoner's Dilemma

The neurobiological basis of psychopathy and immoral behavior can also be examined in the context of the Prisoner's Dilemma model. A recent fMRI study examined the relationship between psychopathy and brain activity during an iterated version of the Prisoner's Dilemma, using the monetary values from Fig. 1 (Rilling et al., 2007). Subjects were scanned while playing the game via computer with a human confederate outside of the scanner. Subjects scoring higher on self-report psychopathy measures (Levenson, Kiehl, & Fitzpatrick, 1995; Lilienfeld & Andrews, 1996) defected more frequently and were less likely to continue cooperating after establishing mutual cooperation with their partner. This could be viewed as evidence of instrumental (unprovoked) aggression. Brain imaging results revealed weaker amygdala activation following situations in which the subject cooperated, but their partner defected, suggesting they may lack the appropriate neural response necessary to learn to avoid such disadvantageous outcomes. When cooperating, subjects scoring higher in psychopathy showed weaker orbitofrontal cortex activation, suggesting that they may be less rewarded by cooperation. This study was conducted on an unselected undergraduate population, and therefore shows that even small individual differences in psychopathy can impact behavior and brain activation in the Prisoner's Dilemma. Results may be even more pronounced in samples of clinically diagnosed psychopaths.

The Neurobiology of Deception

While the majority of studies have observed reduced functioning in particular areas in immoral individuals, it also appears that some psychopaths may have superior functioning in areas that may be necessary for adaptive strategies. It has been argued that psychopaths have evolved advanced mechanisms for taking advantage of others in society (Mealey, 1995). An example of such a mechanism is deception, or lying. There is neurobiological evidence that antisocial individuals who pathologically lie show different brain structure and functioning. A recent study by Yang et al. (2005) found a 22–26% increase in white matter volumes in the prefrontal cortex in pathological liars when compared to antisocial and normal controls. This suggests that these individuals may actually have *increased* cognitive capacity for lying. Lying can be a complex behavior which requires several tasks to be managed at one time, including suppressing the truth, remembering details of the lie and information that the individual being lied to knows, continually assessing the believability of the lie, and modifying behavior in response to the reaction of the receiver. Increased prefrontal white matter may be indicative of increased connectivity between regions necessary to manage the complexity of lying, and thus may be an adaptation that has evolved in some antisocial individuals.

Functional imaging studies of deception have revealed that when individuals lie, compared to when they tell the truth, *increased* activation is observed in the prefrontal cortex (Kozel, Padgett, & George, 2004; Langleben et al., 2005), anterior cingulate (Nunez, Casey, Egner, Hare, & Hirsch, 2005; Spence et al., 2001), temporal cortex (Lee et al., 2002; Mohamed et al., 2006), and supramarginal gyrus (Lee et al., 2005; Phan et al., 2005). It might be the case that individuals who have become skilled at lying, conning, and deceiving others actually show superior functioning in some of these areas.

Cheater Detection

In a world where a certain percentage of the population engages in high rates of cheating behavior, it is adaptive for those who typically cooperate to develop mechanisms for detecting and avoiding interactions with cheaters. We have evolved to sense very subtle signals of insincerity and untrustworthiness. In a recent study, Verplaetse, Vanneste, and Braeckman (2007) found that participants were able to predict whether another individual would cooperate or cheat based on a single, still-shot photograph of the individual during the moment of decision-making. This suggests that very small changes in facial expressions can serve as cues to alert us of potential cheaters. However, our ability to detect cheaters is not 100% accurate. In some instances in which encountering cheaters could be particularly harmful, we may adopt additional strategies for protecting our security. For example, businesses sometimes administer integrity tests to potential employees in attempt to detect cheaters before damage is done. Interestingly, scores on several such integrity tests have been shown to correlate negatively with psychopathy scores (i.e. high integrity associated with low psychopathy) (Connelly, Lilienfeld, & Schmeelk, 2006).

The ability to recognize potential cheaters also appears to have a neurobiological basis. The amygdala, which is involved in the fundamental processing of stimuli which is threatening to an individual, is also involved in making judgments about the trustworthiness of people. Lesion studies have shown that patients with bilateral amygdala damage have difficulty discriminating between trustworthy and untrustworthy-looking faces, and tend to judge people as more trustworthy and approachable than normal controls (Adolphs, Tranel, & Damasio, 1998). In addition, functional imaging studies have shown that the amygdala, as well as the superior temporal sulcus, orbitofrontal cortex, and the insula are active when viewing an untrustworthy face (Winston, Strange, O'Doherty, & Dolan, 2002). This suggests that the identification of cheaters in our society has become an integral part of the threat-response system in the brain.

Upon the identification of cheaters, another subset of social emotions may have evolved to deter us from future interactions with these individuals. Emotions such as indignation, contempt, outrage, and disgust can drive us to engage in acts of moralistic aggression and vengeance. We punish cheaters by threatening, harming, or exiling them in attempt to deter them from violating social norms in the future. A recent brain imaging study has shown that punishing cheaters is rewarding, as seen though activation in the dorsal striatum, an area often implicated in the processing of reward (Quervain et al., 2004). Using the Prisoner's Dilemma, Singer, et al. (2006) showed that after playing an iterated version of PD with a fair (cooperative) and unfair (uncooperative) partner, males showed an empathic response when watching the fair partner receive a painful stimulus, but showed reduced empathic response and activation in reward-related areas, including the ventral striatum / nucleus accumbens, when watching the unfair partner receive the same stimulus. This is yet another example of how mechanisms might have evolved to help us protect ourselves from individuals who seek to take advantage of us.

As human societies became more complex, higher-order cognitive processes likely became more important in dealing with more complex moral dilemmas and for regulating the expression of moral emotions. An example of the conflict between strong emotional responsiveness and cognitive control can be observed in studies using the Ultimatum Game. In the Ultimatum game, two players are given the opportunity to split a sum of money. One player makes an offer and the other player can choose to accept the offered amount or reject the offer, in which case neither player makes any money. It has repeatedly been shown that about half the time, people will reject unfair offers (Bolton & Zwick, 1995), giving up their own monetary reward in order to punish the player who made the offer. Objecting to unfairness may be a fundamental adaptive mechanism by which individuals assert and maintain a social reputation as one who will not tolerate non-cooperative behavior (Nowak, Page, & Sigmund, 2000). In an fMRI study using the Ultimatum Game (Sanfey, Rilling, Aronson, Nystrom, & Cohen, 2003), both the anterior insula and the dorsolateral prefrontal cortex were found to be active upon the receipt of an unfair offer; anterior insula activation was found to be considerably higher in trials in which the person decided to reject the offer, suggesting that when emotions of anger and disgust (which are linked to anterior insula activation) were particularly high, the subject decided to reject the offer, overriding the cognitive evaluation (i.e. dorsolateral PFC activation) that one should accept the money. In another study implementing the Ultimatum Game, Koenigs and Tranel (2007) found that patients with damage to the ventromedial prefrontal cortex, which is critical in the regulation of emotional reactions, were more likely to reject unfair offers. Based on these studies, it appears that the prefrontal cortex is important in modulating the strong aversive emotional response that has evolved in response to the cheating behaviors in others. This can be especially important in cases where attempts to punish a cheater may be costly to the individual.

In summary, the detection of cheaters involves the activation of threat regions such as the amygdala. The generation of an aversive emotional response to cheating behavior involves regions such as the insula, and can be the driving force for vengeance and moralistic aggression. When cheaters are punished, reward areas such as the ventral striatum become active. In situations in which vengeance may be costly to the individual, inhibitory action of the prefrontal cortex may serve to help regulate the emotional response. Taken together, these studies demonstrate that there is a neurobiological basis to the adaptive strategies that cooperators use to protect themselves from cheating.

Neuroethics

The continual struggle to gain fitness advantage appears to be present in both moral and immoral life strategies. Individuals who primarily engage in moral behavior attempt to improve methods for detecting and punishing cheaters so as to avoid personal harm. Concurrently, cheaters attempt to improve methods for deceiving others and escaping detection. In modern societies, it could be argued that the main interface of these competing evolutionary strategies is in the criminal justice system. The evidence that there appears to be a neurobiological basis to immoral behavior presents a new set of challenges to the criminal justice system. For example, if psychopaths lack a core moral sense, as observed by reductions in activation of the moral neural circuit, are they to blame for their actions?

The ethical issues that have emerged from advances in neuroscience have resulted in a new field termed "neuroethics" (Marcus, 2002). Farah (2005) points out that we would not blame Phineus Gage for his bad behavior resulting from physical damage to his ventromedial prefrontal cortex. If we can detect differences in the brains of antisocial individuals, would this not be analogous? Such questions are becoming increasingly important for society and the law. Brain imaging evidence has already been implemented in over 130 court cases (Feigenson, 2006) and its use is likely to continue. Brain imaging evidence may have implications for cases involving the insanity defense, determining criminal intent, detecting deception by witnesses or defendants, or in determining the appropriateness of punishments such as the death penalty. In the future, neurobiological evidence, in combination with other risk factors, may eventually be able to aid in the prediction of future criminal behavior, which could have significant implications for prevention and treatment, but also would raise civil liberties issues regarding the unacceptability of false positives (i.e. classifying non-criminals as criminal). Ultimately, our explicit knowledge of the neurobiological bases of moral and immoral behavior could potentially change the very definition of what is considered moral and immoral.

Summary

This chapter has attempted to explain how both moral and immoral behavior could be viewed as evolutionary strategies with corresponding neurobiological mechanisms. With regard to immoral behavior, psychopathy may be an adaptive strategy designed to thrive in an interpersonal environment dominated by social cooperators. The Prisoner's Dilemma was used to model cooperation and non-cooperation, illustrating the positive and negative social emotions that serve to guide mutually cooperative, moral behavior, as well as how a lack of these emotions allows some individuals to pursue cheating as a viable strategy. The neurobiological evidence implicating several common structures in both moral and immoral behavior was reviewed. Several structures, including the medial and ventral prefrontal cortex, the amygdala, and the angular gyrus have been commonly implicated in both moral decision-making and in immoral behavior, suggesting these areas might be key regions involved in morality. Additional evolutionary and neurobiological processes associated with deception and cheater detection were also discussed, demonstrating the continual struggle between the two strategies. Deception appears to involve good connectivity in the prefrontal cortex, whereas detecting deception appears to involve regions of the brain's threat response system, including the amygdala and insula. Finally, the neuroethical implications of the evolutionary and neurobiological theories of moral and immoral behavior were presented. The integration of such evolutionary and neurobiological perspectives carries great potential for furthering our understanding of both moral and immoral behavior.

References

Adolphs, R., Tranel, D., & Damasio, A. R. (1998). The human amygdala in social judgement. *Nature, 393*, 470–474.

Amen, D. G., Stubblefield, M., Carmichael, B., & Thisted, R. (1996). Brain SPECT findings and aggressiveness. *Annals of Clinical Psychiatry, 8*, 129–137.

Anderson, S. W., Bechara, A., Damasio, H., Tranel, D., & Damasio, A. R. (1999). Impairment of social and moral behavior related to early damage in human prefrontal cortex. *Nature Neuroscience, 2*, 1031–1037.

Barkataki, I., Kumari, V., Das, M., Taylor, P., & Sharma, T. (2006). Volumetric structural brain abnormalities in men with schizophrenia or antisocial personality disorder. *Behavioral Brain Research, 15*, 239–247.

Barr, K. N., & Quinsey, V. L. (2004). Is psychopathy a pathology or a life strategy? Implications for social policy. In C. Crawford & C. Salmon (Eds.), *Evolutionary psychology, public policy, and personal decisions* (pp. 293–317). Hillsdale, NJ: Erlbaum.

Bechara, A. (2004). The role of emotion in decision-making: Evidence from neurological patients with orbitofrontal damage. *Brain and Cognition, 55*, 30–40.

Berthoz, S., Grezes, J., Armony, J. L., Passingham, R. E., & Dolan, R. J. (2006). Affective response to one's own moral violations. *NeuroImage, 31*, 945–950.

Birbaumer, N., Viet, R., Lotze, M., Erb, M., Hermann, C., Grodd, W., et al. (2005). Deficient fear conditioning in psychopathy: A functional magnetic resonance imaging study. *Archives of General Psychiatry, 62*, 799–805.

Blair, R. J. (2005). Applying a cognitive neuroscience perspective to the disorder of psychopathy. *Development and Psychopathology, 17*, 865–891.

Blair, R. J. (2006). Subcortical brain systems in psychopathy. In C. J. Patrick (Ed.), *Handbook of Psychopathy* (pp. 296–312). New York: Guilford.

Bolton, G. E., & Zwick, R. (1995). Anonymity versus punishment in ultimatum bargaining. *Games and Economic Behavior, 10*, 95–121.

Borg, J. S., Hynes, C., Van Horn, J., Grafton, S., & Sinnott-Armstrong, W. (2006). Consequences, action, and intention as factors in moral judgments: an fMRI investigation. *Journal of Cognitive Neuroscience, 18*, 803–817.

Cleckley, H. (1976). *The mask of sanity* (5th ed.). St. Louis, MO: Mosby.

Coccaro, E. F., Kavoussi, R. J., & McNamee, B. (2000). Central neurotransmitter function in criminal aggression. In D. H. Fishbein (Ed.), *The science, treatment, and prevention of antisocial behaviors* (pp. 6-1-6-16). Kingston, NJ: Civic Research Institute.

Connelly, B. S., Lilienfeld, S. O., & Schmeelk, K., M. (2006). Integrity tests and morality: Associations with ego development, moral reasoning, and psychopathic personality. *International Journal of Selection and Assessment, 14*, 82–86.

Crawford, C., & Salmon, C. (2002). Psychopathology or adaptation? Genetic and evolutionary perspectives on individual differences in psychopathology. *Neuro Endocrinology Letters, 23*(Suppl. 4), 39–45.

Damasio, A. R. (1994). *Descartes' error: Emotion, reason, and the human brain.* New York: GP Putnam's Sons.
Dolan, M., Deakin, J. F. W., Roberts, N., & Anderson, I. M. (2002). Quantitative frontal and temporal structural MRI studies in personality-disordered offenders and control subjects. *Psychiatry Research Neuroimaging, 116*, 133–149.
Farah, M. J. (2005). Neuroethics: The practical and the philosophical. *Trends in Cognitive Science, 9*, 34–40.
Feigenson, N. (2006). Brain imaging and courtroom evidence; on the admissibility and persuasiveness of fMRI. *International Journal of Law in Context, 2*, 233–255.
Finger, E. C., Marsh, A. A., Kamel, N., Mitchell, D. G. V., & Blair, R. J. (2006). Caught in the act: The impact of audience on the neural response to morally and socially inappropriate behavior. *NeuroImage, 33*, 414–421.
Gerra, G., Calbiani, B., Zaimovic, A., Sartori, R., Ugolotti, G., Ippolito, L., et al. (1998). Regional cerebral blood flow and comorbid diagnosis in abstinent opioid addicts. *Psychiatry Research Neuroimaging, 26*, 117–126.
Goethals, I., Audenaert, K., Jacobs, F., Van den Eynde, F., Bernagie, K., Kolindou, A., et al. (2005). Brain perfusion SPECT in impulsivity-related personality disorders. *Behavioral Brain Research, 157*, 187–192.
Greene, J. D., & Haidt, J. (2002). How (and where) does moral judgment work? *Trends in Cognitive Science, 6*, 517–523.
Greene, J. D., Nystrom, L. E., Engell, A. D., Darley, J. M., & Cohen, J. (2004). The neural bases of cognitive conflict and control in moral judgment. *Neuron, 44*, 389–400.
Greene, J. D., Sommerville, R. B., Nystrom, L. E., Darley, J. M., & Cohen, J. (2001). An fMRI investigation of emotional engagement in moral judgement. *Science, 293*, 2105–2108.
Halpern, C. T., Campbell, B., Agnew, C. R., Thompson, V., & Udry, J. R. (2002). Associations between stress reactivity and sexual and nonsexual risk taking in young adult human males. *Hormones and Behavior, 42*, 387–398.
Hare, R. D. (1991). *Manual for the hare psychopathy checklist-revised.* Toronto: Multi-Health Systems.
Hare, R. D. (2003). *Hare psychopathy checklist-revised (PCL-R)* (2nd ed.) Toronto: Multi-Health Systems, Inc.
Harenski, C. L., & Hamann, S. (2006). Neural correlates of regulating negative emotions related to moral violations. *NeuroImage, 30*, 313–324.
Harpending, H. C., & Draper, P. (1988). Antisocial behavior and the other side of cultural evolution. In T. E. Moffitt & S. Mednick (Eds.), *Biological contributions to crime causation* (pp. 293–307). Dordrecht, The Netherlands: Martinus Nijhoff.
Harpending, H. C., & Sobus, J. (1987). Sociopathy as an adaptation. *Ethology and Sociobiology, 8*, 63s–72s.
Harris, G. T., Rice, M. E., Hilton, N. Z., Lalumiere, M. L., & Quinsey, V. L. (2007). Coercive and precocious sexuality as a fundamental aspect of psychopathy. *Journal of Personality Disorders, 21*(1), 1–27.
Heekeren, H. R., Wartenburger, I., Schmidt, H., Prehn, K., Schwintowski, H. P., & Villringer, A. (2005). Influence of bodily harm on neural correlates of semantic and moral decision-making. *NeuroImage, 30*, 313–324.
Heekeren, H. R., Wartenburger, I., Schmidt, H., Schwintowski, H. P., & Villringer, A. (2003). An fMRI investigation of emotional engagement in moral judgment. *Neuroreport, 14*, 1215–1219.
Hirono, N., Mega, M. S., Dinov, I. D., Mishkin, F., & Cummings, J. L. (2000). Left frontotemporal hypoperfusion is associated with aggression in patients with dementia. *Archive of Neurology, 57*, 861–866.
Johnson, M. K., Raye, C. R., Mitchell, K. J., Touryan, S. R., Greene, E. J., & Nolen-Hoeksema, S. (2006). Dissociating the medial frontal and posterior cingulate activity during self-reflection. *Social, Cognitive, and Affective Neuroscience, 1*, 64.

Juhasz, C., Behen, M. E., Muzik, O., Chugani, D. C., & Chugani, H. T. (2001). Bilateral medial prefrontal and temporal neocortical hypometabolism in children with epilepsy and aggression. *Epilepsia, 42*, 991–1001.

Kiehl, K. A. (2006). A cognitive neuroscience perspective on psychopathy: Evidence for paralimbic system dysfunction. *Psychiatry Research, 142*, 107–128.

Kiehl, K. A., Smith, A. M., Hare, R. D., Mendrek, A., Forster, B. B., & Brink, J. (2001). Limbic abnormalities in affective processing by criminal psychopaths as revealed by functional magnetic resonance imaging. *Biological Psychiatry, 50*, 677–684.

Kiehl, K. A., Smith, A. M., Mendrek, A., Forster, B. B., Hare, R. D., & Liddle, P. F. (2004). Temporal lobe abnormalities in semantic processing by criminal psychopaths as revealed by functional magnetic resonance imaging. *Psychiatry Research, 130*, 27–42.

Koenigs, M., & Tranel, D. (2007). Irrational economic decision-making after ventromedial prefrontal damage: Evidence from the ultimatum game. *Journal of Neuroscience, 27* (4), 951–956.

Koenigs, M., Young, L., Adolphs, R., Tranel, D., Cushman, F., Hauser, M., et al. (2007). Damage to the prefrontal cortex increases utilitarian moral judgements. *Nature, 446* (7138), 908–911.

Kozel, F. A., Padgett, T. M., & George, M. S. (2004). A replication study of the neural correlates of deception. *Behavioral Neuroscience, 118*, 852–856.

Kruesi, M. J. P., Casanova, M. F., Mannheim, G., & Johnson-Bilder, A. (2004). Reduced temporal lobe volume in early onset conduct disorder. *Psychiatry Research Neuroimaging, 132*, 1–11.

Kuruoglu, A. C., Arikan, Z., Vural, G., & Karatas, M. (1996). Single photon emission computerised tomography in chronic alcoholism: Antisocial personality disorder may be associated with decreased frontal profusion. *British Journal of Psychiatry, 169*, 348–354.

Laakso, M. P., Gunning-Dixon, F., Vaurio, O., Repo-Tiihonen, E., Soininen, H., & Tiihonen, J. (2002). Prefrontal volumes in habitually violent subjects with antisocial personality disorder and type 2 alcoholism. *Psychiatry Research Neuroimaging, 114*, 95–102.

Lalumiere, M. L., & Quinsey, V. L. (1996). Sexual deviance, antisociality, mating effort, and the use of sexually coersive behaviors. *Personality & Individual Differences, 21*, 33–48.

Langleben, D. D., Loughead, J. W., Bilker, W. B., Ruparel, K., Childress, A. R., Busch, S. I., et al. (2005). Telling truth from lie in individual subjects with fast event-related fMRI. *Human Brain Mapping, 26*, 262–272.

Larden, M., Melin, L., Holst, U., & Langstrom, N. (2006). Moral judgment, cognitive distortions and empathy in incarcerated delinquent and community control adolescents. *Psychology, Crime & Law, 12*, 453–462.

Lee, T. M. C., Liu, H., Chan, C. C. H., Ng, Y., Fox, P. T., & Gao, J. (2005). Neural correlates of feigned memory impairment. *NeuroImage, 28*, 305–313.

Lee, T. M. C., Liu, H. L., Tan, L. H., Chan, C. C. H., Mahankali, S., Feng, C. M., et al. (2002). Lie detection by functional magnetic resonance imaging. *Human Brain Mapping, 15*, 157–164.

Levenson, M. R., Kiehl, K. A., & Fitzpatrick, C. M. (1995). Assessing psychopathic attributes in a noninstitutionalized population. *Journal of Personality and Social Psychology, 68*, 151–158.

Lilienfeld, S. O., & Andrews, B. P. (1996). Development and preliminary validation of a self-report measure of psychopathic personality traits in noncriminal populations. *Journal of Personality Assessment, 66*, 488–524.

Luo, Q. A., Nakic, M., Wheatley, T., Richell, R. A., Martin, A., & Blair, R. J. (2006). The neural basis of implicit moral attitude – An IAT study using event-related fMRI. *NeuroImage, 30*, 1449–1457.

Maratos, E. J., Dolan, R. J., Morris, J. S., Henson, R. N. A., & Rugg, M. D. (2001). Neural activity associated with episodic memory for emotional context. *Neuropsychologia, 39*, 910–920.

Marcus, D. (Ed.). (2002). *Neuroethics: Mapping the field, proceedings of the dana foundation conference*: University of Chicago Press.

Mayberg, H. S., Liotti, M., Brannan, S. K., McGinnis, S., Mahurin, R. K., & Jerabek, P. A. (1999). Reciprocal limbic-cortical function and negative mood: Converging PET findings in depression and normal sadness. *American Journal of Psychiatry, 156*, 675–682.

Mealey, L. (1995). The sociobiology of sociopathy: An integrated evolutionary model. *Behavioural and Brain Sciences, 18*, 523–599.
Mendez, M. F. (2006). What frontotemporal dementia reveals about the neurobiological basis of morality. *Medical Hypotheses, 67*, 411–418.
Mendez, M. F., Chen, A. K., Shapira, J. S., & Miller, B. L. (2005). Acquired sociopathy and frontotemporal dementia. *Dementia and Geriatric Cognitive Disorders, 20*, 99–104.
Mohamed, F. B., Faro, S. H., Gordon, N. J., Platek, S. M., Ahmad, H., & Williams, J. M. (2006). Brain mapping of deception and truth telling about an ecologically valid situation: Functional MR imaging and polygraph investigation–initial experience. *Radiology, 238*, 679–688.
Moll, J., Oliveira-Souza, R., & Eslinger, P. J. (2003). Morals and the human brain: A working model. *Neuroreport, 14*, 299–305.
Moll, J., Oliveira-Sousa, R., Bramati, I. E., & Grafman, J. (2002a). Functional networks in emotional moral and nonmoral social judgments. *NeuroImage, 16*, 696–703.
Moll, J., Oliveira-Souza, R., Eslinger, P. J., Bramati, I. E., Mourao-Miranda, J., Andreiuolo, P. A., et al. (2002b). The neural correlates of moral sensitivity: A functional magnetic resonance imaging investigation of basic and moral emotions. *The Journal of Neuroscience: The Official Journal of the Society for Neuroscience, 22*, 2730–2736.
Moll, J., Oliveira-Souza, R., Moll, F. T., Ignacio, F. A., Bramati, I. E., Caparelli-Daquer, E. M., et al. (2005). The moral affiliations of disgust: A functional MRI study. *Cognitive Behavioral Neurology, 18*, 68–78.
Müller, J. L., Sommer, M., Wagner, V., Lange, K., Taschler, H., Roder, C. H., et al. (2003). Abnormalities in emotion processing within cortical and subcortical regions in criminal psychopaths: Evidence from a functional magnetic resonance imaging study using pictures with emotional content. *Psychiatry Research Neuroimaging, 54*, 152–162.
Nesse, R. M. (1990). Evolutionary explanations of emotions. *Human Nature, 1*, 261–289.
Nowak, M. A., Page, K. M., & Sigmund, K. (2000). Fairness versus reason in the ultimatum game. *Science, 289*, 1773–1775.
Nunez, J. M., Casey, B. J., Egner, T., Hare, T., & Hirsch, J. (2005). Intentional false responding shares neural substrates with response conflicts and cognitive control. *NeuroImage, 25*, 267–277.
Ochsner, K. N., Beer, J. S., Robertson, E. R., Cooper, J. C., Gabrieli, J. D. E., Kihsltrom, J. F., et al. (2005). The neural correlates of direct and reflected self-knowledge. *NeuroImage, 28*, 797–814.
Ochsner, K. N., Bunge, S. A., Gross, J. J., & Gabrieli, J. D. E. (2002). Rethinking feelings: An fMRI study of the cognitive regulation of emotion. *Journal of Cognitive Neuroscience, 14*, 1215–1229.
Ochsner, K. N., Knierim, K., Ludlow, D., Hanelin, J., Ramachandran, T., & Mackey, S. (2004). Reflecting upon feelings: An fMRI study of neural systems supporting the attribution of emotion to self and other. *Journal of Cognitive Neuroscience, 16*, 1748–1772.
Oder, W., Goldenberg, G., Spatt, J., Podreka, I., Binder, H., & Deecke, L. (1992). Behavioural and psychosocial and regional cerebral blood flow: A SPECT study. *Journal of Neurology, Neurosurgery and Psychiatry, 55*, 475–480.
Oliveira-Sousa, R., & Moll, J. (2000). The moral brain: Functional MRI correlates of moral judgment in normal adults. *Neurology, 54*, 252.
Phan, K. L., Magalhaes, A., Ziemlewicz, T. J., Fitzgerald, D. A., Green, C., & Smith, W. (2005). Neural correlates of telling lies: A functional magnetic resonance imaging study at 4 tesla. *Academic Radiology, 12*, 164–172.
Prichard, J. (1835). *A treatise on insanity and other disorders affecting the mind*. London: Sherwood, Gilbert, and Piper.
Quervain, D. J. F., Fischbacher, U., Treyer, V., Schellhammer, M., Schnyder, U., Buck, A., et al. (2004). The neural basis of altruistic punishment. *Science, 305*, 1254–1258.
Raine, A. (1993). *The psychopathology of crime: Criminal behavior as a clinical disorder*. San Diego, CA: Academic Press.

Raine, A., Buchsbaum, M. S., & Lacasse, L. (1997). Brain abnormalities in murderers indicated by positron emission tomography. *Biological Psychiatry, 42*, 495–508.

Raine, A., Buchsbaum, M. S., Stanley, J., Lottenberg, S., Abel, L., & Stoddard, J. (1994). Selective reductions in prefrontal glucose metabolism in murderers. *Biological Psychiatry, 36*, 365–373.

Raine, A., Lencz, T., Bihrle, S., LaCasse, L., & Colletti, P. (2000). Reduced prefrontal gray matter volume and reduced autonomic activity in antisocial personality disorder. *Archives of General Psychiatry, 57*, 119–127.

Raine, A., Meloy, J. R., Bihrle, S., Stoddard, J., Lacasse, L., & Buchsbaum, M. S. (1998). Reduced prefrontal and increased subcortical brain functioning assessed using positron emission tomography in predatory and affective murderers. *Behavioral Sciences & the Law, 16*, 319–332.

Raine, A., Park, S., Lencz, T., Bihrle, S., Lacasse, L., Widom, C. S., et al. (2001). Reduced right hemisphere activation in severely abused violent offenders during a working memory task: An fMRI study. *Aggressive Behavior, 27*, 111–129.

Raine, A., & Yang, Y. (2006). The neuroanatomical bases of psychopathy: A review of brain imaging findings. In C. J. Patrick (Ed.), *Handbook of psychopathy* (pp. 278–295). New York: Guilford.

Rilling, J. K., Glenn, A. L., Jairam, M. R., Pagnoni, G., Goldsmith, D. R., Elfenbein, H. A., et al. (2007). Neural correlates of social cooperation and non-cooperation as a function of psychopathy. *Biological Psychiatry, 61*, 1260–1271.

Rilling, J. K., Gutman, D. A., Zeh, T. R., Pagnoni, G., Berns, G. S., & Kilts, C. D. (2002). A neural basis for social cooperation. *Neuron, 35*, 395–405.

Robertson, D., Snarey, J., Ousley, O., Harenski, K., Bowman, F. D., Gilkey, R., et al. (2007). The neural processing of moral sensitivity to issues of justice and care. *Neuropsychologia, 45*, 755–766.

Rolls, E. T., Hornak, J., Wade, D., & McGrath, J. (1994). Emotion-related learning in patients with social and emotional changes associated with frontal lobe damage. *Journal of Neurology, Neurosurgery and Psychiatry, 57*, 1518–1524.

Sanfey, A. G., Rilling, J. K., Aronson, J. A., Nystrom, L. E., & Cohen, J. D. (2003). The neural basis of economic decision-making in the ultimatum game. *Science, 300*, 1755–1758.

Schneider, F., Habel, U., Kessler, C., Posse, S., Grodd, W., & Muller-Gartner, H. W. (2000). Functional imaging of conditioned aversive emotional responses in antisocial personality disorder. *Neuropsychobiology, 42*, 192–201.

Schulz, W. (1998). Predictive reward signal of dopamine neurons. *Journal of Neurophysiology, 80*, 1–27.

Seidenwurm, D., Pounds, T. R., Globus, A., & Valk, P. E. (1997). Abnormal temporal lobe metabolism in violent subjects: Correlation of imaging and neuropsychiatric findings. *American Journal of Neuroradiology, 18*, 625–631.

Shamay-Tsoory, S. G., Tomer, R., Berger, B. D., Goldsher, D., & Aharon-Peretz, J. (2005). Impaired "affective theory of mind" is associated with right ventromedial prefrontal damage. *Cognitive Behavioral Neurology, 18*, 55–67.

Singer, T., Seymour, B., O'Doherty, J. P., Stephan, K. E., Dolan, R. J., & Frith, C. D. (2006). Empathic neural responses are modulated by the perceived fairness of others. *Nature, 439*, 466–469.

Soderstrom, H., Hultin, L., Tullberg, M., Wikkelso, C., Ekholm, S., & Forsman, A. (2002). Reduced frontotemporal perfusion in psychopathic personality. *Psychiatry Research: Neuroimaging, 114*, 81–94.

Soderstrom, H., Tullberg, M., Wikkelso, C., Ekholm, S., & Forsman, A. (2000). Reduced regional cerebral blood flow in non-psychotic violent offenders. *Psychiatry Research: Neuroimaging, 98*, 29–41.

Spence, S. A., Farrow, T. F., Herford, A. E., Wilkinson, I. D., Zheng, Y., & Woodruff, P. W. (2001). Behavioral and functional anatomical correlates of deception in humans. *Neuroreport, 12*, 2349–2353.

Sterzer, P., Stadler, C., Krebs, A., Kleinschmidt, A., & Poustka, F. (2005). Abnormal neural responses to emotional visual stimuli in adolescents with conduct disorder. *Biological Psychiatry, 57*, 7–15.

Takahashi, H., Yahata, N., Koeda, M., Matsuda, T., Asai, K., & Okubo, Y. (2004). Brain activation associated with evaluative processes of guilt and embarrassment: an fMRI study. *NeuroImage, 23*, 967–974.

Thompson, P. M., Cannon, T. D., Narr, K. L., Erp, T. van, Poutanen, V. P., Huttunen, M., et al. (2001). Genetic influences on brain structure. *Nature Neuroscience, 4*, 1253–1258.

Tiihonen, J., Hodgins, S., & Vaurio, O. (2000). Amygdaloid volume loss in psychopathy. *Society for Neuroscience Abstracts*, 20017.

Trivers, R. (1971). The evolution of reciprocal altruism. *The Quarterly Review of Biology, 46*, 35–56.

Verplaetse, J., Vanneste, S., & Braeckman, J. (2007). You can judge a book by its cover: The sequel. A kernel of truth in predictive cheating detection. *Evolution and Human Behavior, 28*, 260–271.

Viet, R., Flor, H., Erb, M., Hermann, C., Lotze, M., Grodd, W., et al. (2002). Brain circuits involved in emotional learning in antisocial behavior and social phobia in humans. *Neuroscience Letters, 328*, 233–236.

Volkow, N. D., & Tancredi, L. R. (1987). Neural substrates of violent behavior: A preliminary study with positron emission tomography. *British Journal of Psychiatry, 151*, 668–673.

Volkow, N. D., Tancredi, L. R., Grant, C., Gillespie, H., Valentine, A., Mullani, N., et al. (1995). Brain glucose metabolism in violent psychiatric patients: A preliminary study. *Psychiatry Research Neuroimaging, 61*, 243–253.

Völlm, B., Richardson, P., Stirling, J., Elliot, R., Dolan, M., Chaudhry, I., et al. (2004). Neurobiological substrates of antisocial and borderline personality disorders: preliminary result of a functional MRI study. *Criminal Behavior and Mental Health, 14*, 39–54.

Winston, J. S., Strange, B. A., O'Doherty, J., & Dolan, R. J. (2002). Automatic and intentional brain responses during evaluation of trustworthiness of faces. *Nature Neuroscience, 5*, 277–283.

Wong, M. T., Lumsden, J., Fenton, G. W., & Fenwick, P. B. (1997). Neuroimaging in mentally abnormal offenders. *Issues of Criminology and Legal Psychology, 27*, 48–58.

Yang, Y., Raine, A., Lencz, T., Bihrle, S., Lacasse, L., & Colletti, P. (2005). Prefrontal white matter in pathological liars. *British Journal of Psychiatry, 187*, 320–325.

Yang, Y., Raine, A., Lencz, T., Bihrle, S., Lacasse, L., & Colletti, P. (2005). Volume reduction in prefrontal gray matter in unsuccessful criminal psychopaths. *Biological Psychiatry, 15*, 1103–1108.

"Extended Attachment" and the Human Brain: Internalized Cultural Values and Evolutionary Implications

Jorge Moll and Ricardo de Oliveira-Souza

In the present essay we propose that an extended form of attachment is an important ingredient of human cooperation. The neural implementation of such ability will be articulated within a general framework for the neurobiological basis of human moral cognition (Moll, Zahn, Oliveira-Souza, Krueger, & Grafman, 2005), which is here extended to human affiliative behaviors in cultural contexts. More specifically, we postulate that ancient mechanisms supporting basic forms of attachment in other species, such as pair-bonding (Insel & Young, 2001), evolved to enable the unique human ability to attach to cultural objects and abstract ideas. This form of attachment, here referred to as "extended attachment", might have played a major role in cooperation and indirect reciprocity during evolution, promoting altruistic behaviors within socio-cultural groups and facilitating out-group moralistic aggression. Based on evidence from comparative and human neuroanatomical data, social psychology and cognitive neuroscience, the following hypotheses are put forth: (1) anatomical and functional reorganization of basic mechanisms of social attachment described in other species resulted in the expression of an extended form of attachment in humans; (2) this ability relies on a specific neural architecture, in which limbic/brainstem networks are directly connected to phylogenetically recent association cortical systems; (3) the extended attachment mechanism provides the basis for the human ability to attach motivational significance to abstract ideas, cultural symbols and beliefs; and (4) extended attachment may help explain high level of cooperation among non-kin observed in human societies (Fig. 1).

Attachment Operationally Defined

The inclination to form affective bonds is one major driving force behind the behavior of social mammals. This motivation for attaching to others is illustrated by the affiliative bonds which are set up between mother and child since an early

J. Moll (✉)
Cognitive and Behavioral Neuroscience Unit, Neuroscience Center LABS-D'Or
Hospital Network Rio de Janeiro, Brazil
e-mail: mollj@neuroscience-rio.org

J. Verplaetse et al. (eds.), *The Moral Brain*, DOI 10.1007/978-1-4020-6287-2_3,

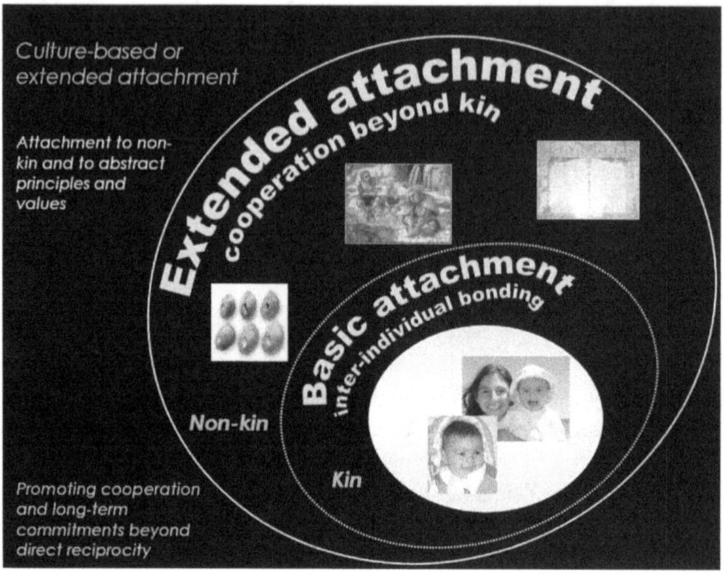

Fig. 1 Schematic representation two types of attachment. "Basic" attachment is typically shared by humans and other social animals, and supports inter-individual attachement based on kinship (e.g., mother-offspring) and sexual pair-bonding. Extended attachment, in contrast, is believed to be uniquely human. It is intrinsically related to culturally-shaped symbols, abstract ideas and values. Extended attachment enables cooperation beyond kinship relations and motivates indirect reciprocity

age, which is present in virtually all classes of mammals providing a mold for the interpersonal affective bonds that the individual establishes in life. For our current purposes, "proximate" or "basic" attachment is operationally defined as a collection of motivated behaviors that bolsters the formation of three basic types of bonding (Insel, 1997): parent-infant, filial, and pair-bonding (Table 1). Attachment provides a basic motivational ingredient for inter-individual bonding and affiliative behaviors. In addition, it is associated with the experience of belongingness (Baumeister, 2005) and, we argue, with the prosocial moral emotions of guilt, pity, gratitude, trust, admiration, and sympathy (Moll, Oliveira-Souza, Zahn, & Grafman, 2007). These

Table 1 Operational criteria for attachment

– The object of attachment is a concrete entity: another human being, a nonhuman living being, or a nonliving surrogate
– Relatively enduring, albeit not necessarily permanent
– Based on parent-infant, filial, and pair-bonding associations
– Benevolent dispositions towards the object of attachment
– Emotional dependence on the object of attachment
– Implies the experience of prosocial moral emotions (other-suffering and other-praising)
– Supports feelings of belongingness and shared social nature
– Bereavement reaction when attachment is lost, leading to the experience of the self-conscious emotions of shame or guilt

emotions act in the reparation of broken social links and in the promotion of helping behaviors, while at the same time they restrain self-serving behaviors (Haidt, 2003; Smith, 1759/1966). Although attachment is typically established among humans or between a human and another living being (e.g., a pet, a Yayá orchid), in special circumstances it may extend to an inanimate surrogate of a living being (e.g., to a ring that belonged to your grandfather) (Harlow & Suomi, 1970). Attachment also implies a special emotional stance in relation to the object of attachment, which may be reciprocated or not. Rupture of attachment bonds lead to frustration, despondency and grief (Clayton, Desmarais, & Winokur, 1968). In the field of developmental psychology, the work of Bowlby on child attachment has been highly influential (Bretherton, 1997). Severe psychopathology is often associated with chronic childhood maltreatment, leading to pathological social fear and avoidance, aggressiveness and other manifestations of poor psychological and emotional development. These developmental problems have been framed within the rubric of the "attachment disorders". Although the developmental literature on attachment has obvious links to the topic of this essay, it should be emphasized that our view on the role of attachment is not efficiently covered by these views on attachment loss (liked to social anxiety and insecurity) or attachment underdevelopment (related to callousness and antisocial behaviors). As will be discussed, attachment plays crucial roles in approaching and bonding behaviors, cultural affiliations and norm-abiding.

The above definition of attachment excludes several stable social interactions that may even be productive and satisfying, but which are not driven by attachment dispositions or by the bereavement that follows their loss. It also excludes feelings associated with hedonic pleasure or dominance, such as those that bridge people to money or to a brand new car. Finally, because attachment is particularly associated with prosocial emotions and sympathy, its relatively enduring character must be emphasized.

A common thread underlying the preceding forms of attachment is that they refer to concrete entities, regardless of whether they are living or inanimate. So far, the obvious human ability to establish enduring bonds with abstract entities has not been considered in most influential formulations on the nature of attachment. Yet, this extension of attachment to embrace abstract entities can be translated into essentially the same set of criteria that stands for primitive attachment. Moreover, as discussed later on, when brain damage impairs the bonds to abstract entities it does so roughly in parallel with the better known forms of primitive attachment. There are thus grounds for supposing that extended attachment is structured upon principles analogous to the ones that govern the bonding of humans to concrete entities.

Scope and Aims

The present essay has two main goals. Firstly, to put forth the hypothesis that humans are biologically endowed with the inclination to become emotionally attached to abstract objects collectively known as "symbols". The meaning of symbols is highly arbitrary and fully comprehensible only to those individuals who belong to the particular group within which symbols have been bred (Robb, 1998). They entail an enormous set of values (e.g., conformity, achievement, power), norms (e.g., waiting in line, etiquette rules, not stealing), principles and ideologies (e.g., equality, freedom, loyalty), as well as esthetic likes, which are internalized by a process of active cultural learning during critical periods of individual development (Vygotsky, 1934/1986). Thus, regardless of the particular content that abstract objects may assume in each society in time and space, they are always symbolic in form. Therefore, one of our main tenets is that humans are hardwired to link motivational values to the elements of culture, and that extended attachment is a crucial mediator of this capacity.

Our second major goal is to advance a set of testable hypotheses concerning the cerebral structures that mediate extended attachment. If extended attachment is a neurobiological derivative of primitive attachment, the neural structures mediating both forms of attachment are likely to be related, both anatomically and functionally. We therefore propose that the motivational engine for extended attachment must be sought for in neuropeptidergic and neurohumoral basal forebrain (limbic) systems operating in concert with isocortical networks which are expanded in the human brain (Allman, Hakeem, & Watson, 2002). Information encoded within these cortico-limbic networks, which is deeply influenced by social contact and experience, are believed represent "culturalized breeds" of basic motivational "wants" and "needs" (Baumeister, 2005).

Culture, Extended Attachment and the Human Brain

The cultural mediation of man's interactions with the social environment is the biological hallmark of our species (Baumeister, 2005; Ehrlich, 2000). Yet, due to a lack of suitable technologies the neural mechanisms that underpin cultural behavior have been unexplored until recently. As a result, research carried on in the 20th century is characterized by the search for commonalities between humans and other species. Only in the past years has it become possible to empirically investigate the subtle ways by which culture interacts with the brain to imprint unique qualities to human cognition and behavior. Extended attachment, we hypothesize, arises from the reorganization of these neural systems, enabling the unique forms by which humans relate to their social world, a speculation that is now amenable to empirical testing.

Human sociability finds parallels in pre-existent neurobehavioral mechanisms that are discernible in other species, particularly, but not exclusively, in the higher primates. For example, a "proto-politics" has been documented in chimpanzees as a set of rules that mediates intra-group interactions (de Waal, 1998). Similarly, rules of fairness and reciprocation can be inferred from the behaviors of tamarin monkeys (Hauser, Chen, Chen, & Chuang, 2003). Although the behavior of virtually every living species unfolds upon a continuum of approach-avoidance dispositions, the behavior of even the most primitive forms of animal life can not be reduced to such elementary reaction-patterns. Broadly speaking, two classes of motivations can be recognized in humans both with far-reaching social implications: one linked to aversion and rejection (which is essential for disputes for territory, mating opportunities, and power hierarchies) and the other linked to approach and affiliation (crucial for maternal-infant bonding, pair-bonding and, at least in humans, instances of cooperation, altruism and reciprocity). What distinguishes humans from other species is, therefore, the ways by which culture interacts with these basic motivations, giving rise to conceptually rich social motivations, values and preferences. This does not mean that culturally-shaped motivations rely mostly on cortical systems; contrary to a commonly held belief, certain limbic structures evolved hand in hand with cortical networks. The septal nuclei, for example, increased in size and complexity as humans ascended the ancestral tree, and are most developed in our species (Joseph, 1996).

In recent years, several studies have begun to explore the relationships that are established between specific brain systems and complex human abilities, such as moral judgment (Oliveira-Souza & Moll, 2000; Koenigs et al., 2007; Moll, Eslinger, & Oliveira-Souza, 2001), moral sentiments (Moll et al., 2002; Moll, Oliveira-Souza, Garrido, et al., 2007; Takahashi et al., 2004), political and ideological inclinations (Knutson, Wood, Spampinato, & Grafman, 2006; Westen, Blagov, Harenski, Kilts, & Hamann, 2006), ethnic attitudes (Lieberman, Hariri, Jarcho, Eisenberger, & Bookheimer, 2005), aesthetical appreciation (Jacobsen, Schubotz, Hofel, & Cramon, 2006), economic interactions (de Quervain et al., 2004; Rilling, Sanfey, Aronson, Nystrom, & Cohen, 2004; Sanfey, Rilling, Aronson, Nystrom, & Cohen, 2003) and altruistic charitable donation (Harbaugh, Mayr, & Burghart, 2007; Moll et al., 2006; Tankersley, Stowe, & Huettel, 2007). Such dispositions are as instrumental in promoting the care of others, cooperation and reciprocity, as in fostering blame, prejudice and group dissolution (de Waal, 1998; Moll, Oliveira-Souza, & Eslinger, 2003; Schulkin, 2000). These studies set the stage for the scientific exploration of the bases of culturalized forms of attachment, represented by altruistic behaviors, anonymous helping and respect for social norms and traditions, and aversion, manifested under the form of prejudice, intolerance and moral outrage. A special emphasis will be given for the concept of extended attachment.

The Neural Basis of Social Attachment and Aversion

Several neuropeptides and neuromediators, especially vasopressin, oxytocin, dopamine and peptide Y, contribute to attachment in mammals. Different peptides promote prosocial behaviors in different ways: for example, neuropeptide Y acts at the ventral striatum and perifornical region and has both anxiety-relieving properties and rewarding effects, while oxytocin facilitates social bonding (Insel & Young, 2001). Among these, oxytocin has received increasingly attention, as it regulates parturition and milk letdown, besides playing an important role in the establishment of interindividual attachment through receptors located in subcortical circuits including the ventral striatum, septal nuclei, amygdala and hypothalamus (Insel & Fernald, 2004; Keverne & Curley, 2004).

Recent experiments provide evidence that primitive attachment also plays a major role in human bonding. Structures of the brain reward system, such as the midbrain ventral tegmental area, hypothalamus and striatum, along with the subgenual cortex and adjacent basal structures were activated when humans looked at their own babies or romantic partners (Aron et al., 2005; Bartels & Zeki, 2004). Other studies provided causal evidence for the effects of certain peptides, such as oxytocin, on human social behavior. In a sequential economic game involving trust, blood oxytocin levels were higher in subjects who received a monetary transfer signaling an intention to trust, in comparison to an unintentional monetary transfer of the same amount from another player (Zak, Kurzban, & Matzner, 2004). In addition, higher oxytocin levels were associated with increased likelihood of reciprocation. Besides the putative role of oxytocin in increasing social attachment, decreasing social anxiety may be as important, as shown by a recent pharmaco-fMRI study in which oxytocin decreased amygdala activation to fearful stimuli (Kirsch et al., 2005). Still in another study, intranasal administration of oxytocin induced more cooperation in an anonymous trust game (Kosfeld, Heinrichs, Zak, Fischbacher, & Fehr, 2005). In this game, the first player chooses to transfer an amount of money (if any) to another player. The amount is multiplied, and the second player may choose how much he/she will transfer back to the first player (i.e., reciprocation). Exogenous oxytocin administration was associated with increased monetary transfers in the trust game by first movers. This effect suggests that oxytocin has causal role in trust and prosocial behavior, possibly by enhancing social attachment (Zak & Fakhar, 2006).

A recent fMRI experiment provided evidence for a direct link between altruistic decision-making and the recruitment of discrete basal forebrain structures. More relevant to our goal here, this study provided the first clear link between the role of limbic networks linked to "primitive" forms of social attachment and attachment to highly abstract societal causes,—such as human rights, child welfare and environmental protection. In this study, subjects were scanned using fMRI while they decided whether to make real donations to or to oppose a number of charities (Moll et al., 2006). Decisions, depending on trial type, could be either financially costly or non-costly to the participant. In other trials, participants were able to receive "pure" monetary rewards (without consequences for the charities). The charities were associated with causes with important societal implications, such as abortion, children's rights, nuclear energy, war and euthanasia. Pure monetary rewards activated the mesolimbic reward system, including the ventral tegmental area and the striatum. Remarkably, decisions to donate also activated the brain reward system. However, they also selectively activated subgenual-septal area, which has been implicated in social attachment. These results thus started to extend our knowledge on the role of fronto-limbic networks in social cooperation from interpersonal economic interactions, as addressed by a number of studies (de Quervain et al., 2004; Delgado, Frank, & Phelps, 2005; King-Casas et al., 2005; Sanfey et al., 2003; Singer, Kiebel, Winston, Dolan, & Frith, 2004), to the realm of culturally shaped societal values and preferences. In sum, decisions to donate to "worthy causes" led to activation of the subgenual cortex and neighboring septal region, structures which provide the main input to the preoptic hypothalamus, which is primarily involved in the control of the secretion of the neuropeptides oxytocin and vasopressin. The finding that the same circuitry which is intimately linked with mechanisms of social bonding and affiliative behaviors is engaged by abstract ideas associated with societal causes strongly supports the concept of an extended form of attachment in humans.

In the same way that neural systems underlying primitive forms of pleasure and social bonding operate in the complex situations associated with human cooperation, neural systems underlying aversive responses related to physical properties of odors and foods seem to have adapted to sustain social disapproval. While morality often promotes cooperation and helping, it often steers hostility among individuals and social groups. Moral attitudes and values powerfully incite a challenge of others' beliefs and ideologies (Allport, 1954; Vogel, 2004). Research has consistently implicated a number of brain regions and circuits in social aversion, including brainstem regions, amygdala, basal forebrain and hypothalamic nuclei, piriform and cingulate cortex, and temporal and frontal connections (Calder, Lawrence, & Young, 2001; Mega, Cummings, Salloway, & Malloy, 1997; Moll, Oliveira-Souza, et al., 2005; Volavka, 1999). The lateral OFC and neighboring agranular insula, in special, have been implicated in interpersonal aversive mechanisms (Bechara, Tranel, & Damasio, 2000; Kringelbach, 2005), including punishment of non-cooperators in economic interactions (de Quervain et al., 2004; Sanfey et al., 2003) and anger responses (Bechara et al., 2000; Blair, Morris, Frith, Perrett, & Dolan, 1999). Brain regions involved in basic and moral disgust (e.g., social disapproval) are largely shared (Moll, Oliveira-Souza, et al., 2005). Accordingly, decisions to oppose charities linked to societal causes, whether at a personal cost or at no cost, were associated with activity in the lateral OFC and anterior insula, supporting the idea that the role of these regions in basic aversive responses such as anger and disgust has been extended to operate at sophisticated cultural levels (Moll et al., 2006).

Acquired brain damage might cause impairments of attachment. Clinico-anatomic observations indicate that abnormalities in attachment result from disruption of discrete brain structures and pathways in the ventromedial frontal and temporal lobes, including the amygdala, and the subgenual-septal-preoptic-anterior hypothalamic continuum and ventral strio-pallidal and brainstem nuclei (Insel & Young, 2001; Moll et al., 2003). Researchers have seldom referred to attachment when describing the manifestations of impaired social behaviors in humans, with the notable exception of "separation anxiety" by child psychologists (Bowlby, 1960; Bretherton, 1997). In this sense, the impairment of the ability to develop attachment or its loss as a result of strategic brain lesions or neurochemical imbalances may underlie sociopathic behaviors. Patients fulfilling the clinical picture of "acquired sociopathy" have been variously described as "detached", "cold", and "aloof" (Damasio, Tranel, & Damasio, 1990; Eslinger, Grattan, Damasio, & Damasio, 1992; Moll, Zahn, et al., 2005). For example, EVR, the patient who rekindled the modern interest for the cerebral correlates of human social behavior (Eslinger & Damasio, 1985), became detached from his closest relatives and peers after an injury of the

VMPFC. Patients like EVR, who also suffered bilateral VMPFC damage, have a disproportional impairment of prosocial moral sentiments (Koenigs et al., 2007; Moll & Oliveira-Souza, 2007), for which the basic drive for attachment is crucial (Moll, Oliveira-Souza, et al., 2007). In patients with fronto-temporal dementia, atrophy of the right temporal pole and right subgenual cortex plays a major role in the loss of empathic concern (Rankin et al., 2006) and is directly related to their interpersonal detachment (Mychack, Kramer, Boone, & Miller, 2001). The role of the temporal pole is further illustrated by the behavioral changes observed in patients undergoing therapeutic temporal lobe resection, and in the "sensory-limbic hyperconnection syndrome" of patients with temporolimbic epilepsy (Bear, 1979). This syndrome includes an inappropriate prolongation of ordinary social interactions (interpersonal viscosity), a feeling of immanence, and a mystical stance towards life (Waxman & Geschwind, 1975).

The aforementioned observations indicate that attachment is often impaired by lesions that fall within a broad, yet circumscribed, region of the cerebral hemispheres and upper brainstem. From a strict neuroanatomical perspective, this region is centered in the basal forebrain and related structures. The basal forebrain is a general term for a collection of fiber pathways and nuclei that compound a longitudinal continuum that extends from the anterior frontal and temporal lobes to the upper brainstem always occupying a position below the plane that cuts through the anterior and posterior commissures. The complex anatomy of this region has been subjected to several studies in the past years and some neural systems with possible functional significance have been identified. The main anatomical structures that comprise the basal forebrain continuum are (i) the primary and accessory olfactory tracts and their central projections, (ii) the collection of nuclei that comprise the amygdaloid complex, (iii) the septal area and nuclei, (iv) the pre-optic and anterior hypothalamus, (v) the pathways that traverse this region and interconnect them, and (vi) the temporo-polar cortex and the orbitomedial and subgenual frontal cortices, (vii) the basal nuclei of Meynert, the extended amygdala, and the ventral (i.e., subcommissural) strio-pallidum.While these structures are not exclusively linked to attachment, they are necessary for the development of interpersonal bonds and for the experience of interpersonal attachment through close anatomical and functional integration with specific prefrontal and temporal isocortical regions (Moll, Zahn, et al., 2005). One major challenge for future studies is to define the part played by discrete circuits within these systems in the promotion of specific aspects of attachment and bonding.

Evolutionary Implications

The evidence reviewed above provides the ground for the existence of a form of attachment that evolved from the primitive mechanisms of social attachment allowing humans to develop and sustain vicarious emotional bonds beyond the sphere of kinship to the realms of abstract values (e.g., self-transcendence, achievement), norms (e.g., paying taxes, respecting cultural traditions) and principles (e.g., generosity, honesty, courage) (Brown, 1991). Extended attachment may have contributed to survival fitness in humans, by strengthening interpersonal bonds, promoting social cohesion and by operating as a motivational factor for the observance of cultural standards.

We speculate that extended attachment may have co-evolved, and at times acting as a driving force, for the capacity for artwork and symbolic thinking, which reached a critical point in the cultural explosion of the Upper Paleolithic around 100,000 years ago. This stage marks one of the major recent transitions in human evolution (Maynard-Smith & Szathmary, 1997). Anthropological evidence suggests that humans invested significant amounts of time and energy, and they still do? in cultural and symbolic activities which did not provide immediate survival benefits. Those included carefully designed and executed symbolic burials, crafting adornments, taking part in communal rituals, painting cave walls, and so on (Ehrlich, 2000). Extended attachment may be a key ingredient that allowed our ancestors to develop emotional bonds to cultural artifacts and ideas, promoting social coherence in endeavors such as collective hunting, building shelter, participating in rituals and other kinds of social exchanges going beyond simple interpersonal reciprocity. Imprinting values to abstract symbols, practices and beliefs rendered humans capable of developing and internalizing values and virtues, helping self-evaluations of performance in social groups of ever-growing complexities. Affiliation to culture-specific shared values, agreed upon by members of the group, may have served as "commitment devices" (Frank, 1988) for all individuals, and rendered humans less dependent on more fleeting states of social status, which govern reciprocal mechanisms of cooperation and punishment among individuals. Extended attachment, thus, could significantly increase social cohesion, essentially in a intra-group fashion; it is possible that this mechanism might also promote inter-group competition. In this sense, extended attachment would not only drive intra-group altruism and cooperation, but also help demarcate out-group boundaries, enhancing group distinctiveness and promoting aggression towards out-groups – a haunting feature has permeated human evolution (Bowles, 2006; Bowles & Gintis, 2004).

Psychologically, extended attachment may be an important feature of self-identity that regulates self-esteem and social relations, as individuals of a given cultural group may safely rely on self-evaluations according to how they stick to in-group values and norms. Self-evaluations may well work as an internal currency which depends, to a great extent, on the capacity to attach to culture-specific social values and symbols, providing important guidelines in complex social contexts and helping individuals to assess the status of their own reputation. Failing to comply with internalized cultural values (say, when egotistically opting for personally advantageous options) may lead to feelings of shame and under-achievement, whereas succeeding in doing so leads to pride and improved reputation. This view is supported by the widespread observation that people often engage in collective causes which go far beyond interpersonal interactions, in the absence of expected gains through direct or indirect reciprocity, and even at substantial personal cost in terms of physical integrity and financial costs, even in conditions of complete anonymity (Fehr & Rockenbach, 2004; Moll et al., 2006).

Anthropological research suggests that cultural stability within cultural groups depends on a constant flow of information transmitted in an explicit or implicit form, which is shared by most members of a given group (Sperber & Hirschfeld, 2004); "culture" thus refers to this widely distributed information in the minds of group members, and its behavioral expressions. Cultural learning is not unbounded, and is highly influenced by cognitive and emotional dispositions, which by themselves are tied to cognitive and emotional domains (Hauser, 2006; Sperber & Hirschfeld, 2004). This view is in line with the our proposal that extended attachment arises from a more basic capacity to feel emotionally bounded to others, a capacity that is highly dependent on limbic circuits, which humans share with several social species. This requires functional integration between limbic circuits and neocortical structures, which represent highly complex social information. These neural networks operate through tight anatomical and functional connectivity, providing the bases for culturally-dependent psychological phenomena such as moral sentiments and values (Moll, Zahn, et al., 2005). As we have recently propose, moral phenomena arise from integration of the following main components: (1) abstract action / event knowledge, for which the prefrontal cortex is essential (Krueger, Moll, Zahn, Heinecke, & Grafman, 2007; Wood & Grafman, 2003); (2) social conceptual and perceptual knowledge, provided by the anterior (Zahn et al., 2007) and posterior (Frith & Frith, 2003) temporal neocortex, respectively; and (3) basic emotional-motivational states, which arise from limbic-paralimbic structures (Stellar, 1954/1994).

Altruistic behaviors, i.e., behaviors that benefit a third party while being costly to the agent, have long been described in the social species (Trivers, 1971). Altruistic behavior was first characterized among kin (Hamilton, 1964) being frequent in several social species, such as bees, ants and gregarious birds. This behavior essentially contributes to the perpetuation of the genes of the agent of the altruistic act, leading to kin selection. Humans, however, often cooperate with non-relatives in ways that cannot be explained by kin selection. Trivers provided a model that explains cooperation among non-relatives which was still compatible with survival fitness: reciprocal altruism (Trivers, 1971). In this model, a small personal risk of helping another is paid off by an increased chance that the organism who received the help will repay the favor, thus increasing the survival fitness of the altruistic agent –"I scratch your back, and you will scratch mine". A further advance for theories of reciprocal altruism was provided by the development of the concept of indirect reciprocity [for a review, see (Nowak & Sigmund, 2005)]. In indirect reciprocity, "I scratch your back and someone else will scratch mine". In other words, the agent performs an altruistic act towards another which will not necessarily repay the favor in kind; instead, the altruist will receive the benefit from someone else. A number of mechanisms have been proposed for this sophisticated from of altruism, which is most distinctive of our species. Humans often incur in significant personal costs to reward "good" behaviors or to punish "bad" behaviors of third-parties. It has been argued that indirect reciprocity is connected with the origins of moral norms, which is essentially dependent on culture. Such norms are often culture-specific, and there is little evidence for their existence in other species (Brown, 1991;

Fehr & Fischbacher, 2004). A number of mechanisms have been proposed to support indirect reciprocity humans, including the tendency to punish non-cooperators and reward cooperators, or "strong reciprocity" (Fehr & Rockenbach, 2004). Much less explored, however, is the issue of which internal motivational mechanisms support such behavioral tendencies, and how they are implemented in our nervous systems. While simple forms of emotional states linked to pleasure / reward and aversion / punishment can be sufficient to explain less complex social behaviors, the highly complex forms of human altruism and moralistic aggression require more sophisticated neurocognitive mechanisms. Such mechanisms depart from the simple pleasure-pain axis, combining cognitive-emotional elements such as outcome prediction, social conceptual knowledge, mental state attribution, etc, to give rise to culturally ubiquitous moral sentiments (Moll, Oliveira-Souza, et al., 2007). The functional anatomy of specific moral sentiments has begun to be unveiled (Moll et al., 2002; Moll, Oliveira-Souza, et al., 2007). Supporting our claim that extended forms of attachment share in part the neural mechanisms involved in proximate forms of attachment, we found that the experience of moral sentiments as diverse as compassion and guilt is associated with common activations in attachment-related regions of the basal forebrain (Moll, Oliveira-Souza, et al., 2007). While these experiments did not involve real decisions, and therefore did not directly test altruistic choices, our study on the neural basis of charitable donations further supported the notion of attachment as an important ingredient of human altruism in a specific cultural context. While both "pure" monetary rewards and decisions to donate activated the mesolimbic reward system, only charitable donations led to consistent activation of the subgenual-septal region, responses that can be interpreted as arising from the experience of attachment to the mission of the charitable organizations. Since the experiment was conducted under complete anonymity, the contribution of reputation (Milinski, Semmann, & Krambeck, 2002a) to decisions to donate can be excluded. Noteworthy, this task did not involve donations to individual persons, but to highly abstract societal causes (e.g., civil liberty, environmental protection, animal rights, medical research, etc). Extended attachment thus provides a possible mechanism by which humans build direct emotional links to culturally-shaped beliefs and symbols and which, in turn, play a determinant role in guiding our choices related to others.

Prospects and Final Remarks

We have provided an argument for the existence of an extended form of attachment in humans and broadly described some candidate neural underpinnings. That extended altruism may increase survival fitness of individuals within a group and of one group over others finds some support in neurobehavioral and cognitive evidence, and can be tested with the tools currently available to evolutionary biologists. Mathematical simulations can potentially model the ability of agents to develop attachments to arbitrary "cultural symbols", and test the hypothesis that individual attachment to such symbols and increased cooperation with members who share beliefs (i.e., which attach to the same sets of cultural symbols) can increase their survival fitness. This approach may provide interesting insights bridging the function of specific neural systems that are especially developed in humans, and the uniquely developed capacity for altruism and social cooperation observed in our species.

Although we propose that extended attachment might have evolved from primitive forms of attachment, with which it seems to share neural underpinnings, many aspects of human sociality cannot be easily accommodated. Based on our correspondence with Jan Verplaetse, we will briefly address some of these points and speculate on possible solutions. Firstly, in many instances the motivational forces supporting cultural acquisition of norms and social behaviors are not related to attachment, in its basic or extended forms. For example, fear of physical punishment or of reputation loss could be a sufficient motive for promotion of prosocial behaviors or for avoidance of socially inappropriate choices. In another example, prediction of future rewards or reputation gains following social cooperation can provide enough motivational force for an individual to behave "altruistically". Accordingly, "image scoring" or reputation monitoring, direct and indirect reciprocity models (Milinski et al., 2002a; Milinski, Semmann, & Krambeck, 2002b; Nowak & Sigmund, 2005; Trivers, 1971; Wedekind & Braithwaite, 2002) rarely mention the need of emotions in supporting cooperation; even if they need to do so, simple emotional states, shared by humans and non-humans alike (such as fear and anger), seem to provide a sufficient repertoire of internal motivations to do the trick. The same applies to the strong reciprocity model: feeling angry towards an unfair agent or having positive feelings after the other agent reciprocates provide the necessary motivational bases for cooperative and altruistic behaviors (Bowles & Gintis, 2004; Quervain et al., 2004). Our claim, however, is not that these mechanisms and the motivations operating behind them are overturned by extended attachment in humans. On the contrary, we suggest that extended attachment may operate as an additional mechanism and in many instances in a synergic way, along with other cooperation mechanisms. For example, the ability to genuinely attach to some arbitrary cultural norm or symbol – to use simple examples, adopting certain formalisms in academic social situations which are felt to create an atmosphere of tradition and soberness, or dressing ragged jeans to symbolize irreverence and autonomy – may tremendously boost one's acceptance within a social group in which such symbols have shared meaning, and therefore boost one's social status. This would facilitate and stabilize indirect reciprocity, for example.

Another relevant issue, which is related to the aspects discussed above, is whether extended attachment would apply seamlessly not only to values related to affiliative feelings (such as helping, empathy and altruism) but also to other values such as achievement and reliability, which might rely on different motivations, such as dominance and social status. If extended attachment indeed evolved from proximate forms of attachment, it would be reasonable to think that it should be more important for affiliative and related positive emotional states. Indeed, available neuroscientific evidence mainly support this view; for example, charitable donations engage the brain regions related to reward and affiliative responses (Moll et al., 2006), and activity in reward-related brain regions is evoked by interpersonal reciprocal exchanges (de Quervain et al., 2004; Rilling et al., 2004; Singer et al., 2006) and by culturally-shaped rewards, such as sports cars (Erk, Spitzer, Wunderlich, Galley, & Walter, 2002). Nevertheless, other lines of investigation do start to provide some evidence that brain regions related to affiliation are also in processing non-pleasant feelings which, we argue, could be related to attachment mechanisms. For example, clinical neuroscience studies consistently show a dysfunction of the subgenual ventral cingulate area in patients with major depression, which is associated with increased feelings of guilt (Drevets, Ongur, & Price, 1998). Direct electric stimulation of the subgenual area, performed in a small sample of patients who were refractory to antidepressant treatment, led to symptom amelioration and, curiously, to a reported feeling of "increased connectedness" (Mayberg et al., 2005). Further, on a recent fMRI study addressing the neural bases of specific moral sentiments we have demonstrated that negatively-valenced emotions such as guilt and compassion activate the brain reward and social attachment regions (Moll, Oliveira-Souza, et al., 2007). These findings may bear important implications to the issue raised by J. Verplaetse of whether the possible role of extended attachment in "attachment moralities" (related to love and empathy) would necessarily apply to "principled moralities" (relying on a feeling of "duty"). Whereas simple introspection seems to suggest that the inner experience of "warm glow" when helping someone in need differs in essential ways from the experience of helping based on a feeling of duty, an inherent human capacity of extended attachment may be important for both cases; accordingly, being able to feel connected and attribute importance to a person in trouble (or to a principle, such as the duty of promoting human well-being) can be a common underlying factor which, we suggest, may be linked to extended attachment. Failing to do so triggers the sentiment guilt, which is linked to reduced attribution of self-value or esteem (Moll, Oliveira-Souza, Garrido, et al., 2007), for which a sense of moral identity appears to be essential (Blasi, 1999). As pointed out by J. Verplaetse, in some cases an ideology or sense of duty may be at odds with a feeling of attachment, in some instances even suppressing empathy (e.g., a soldier that inflicts pain to other humans on behalf of his commitment to his country). But a mechanism of extended attachment (and guilt avoidance) may well be at work for someone to feel committed to a given moral duty (e.g., the soldier feels attached to his country and to the values associated with it). These are, of course, wild speculations at this point, which need to be addressed by empirical research. Our prediction is that brain attachment systems will be engaged not only in response to "warm glow", but also by feelings of commitment towards a large set of cultural entities. Fortunately, conceptual frameworks and methods currently available in cognitive psychology and neuroscience now enable researchers to address these crucial questions.

Finally, would extended attachment be a plausible mechanism of evolutionary fitness? As discussed above, extended attachment may promote both direct and indirect reciprocity, operating as an important commitment device and therefore promoting costly prosocial behaviors which may payoff through improved social image or reputation building in the long term. Recent experimental work in humans indeed indicate that individuals who contribute more to public goods and act as volunteers, even bearing the costs involved in doing so, seem to do better in the long run (Milinski et al., 2002a). As discussed, extended attachment may provide a powerful cognitive and motivational ingredient for social cohesion and commitment, a claim may be validated or refuted by future experimental work and theoretical models.

References

Allman, J., Hakeem, A., & Watson, K. (2002). Two phylogenetic specializations in the human brain. *Neuroscientist, 8*, 335–346.

Allport, G. W. (1954). *The nature of prejudice*. Boston: Beacon Press.

Aron, A., Fisher, H., Mashek, D. J., Strong, G., Li, H., & Brown, L. L. (2005). Reward, motivation, and emotion systems associated with early-stage intense romantic love. *Journal of Neurophysiology, 94*, 327–337.

Bartels, A., & Zeki, S. (2004). The neural correlates of maternal and romantic love. *NeuroImage, 21*, 1155–1166.

Baumeister, R. F. (2005). *The cultural animal*. New York: Oxford University Press.

Bear, D. M. (1979). Temporal lobe epilepsy: A syndrome of sensory-limbic hyperconnection. *Cortex, 15*, 357–384.

Bechara, A., Tranel, D., & Damasio, H. (2000). Characterization of the decision-making deficit of patients with ventromedial prefrontal cortex lesions. *Brain, 123*, 2189–2202.

Blair, R. J., Morris, J. S., Frith, C. D., Perrett, D. I., & Dolan, R. J. (1999). Dissociable neural responses to facial expressions of sadness and anger. *Brain, 122*, 883–893.

Blasi, A. (1999). Emotions and moral motivation. *Journal for the Theory of Social Behaviour, 29*, 1–19.

Bowlby, J. (1960). Separation anxiety. *International Journal of Psychoanalysis, 41*, 89–113.

Bowles, S. (2006). Group competition, reproductive leveling, and the evolution of human altruism. *Science, 314*, 1569–1572.

Bowles, S., & Gintis, H. (2004). The evolution of strong reciprocity: Cooperation in heterogeneous populations. *Theoretical Population Biology, 65*, 17–28.

Bretherton, I. (1997). Bowlby's legacy to developmental psychology. *Child Psychiatry and Human Development, 28*, 33–43.

Brown, D. E. (1991). *Human universals*. New York: McGraw-Hill.

Calder, A. J., Lawrence, A. D., & Young, A. W. (2001). Neuropsychology of fear and loathing. *Nature Reviews Neuroscience, 2*, 352–363.

Clayton, P. J., Desmarais, L., & Winokur, G. (1968). A study of normal bereavement. *American Journal of Psychiatry, 125*, 64–74.

Damasio, A. R., Tranel, D., & Damasio, H. (1990). Individuals with sociopathic behavior caused by frontal damage fail to respond autonomically to social stimuli. *Behavioural Brain Research, 41*, 81–94.

Delgado, M. R., Frank, R. H., & Phelps, E. A. (2005). Perceptions of moral character modulate the neural systems of reward during the trust game. *Nature Neuroscience, 8*, 1611–1618.

de Quervain, D. J.-F., Fischbacher, U., Treyer, V., Schellhammer, M., Schnyder, U., Buck, A., et al. (2004). The neural basis of altruistic punishment. *Science, 305*, 1254–1258.

de Waal, F. (1998). *Chimpanzee politics power and sex among apes*. Baltimore: Johns Hopkins University Press.

Drevets, W. C., Ongur, D., & Price, J. L. (1998). Neuroimaging abnormalities in the subgenual prefrontal cortex: Implications for the pathophysiology of familial mood disorders. *Molecular Psychiatry, 3*, 220–226.

Ehrlich, P. R. (2000). *Human natures: Genes, cultures, and the human prospect*. Washington, DC: Island Press.

Erk, S., Spitzer, M., Wunderlich, A. P., Galley, L., & Walter, H. (2002). Cultural objects modulate reward circuitry. *Neuroreport, 13*, 2499–2503.

Eslinger, P. J., & Damasio, A. R. (1985). Severe disturbance of higher cognition after bilateral frontal lobe ablation: Patient EVR. *Neurology, 35*, 1731–1741.

Eslinger, P. J., Grattan, L. M., Damasio, H., & Damasio, A. R. (1992). Developmental consequences of childhood frontal lobe damage. *Archives Neurology, 49*, 764–769.
Fehr, E., & Fischbacher, U. (2004). Social norms and human cooperation. *Trends in Cognitive Science, 8*, 185–190.
Fehr, E., & Rockenbach, B. (2004). Human altruism: Economic, neural, and evolutionary perspectives. *Current Opinion in Neurobiology,14*, 784–790.
Frank, R. H. (1988). *Passions within reason: The strategic role of the emotions*. New York: W W Norton & Co Ltd.
Frith, U., & Frith, C. D. (2003). Development and neurophysiology of mentalizing. *Philosophical Transactions of the Royal Society of London. Series B, Biological Sciences, 358*, 459–473.
Haidt, J. (2003). The moral emotions. In R. J. Davidson, K. R. Scherer, & H. H. Goldsmith (Eds.), *Handbook of affective sciences* (pp. 852–870). Oxford: Oxford University Press.
Hamilton, W. D. (1964). The genetic evolution of social behavior. *Journal of Theoretical Biology, 7*, 17–18.
Harbaugh, W. T., Mayr, U., & Burghart, D. R. (2007). Neural responses to taxation and voluntary giving reveal motives for charitable donations. *Science, 316*, 1622–1625.
Harlow, H. F., & Suomi, S. J. (1970). Nature of love-simplified. *American Psychologist, 25*, 161–168.
Hauser, M. D. (2006). *Moral minds: How nature designed our universal sense of right and wrong*. New York: Ecco.
Hauser, M. D., Chen, M. K., Chen, F., & Chuang, E. (2003). Give unto others: Genetically unrelated cotton-top tamarin monkeys preferentially give food to those who altruistically give food back. *Proceedings. Biological Sciences, 270*, 2363–2370.
Insel, T. R. (1997). A neurobiological basis of social attachment. *American Journal of Psychiatry, 154*, 726–735.
Insel, T. R., & Fernald, R. D. (2004). How the brain processes social information: Searching for the social brain. *Annual Review of Neuroscience, 27*, 697–722.
Insel, T. R., & Young, L. J. (2001). The neurobiology of attachment. *Nature Reviews Neuroscience, 2*, 129–136.
Jacobsen, T., Schubotz, R. I., Hofel, L., & Cramon, D. Y. von. (2006). Brain correlates of aesthetic judgment of beauty. *NeuroImage, 29*, 276–285.
Joseph, R. (1996). *Neuropsychiatry, neuropsychology and clinical neuroscience*. Baltimore: Williams & Wilkins.
Keverne, E. B., & Curley, J. P. (2004). Vasopressin, oxytocin and social behaviour. *Current Opinion in Neurobiology, 14*, 777–783.
King-Casas, B., Tomlin, D., Anen, C., Camerer, C. F., Quartz, S. R., & Montague, P. R. (2005). Getting to know you: Reputation and trust in a two-person economic exchange. *Science, 308*, 78–83.
Kirsch, P., Esslinger, C., Chen, Q., Mier, D., Lis, S., Siddhanti, S., et al. (2005). Oxytocin modulates neural circuitry for social cognition and fear in humans. *Journal of Neuroscience, 25*, 11489–11493.
Knutson, K. M., Wood, J. N., Spampinato, M. V., & Grafman, J. (2006). Politics on the brain: An fMRI Investigation. *Social Neuroscience, 1*, 25–40.
Koenigs, M., Young, L., Adolphs, R., Tranel, D., Cushman, F., Hauser, M., et al. (2007). Damage to the prefrontal cortex increases utilitarian moral judgements. *Nature, 446*, 908–911.
Kosfeld, M., Heinrichs, M., Zak, P. J., Fischbacher, U., & Fehr, E. (2005). Oxytocin increases trust in humans. *Nature, 435*, 673–676.
Kringelbach, M. L. (2005). The human orbitofrontal cortex: Linking reward to hedonic experience. *Nature Reviews Neuroscience, 6*, 691–702.
Krueger, F., Moll, J., Zahn, R., Heinecke, A., & Grafman J. (2007, Oct). Event frequency modulates the processing of daily life activities in human medial prefrontal cortex. *Cerebral Cortex, 17* (10), 2346–2353.

Lieberman, M. D., Hariri, A., Jarcho, J. M., Eisenberger, N. I., & Bookheimer, S. Y. (2005). An fMRI investigation of race-related amygdala activity in African-American and Caucasian-American individuals. *Nature Neuroscience, 8*, 720–722.

Mayberg, H. S., Lozano, A. M., Voon, V., McNeely, H. E., Seminowicz, D., Hamani, C., et al. (2005). Deep brain stimulation for treatment-resistant depression. *Neuron, 45*, 651–660.

Maynard-Smith, J., & Szathmary, E. (1997). *The major transitions in evolution*. New York: Oxford University Press.

Mega, M., Cummings, J., Salloway, S., & Malloy, P. (1997). The limbic system: An anatomic, phylogenetic, and clinical perspective. *Journal of Neuropsychiatry Clinical Neuroscience, 9*, 315–330.

Milinski, M., Semmann, D., & Krambeck, H. J. (2002a). Donors to charity gain in both indirect reciprocity and political reputation. *Proceedings. Biological Sciences, 269*, 881–883.

Milinski, M., Semmann, D., & Krambeck, H. J. (2002b). Reputation helps solve the 'tragedy of the commons'. *Nature, 415*, 424–426.

Moll, J., & Oliveira-Souza, R. (2007). Moral judgments, emotions and the utilitarian brain. *Trends in Cognitive Science*.

Moll, J., Oliveira-Souza, R., & Eslinger, P. J. (2003). Morals and the human brain: A working model. *Neuroreport, 14*, 299–305.

Moll, J., Oliveira-Souza, R., Eslinger, P. J., Bramati, I. E., Mourao-Miranda, J., Andreiuolo, P. A., et al. (2002). The neural correlates of moral sensitivity: A functional magnetic resonance imaging investigation of basic and moral emotions. *Journal of Neuroscience, 22*, 2730–2736.

Moll, J., Oliveira-Souza, R., Garrido, G. J., Bramati, I. E., Caparelli-Daquer, E. M. A., Paiva, M. M. F., et al. (2007). The self as a moral agent: Linking the neural bases of social agency and moral sensitivity. *Social Neuroscience, 2*, 336–352.

Moll, J., & Oliveira-Souza, R. (2007). Moral judgments, emotions and the utilitarian brain. *Trends in Cognitive Sciences, 11*(8), 319–321.

Moll, J., Eslinger, P. J., & Oliveira-Souza, R. (2001). Frontopolar and anterior temporal cortex activation in a moral judgment task: Preliminary functional MRI results in normal subjects. *Arq Neuropsiquiatr, 59*, 657–664.

Moll, J., Krueger, F., Zahn, R., Pardini, M., Oliveira-Souza, R., & Grafman, J. (2006). Human fronto-mesolimbic networks guide decisions about charitable donation. *Proceedings of the National Academy of Sciences of the United States of America, 103*, 15623–15628.

Moll, J., Oliveira-Souza, R., Tovar-Moll, F., Ignacio, F. A., Bramati, I., Caparelli-Daquer, E. M., et al. (2005). The moral affiliations of disgust: A functional MRI study. *Cognitive and Behavioral Neurology, 18*(1), 68–78.

Moll, J., Oliveira-Souza, R., Zahn, R., & Grafman, J. (2007). The cognitive neuroscience of moral emotions. In W. Sinnott-Armstrong (Ed.), *Moral Psychology, Volume 3: Morals and the Brain*. Cambridge, MA: MIT Press.

Moll, J., Zahn, R., Oliveira-Souza, R., Krueger, F., & Grafman, J. (2005). Opinion: The neural basis of human moral cognition. *Nature Reviews Neuroscience, 6*, 799–809.

Mychack, P., Kramer, J. H., Boone, K. B., & Miller, B. L. (2001). The influence of right frontotemporal dysfunction on social behavior in frontotemporal dementia. *Neurology, 56*, 11–15.

Nowak, M. A., & Sigmund, K. (2005). Evolution of indirect reciprocity. *Nature, 437*(7063), 1291–1298.

Oliveira-Souza, R., & Moll, J. (2000). The moral brain: A functional MRI study of moral judgment. Paper presented at the Neurology Conference.

Rankin, K. P., Gorno-Tempini, M. L., Allison, S. C., Stanley, C. M., Glenn, S., Weiner, M. W., et al. (2006). Structural anatomy of empathy in neurodegenerative disease. *Brain, 129*, 2945–2956.

Rilling, J. K., Sanfey, A. G., Aronson, J. A., Nystrom, L. E., & Cohen, J. D. (2004). Opposing BOLD responses to reciprocated and unreciprocated altruism in putative reward pathways. *Neuroreport, 15*, 2539–2543.

Robb, J. E. (1998). The archaeology of symbols. *Annual Review of Anthropology, 27*, 329–346.
Sanfey, A. G., Rilling, J. K., Aronson, J. A., Nystrom, L. E., & Cohen, J. D. (2003). The neural basis of economic decision-making in the Ultimatum Game. *Science, 300*, 1755–1758.
Schulkin, J. (2000). *Roots of social sensitivity and neural function*. Cambridge, MA: MIT Press.
Singer, T., Kiebel, S. J., Winston, J. S., Dolan, R. J., & Frith, C. D. (2004, Feb). Brain responses to the acquired moral status of faces. *Neuron, 41*(4), 653–662.
Singer, T., Seymour, B., O'Doherty, J. P., Stephan, K. E., Dolan, R. J., & Frith, C. D. (2006). Empathic neural responses are modulated by the perceived fairness of others. *Nature, 439*, 466–469.
Smith, A. (1759/1966). *The theory of moral sentiments* (6th ed.). New York: Kelly.
Sperber, D., & Hirschfeld, L. A. (2004). The cognitive foundations of cultural stability and diversity.*Trends in Cognitive Science, 8*, 40–46.
Stellar, E. (1954/1994). The physiology of motivation. 1954. *Psychological Review, 101*, 301–311.
Takahashi, H., Yahata, N., Koeda, M., Matsuda, T., Asai, K., & Okubo, Y. (2004). Brain activation associated with evaluative processes of guilt and embarrassment: An fMRI study. *NeuroImage, 23*, 967–974.
Tankersley, D., Stowe, C. J., & Huettel, S. A. (2007). Altruism is associated with an increased neural response to agency. *Nature Neuroscience, 10*, 150–151.
Trivers, R. L. (1971). The evolution of reciprocal altruism. *The Quarterly Review of Biology, 46*, 35–57.
Vogel, G. (2004). Behavioral evolution. The evolution of the golden rule. *Science, 303*, 1128–1131.
Volavka, J. (1999). The neurobiology of violence: An update. *Journal of Neuropsychiatry Clinical Neuroscience, 11*, 307–314.
Vygotsky, L. S. (1934/1986). *Thought and language*. Cambridge, MA: The MIT Press.
Waxman, S. G., & Geschwind, N. (1975). The interictal behavior syndrome in temporal lobe epilepsy. *Archives of General Psychiatry, 32*, 1580–1586.
Wedekind, C., & Braithwaite, V. A. (2002). The long-term benefits of human generosity in indirect reciprocity. *Current Biology, 12*, 1012–1015.
Westen, D., Blagov, P. S., Harenski, K., Kilts, C., & Hamann, S. (2006). Neural bases of motivated reasoning: An FMRI study of emotional constraints on partisan political judgment in the 2004 U.S. presidential election. *Journal of Cognitive Neuroscience, 18*, 1947–1958.
Wood, J. N., & Grafman, J. (2003). Human prefrontal cortex: Processing and representational perspectives. *Nature Reviews Neuroscience, 4*, 139–147.
Zahn, R., Moll, J., Krueger, F., Huey, E. D., Garrido, G., & Grafman, J. (2007). Social concepts are represented in the superior anterior temporal cortex. *Proceedings of the National Academy of Sciences of the United States of America, 104*, 6430–6435.
Zak, P. J., & Fakhar, A. (2006). Neuroactive hormones and interpersonal trust: International evidence. *Economics and Human Biology, 4*, 412–429.
Zak, P. J., Kurzban, R., & Matzner, W. T. (2004). The neurobiology of trust. *Annals of the New York Academy of Sciences, 1032*, 224–227.

Neuro-Cognitive Systems Involved in Moral Reasoning

James Blair

The Development of Morality

Before considering the systems involved in moral reasoning, it is worth briefly considering what morality is. Two broad viewpoints on morality effectively exist. According to one, moral reasoning is mediated by some form of unitary system for processing social rule related information. Early adherents of this viewpoint included Kohlberg who considered the moral reasoning developed from conventional reasoning (Colby & Kohlberg, 1987). More recent adherents include Moll who has defined morality as: "the sets of customs and values that are embraced by a cultural group to guide social conduct" (Moll, Zahn, Oliveira-Souza, Krueger, & Grafman, 2005, p. 799).

An alternative viewpoint is that there are multiple systems which mediate largely separable forms of social rule related processing that are lumped together within a culture's concept of morality. One of the earliest of these viewpoints was provided by the domain theorists (e.g., Nucci, 1982; Turiel, Killen, & Helwig, 1987; Smetana, 1993). They principally distinguished between moral (harm and justice based) as opposed to conventional (social disorder based) transgressions. More recent adherents to such a view include both Blair and Haidt (e.g., Haidt & Joseph, 2004; Blair, Marsh, Finger, Blair, & Luo, 2006). For example, Haidt and Graham (2007) have argued that there are five psychological foundations of morality, which they term harm, reciprocity, purity, hierarchy and ingroup. Blair has similarly argued for multiple systems for morality. Most of these overlap with Haidt's (care-based: harm, reciprocity, disgust-based: purity, social convention: hierarchy) but one does not (affect-free morality).

In this chapter, I will initially briefly consider the idea of morality as mediated by a unitary system before considering the multiple moralities approach.

J. Blair (✉)
Department of Health and Human Services, Mood and Anxiety Program National Institute of Mental Health, National Institutes of Health, MD, Bethesda, USA
e-mail: JamesBlair@mail.nih.gov

J. Verplaetse et al. (eds.), *The Moral Brain*, DOI 10.1007/978-1-4020-6287-2_4,

Morality as a Unitary System

A variety of positions have considered that morality is a unitary entity, effectively assuming that all forms of norm recruit a specific neuro-cognitive architecture. For example, several recent reviews have provided a description of neural regions frequently activated in fMRI studies of morality (Moll, Zahn, et al., 2005; Raine & Yang, 2006). However, the functional contributions of these regions with respect to moral reasoning remain relatively under-specified. In one of the more detailed accounts, Moll and colleagues suggest the existence of a three component system. These components include: "structured event knowledge, which corresponds to context dependent representations of events and event sequences in the PFC; social perceptual and functional features, represented as context-independent knowledge in the anterior and posterior temporal cortex; and central motive and emotional states, which correspond to context-independent activation in limbic and paralimbic structures." The suggestion is that "component representations interact and give rise to event–feature–emotion complexes through binding mechanisms" (Moll, Zahn, et al., 2005, p. 806). It is argued that these components can be selectively disrupted, giving rise to specific impairments, but as the model is currently articulated (Moll, Zahn, et al., 2005), these components are responding to all forms of norms.

However, unitary morality viewpoints face significant difficulties. One major problem is the existence of the moral/conventional distinction (Nucci & Turiel, 1978; Turiel et al., 1987; Smetana, 1993). Moral transgressions (e.g., one person hitting another) are defined by their consequences for the rights and welfare of others. Social conventional transgressions are defined as violations of the behavioral uniformities that structure social interactions within social systems (e.g., dressing in opposite gender clothing). Healthy individuals distinguish conventional and moral transgressions in their judgments from the age of 39 months (Smetana, 1981) and across cultures (Song, Smetana, & Kim, 1987). In particular, moral transgressions are judged less *rule contingent* than conventional transgressions; i.e., individuals are less likely to state that moral, rather than conventional, transgressions are permissible in the absence of prohibiting rules (Turiel et al., 1987).

A unitary processing position such as Moll's could suggest that the three component system responds differently to moral and conventional transgressions (though the details of the difference in response would have to be specified). However, such a position faces difficulty in the face of data that individuals with psychopathy appear intact regarding the processing of conventional transgressions but have marked impairment processing moral transgressions (Blair, 1995; Blair & Cipolotti, 2000). Why should a neuro-psychiatric population face difficulty with one type of processing, but not another type of processing, within a unitary system?

Developmental unitary morality positions suggest that there is a unitary method of social rule learning; e.g., primarily through verbal based cultural transmission (Shweder, Mahapatra, & Miller, 1987) or punishment (Trasler, 1978). The problem for such positions is explaining how a unitary method of social rule learning could give rise to a situation where children process moral transgressions differently from conventional transgressions. Moreover, it appears clear that the social consequences of moral and conventional transgressions are fundamentally different; caregivers are more likely to refer transgressors of moral rules to the consequences of their actions for the victim and transgressors of conventional rules to the rules themselves or the sanctions against prohibition (Nucci & Nucci, 1982; Grusec & Goodnow, 1994); i.e., the central assumption of such positions appears to be incorrect.

In addition to the moral/conventional distinction data, it is worth noting recent data provided by Haidt and colleagues (Haidt & Graham, 2007). These authors have reported individual differences in peoples' conceptions of what is relevant to their moral judgments. They observed that political liberals regard issues relating to others' welfare and reciprocity as significantly more relevant than political conservatives. Political conservatives regarded issues of disgust and hierarchy as significantly more relevant than political liberals (Haidt & Graham, 2007). Again such data are not obviously predicted by unitary morality viewpoints.

Multiple Moralities

An early view considering that there might be separable systems mediating different aspects of social rule information was provided by Turiel and colleagues (Nucci & Turiel, 1978; Turiel et al., 1987; Smetana, 1993). According to this viewpoint, the child generates "understandings of the social world by forming intuitive theories regarding experienced social events" (Turiel et al., 1987, p. 170). The differing social consequences of moral and conventional transgressions result in the child's development of separate theories (termed within this literature as domains) for moral and conventional transgressions.

A major difference between the domain theorists and more recent multiple morality positions is the level of emphasis on emotion. For example, Turiel and colleagues (1987) wrote that while they did not "exclude the emotional features of individuals' moral and social orientations" they regarded that "children's judgments and conceptual transformations . . . are a central aspect of moral and social convention" (Turiel et al., 1987, p. 170). In contrast, the more recent multiple morality positions regard emotional responding as central forms of processing (e.g., Blair, 1995; Haidt, 2001; see also Nichols, 2004; Prinz, 2007).

The problem for positions such as Turiel's suggesting the importance of conceptual transformations and abstract reasoning is that two major predictions derived from them are not supported. These are: First, populations with impaired abstract reasoning should fail to develop the moral/conventional distinction. However, no data has been presented in support of this position. Moreover, at least one population who might be considered to be impaired in both conceptual transformations and abstract reasoning, children with autism, pass the moral/conventional distinction (Blair, 1996; Steele, Joseph, & Tager-Flusberg, 2003; Leslie, Knobe, & Cohen, 2006; Rogers, Viding, James Blair, Frith, & Happe, 2006). Second, populations who show impairment in the development of the moral/conventional distinction should show impairment in conceptual transformations and abstract reasoning. Individuals with psychopathy show significantly less of a moral/conventional distinction than healthy individuals (Blair, 1995, 1997). However, no data suggests impairment in either conceptual transformations or abstract reasoning in this population.

More recently, a greater number of at least partially independent systems for moral reasoning have been suggested (Blair, 1995; Haidt, 2001; Blair, Marsh, Finger, et al., 2006; Haidt, 2007). Blair has suggested the existence of at least four partially separable neuro-cognitive systems that mediate different aspects of social rule related processing which can be considered moral. Haidt and Joseph concluded that there are five psychological systems, each with its own evolutionary history, that give rise to moral intuitions across cultures (Haidt & Joseph, 2004).

In some respects, the main differences between Blair and Haidt's positions are in emphasis. Haidt and Joseph take a evolutionary psychology perspective and consider that each moral system has its own evolutionary history. In contrast, Blair considers that at least some of these systems (the care- and disgust-based moralities), reflect the functioning of more general emotional learning mechanisms. These emotional learning mechanisms certainly will have an evolutionary history. However, it is unclear whether that history solely relates to morality. In addition, Haidt is relatively uninterested in the neural and cognitive architectures underpinning these types of process. In contrast, this is of central relevance to Blair.

A second difference concerns what Haidt and Joseph (2004) termed the "ingroup system". They argue that "the long history of living in kin-based groups of a few dozen individuals . . . has led to special social-cognitive abilities, backed up by strong social emotions related to recognizing, trusting, and cooperating with members of one's co-residing ingroup, while being wary and distrustful of members of other groups." They consider that these "special social-cognitive abilities" allow virtues of "loyalty, patriotism, and heroism". The details of this system have yet to be specified. Blair has no similar speculation within his framework.

A third difference concerns what can be considered non-affect based reasoning. From Haidt's perspective all morality is based on affect. However, this may not be accurate. Recent work has examined neural responses while individuals reason about Trolley problems (Greene, Sommerville, Nystrom, Darley, & Cohen, 2001; Greene, Nystrom, Engell, Darley, & Cohen, 2004). In these problems and their variants, a circumstance (e.g., a run-away trolley) is described as about to cause substantial loss of life (five people will be killed if it proceeds on its present course). Individual variants of the problem provide the participant with different options for saving the five though these typically have implications for at least one other individual. For example, one option is to hit a switch that will turn the trolley onto an alternate set of tracks where it will kill one person instead of five. Another option is to push a stranger off the bridge onto the tracks below. This person will die but his body will stop the trolley from reaching the five. Most people choose to save the life of the five, if the option to do so involves hitting the switch. However, most people choose *not* to save the five, if the option involves pushing another individual on to some tracks. The big difference between whether one saves the five or the one appears to be the salience of the suffering of the one (see section below: care-based morality). If the individual victim's distress is salient (i.e., because you push him off the bridge), regions implicated in emotional responding (e.g., ventromedial prefrontal cortex) are significantly activated and the one is saved at the expense of the five. In contrast, if the individual victim's distress is not salient (i.e., because you only push a button), regions implicated in emotional responding are not significantly activated and the five are saved at the expense of the one (Greene et al., 2001). Interestingly, regions implicated when reasoning in situations where the victim's distress is not salient, dorsolateral prefrontal cortex and parietal cortex, are also implicated in mathematical processing (Feigenson, Dehaene, S., & Spelke, 2004). It appears plausible that in situations where there is no salient victim, participants may simply turn this "moral" reasoning task into a mathematics problem: i.e., which is bigger five or one?

In this next section of the chapter, we will consider Haidt's and Blair's positions with respect to four systems for morality: care-based (harm in Haidt's terminology), reciprocity, social convention (hierarchy in Haidt's terminology) and disgust-based (purity in Haidt's terminology).

The Development of Care-Based Morality

The moral domain, as considered by the domain theorists (Nucci & Turiel, 1978; Turiel et al., 1987; Smetana, 1993), involved both care-based and reciprocity reasoning. In contrast, both Blair and Haidt have wished to distinguish between processing involved in care-based and reciprocity reasoning (Blair, 1995; Haidt & Joseph, 2004; Blair, Marsh, Finger, et al., 2006). These two forms of processing are distinguished here.

According to Haidt and Graham (2007), "the long history of mammalian evolution has shaped maternal brains to be sensitive to signs of suffering in one's own offspring. In many primate species, particularly humans, this sensitivity has extended beyond the mother-child relationship, so that all normally developed individuals dislike seeing suffering in others, and have the potential to feel the emotion of compassion in response."

However, there are several problems with Haidt's position. First, it remains unspecified how and why "this sensitivity... extended beyond the mother-child relationship". Given that this is primarily a evolutionary psychological model, this is a serious issue. Second, many social animals, beyond primates, find the distress of conspecifics aversive (Church, 1959; Rice & Gainer, 1962; Masserman, Wechkin, & Terris, 1964; Rice, 1965). For example, if a rat learns that pressing a bar will lower another, suspended rat to the ground (a distressing experience for the suspended rat), the rat will press the lever to allow the distressed rat to reach the ground (Rice & Gainer, 1962). Third, no detail is provided regarding the functioning of the systems at the neural and cognitive levels.

Blair has been less interested in developing an evolutionary psychology account and instead concentrated on developing a position specified at the cognitive and neural levels. It is worth remembering that the transfer of valence information between conspecifics is seen in many species (Mineka & Zinbarg, 2006). It is possible that the evolutionary pressures were on these capacities, important for learning which foods to approach and avoid, which objects are threatening) and that moral rules are the verbalizable byproduct of this learning when it related to the behavior of conspecifics.

The original position was the Violence Inhibition Mechanism model (Blair, 1995) which has been more recently considerably extended as the Integrated Emotion Systems (IES) model (Blair, 2004; Blair, Marsh, Finger, et al., 2006). Notably, the IES model is a model of emotional processing; i.e., an evolutionary psychological account of its existence should focus on why specific aspects of emotional processing have evolved to exist. Even aspects regarding learning on the basis of the emotional signals of others are not morality specific. Thus, fearful expressions allow rapid communication of valence information across conspecifics; i.e., the infant monkey learns to avoid objects that caregiver monkeys have shown fear reactions to (Mineka & Zinbarg, 2006).

There are two main components of the IES model that are particularly relevant here: an emotional learning system mediated by the amygdala and a system necessary for "decision making" on the basis of reinforcement expectations mediated by medial orbital frontal cortex.

The emotional learning system allows conditioned stimuli (CSs; including representations of moral transgressions) to be associated with unconditioned stimuli (USs; including the victim's distress cues). This proposed emotional learning system has functional similarities with the earlier ideas on the functioning of the VIM; it allows the individual to learn about both the "goodness" and "badness" of objects on the basis of moral socialization. However, the IES reformulation concludes that the functioning of this system should be understood in terms of the functioning of the amygdala. The amygdala is crucial for the formation of stimulus-reinforcement associations generally; i.e., it allows previously neutral objects to come to be valued as either good or bad according to whether they were associated with punishment or reward (LeDoux, 1998; Everitt, Cardinal, Parkinson, & Robbins, 2003).

This position predicts that individuals who show reduced processing of the distress of others and/ or problems with stimulus-reinforcement learning (i.e., impaired amygdala functioning) should show, developmentally, reduced indications of appropriate moral socialization. Individuals with psychopathy show reduced autonomic responses to the distress of others (Aniskiewicz, 1979; Blair, Jones, Clark, & Smith, 1997) and reduced recognition of sad and fearful expressions (Blair, Colledge, Murray, & Mitchell, 2001; Dolan & Fullam, 2006). In addition, they show impairment in measures of stimulus-reinforcement learning (Birbaumer et al., 2005). An indication of appropriate moral socialization is the development of the moral/conventional distinction (Blair, Marsh, Finger, et al., 2006). Individuals with psychopathy make significantly less of a moral/conventional distinction than comparison individuals (Blair, 1995). Moreover, data suggest that socialization (parenting style) has significantly less of an impact on the level of antisocial behavior in children showing the emotional impairment associated with psychopathy relative to healthy children (Wootton, Frick, Shelton, & Silverthorn, 1997; Oxford, Cavell, & Hughes, 2003).

Kochanska and others have found that fearful children show higher levels of moral development/ conscience using a variety of measures (Asendorpf & Nunner-Winkler, 1992; Kochanska, De Vet, Goldman, Murray, & Putman, 1994; Rothbart, Ahadi, & Hershey, 1994; Kochanska, 1997). Yet we do not frighten our children into being socialized; empathy induction is a significantly more effective socialization strategy than punishment (Hoffman, 1970; Brody & Shaffer, 1982). However, fearfulness can be considered an index of the integrity of the amygdala (Blair, 2003). As such it indexes the integrity of the neural system necessary for responding appropriately to distress cues and moral socialization.

The representation of reinforcement expectancies and decision making. One major addition within the IES over the earlier VIM model is the attempt to provide an account of a particular type of decision making relevant to emotion and, by extension, morality (Blair, 2004). The suggestion is that there is a system which receives information on the expected reinforcement associated with a particular object/action (aversion induced by a potential victim's distress or reward induced by observed happiness). This reinforcement information then allows decisions regarding whether to approach or avoid the object or enact or not the behavior. Expectations of reward will encourage approaches towards the object/action while expectations of punishment will encourage avoidance. This representation of reinforcement information is thought to be mediated by ventromedial prefrontal cortex (Blair, 2004). Considerable data supports a role for ventromedial frontal cortex in the representation of reinforcement information (Knutson, Fong, Adams, Varner, & Hommer, 2001; Paulus, Hozack, Frank, & Brown, 2002; Gottfried & Dolan, 2004; Rogers et al., 2004; Blair, Marsh, Morton, et al., 2006; Valentin, Dickinson, & O'Doherty, 2007).

A series of studies have examined neural activity in response to care-based morality. These studies have investigated responses to trolley problems (Greene et al., 2001, 2004), passive responding to scenes of moral transgressions (Moll, Oliveira-Souza, Eslinger, et al., 2002; Harenski & Hamann, 2006), judging descriptions of behaviors as moral or immoral (Moll, Oliveira-Souza, Bramati, & Grafman, 2002; Heekeren, Wartenburger, Schmidt, Schwintowski, & Villringer, 2003; Heekeren et al., 2005; Borg, Hynes, Van Horn, Grafton, & Sinnott-Armstrong, 2006) or during performance of a morality implicit association task (Luo et al., 2006). Many of these studies report amygdala activity (Moll, Oliveira-Souza, Eslinger, et al., 2002; Greene et al., 2004; Heekeren et al., 2005; Harenski & Hamann, 2006; Luo et al., 2006); all of the picture-based studies (Moll, Oliveira-Souza, Eslinger, et al., 2002; Harenski & Hamann, 2006; Luo et al., 2006) though only 2/6 text-based studies (Greene et al., 2004; Heekeren et al., 2005). All report vmPFC activity. In short, the neuro-imaging literature on care-based morality consistently agrees on the important role of vmPFC and less consistently agrees on the role of the amygdala.

It should be noted that the presence of a victim in distress need not automatically activate this circuitry. While some do consider that the amygdala is automatically activated by exposure to distress cues, for example the fearful expressions of others (Vuilleumier, Armony, Driver, & Dolan, 2001; Dolan & Vuilleumier,

2003), repeated studies have demonstrated amygdala activity to fearful expressions, and other emotional stimuli, is under considerable attentional control (Pessoa, McKenna, Gutierrez, & Ungerleider, 2002; Pessoa, Padmala, & Morland, 2005; Blair et al., 2007; Mitchell et al., 2007). According to a dominant view on attention, stimuli compete for representation and those that "win" this competition are those that are attended to (Desimone & Duncan, 1995). Stimuli can be primed to win this representational competition by top–down control (i.e., by frontally mediated executive attentional mechanisms) as well as bottom-up mechanisms (i.e., as a function of properties of the visual system). If representations necessary for task performance are primed, other representations, including emotional representations, will be less strongly activated; due to suppression from the task relevant representations. This will lead to reduced amygdala activation to the emotional representations (cf. Pessoa et al., 2002; Mitchell et al., 2007). Similarly, it has been shown that participants can direct their attention away from the emotional components of emotional scenes in emotional reappraisal paradigms (Ochsner & Gross, 2005). This again can reduce amygdala activation.

In short, whether an individual represents the distress of another individual will depend on a variety of factors, including the degree to which they are attending to, priming the representations of other stimuli in the environment and/or whether they are engaged in any attentional appraisal of the environment. Some recent work by Leslie and colleagues can be understood in this light (Leslie, Mallon, & Dicorcia, 2006). They demonstrated that children, including children with autism, did not consider it wrong to commit an action that resulted in another's distress, if the other individual's distress was unreasonable (e.g., they were crying because, although they had already eaten a cookie, they were not being allowed to eat the protagonist's cookie). The argument here is that under these circumstances the individual's attentional focus on an "unreasonable victim" is very different from on a "reasonable victim"; the unreasonableness of the individual's behavior is the focus of attention rather than their distress.

Reciprocity

While the early domain theorists considered conceptions of justice/reciprocity as part of the moral domain, both Blair and Haidt have considered the systems mediating reciprocity can be considered at least partially independent of those mediating care-based morality. Haidt has considered that "the long history of alliance formation and cooperation among unrelated individuals in many primate species has led to the evolution of a suite of emotions that motivate reciprocal altruism". However, the details of the processing involved have not been specified by either Haidt or Blair. Preliminary data that would allow a neuro-cognitive account is only beginning to emerge (Robertson et al., 2007). As such reciprocity will not be considered further here.

Disgust-Based Morality

Haidt has emphasized disgust as core in a system he has termed "purity" (Haidt, 2001; Nichols, 2002; Haidt & Joseph, 2004; Haidt & Graham, 2007). According to Haidt: "In many cultures, disgust goes beyond such contaminant-related issues and supports a set of virtues and vices linked to bodily activities in general, and religious activities in particular" (Haidt & Graham, 2007). Certainly, stimuli associated with disgust can reinforce moral intuitions. In an interesting study, Wheatley and Haidt (2005) gave highly hypnotizable participants a posthypnotic suggestion to feel a flash of disgust whenever they read an arbitrary word. The participants then rated moral transgressions that either did or did not include the disgust-inducing word. Moral transgression statements containing the word associated with disgust were regarded as more severe than those that did not contain these words (Wheatley & Haidt, 2005).

Haidt has not specified in detail the neuro-cognitive architecture that might mediate and allow the development disgust-based (purity) morality. However, it is important to note that disgusted expressions, like fearful, sad and happy expressions, are reinforcers. Usually, they provide information about foods (Rozin, Haidt, & McCauley, 1993). In particular, they allow the rapid transmission of taste aversions; the observer is warned not to approach the food that the expresser is displaying the disgust reaction to. However, they are clearly also used to convey distaste at another individual's actions. By doing so they allow the development of a disgust-based morality where the emotional force behind the proscribed actions is not anger but disgust.

Disgusted expressions have been shown to engage the insula and putamen (Phillips et al., 1997, 1998; Sprengelmeyer, Rausch, Eysel, & Przuntek, 1998). The insula is also very important for taste aversion learning (Cubero, Thiele, & Bernstein, 1999). Interestingly, recent fMRI work has indicated that "purity" based transgressions (transgressions associated with "indignation" based on the terminology of the authors) are indeed associated with insula (as well as orbital and inferior frontal cortex) activity (Moll, Oliveira-Souza, et al., 2005). Given that disgust-based learning recruits regions currently not thought to be dysfunctional in psychopathy, it is plausible that disgust-based morality may be intact in individuals with psychopathy.

Social Convention

The domain theorists were the first to consider that systems processing social conventional rules might be dissociable from those processing moral (care or justice based) rules (e.g., Nucci, 1982; Turiel et al., 1987; Smetana, 1993). Conventional transgressions are social order based (e.g., talking in class, dressing in opposite sex clothes). No-one is hurt by the transgression but the action is proscribed. The domain theorists, particularly Turiel, stressed the importance of conceptual transformations and abstract reasoning in the development of these separate domains (Turiel et al., 1987). This position was critiqued above.

Kagan argued that morality was distinguished from convention because only morality was associated with emotional responding (Kagan & Lamb, 1987). However, there are emotional responses associated with conventional transgressions, in particular anger. For example, a child continuously talking in the classroom is likely to lead to the teacher being angry. Similarly, the talking child presumably finds that the teacher lacks sufficient authority; he/she has a low expectation of the teacher's anger. In short, the moral/conventional distinction is not a distinction between affect and non-affect based rules.

Haidt has proposed the existence of a "hierarchy" system and argues that "the long history of living in hierarchically-structured ingroups, where dominant males and females get certain perquisites but are also expected to provide certain protections or services, has shaped human (and chimpanzee, and to a lesser extent bonobo) brains to help them flexibly navigate in hierarchical communities (Haidt & Joseph, 2004; Haidt & Graham, 2007). However, the details of this view from a neuro-cognitive perspective remain underspecified.

Blair has proposed a neuro-cognitive model of social conventional reasoning; the Social Response Reversal (SRR) system model (Blair & Cipolotti, 2000; Blair, 2004). SRR is thought to be initiated by aversive social cues (particularly, but not limited to, angry expressions) or expectations of such cues (as would be engendered by representations previously associated with such cues; i.e., representations of actions that make other individuals angry). This system is considered to (1) guide the individual away from committing conventional transgressions (particularly in the presence of higher status individuals) and; (2) orchestrate a response to witnessed conventional transgressions (particularly when these are committed by lower status individuals) (Blair & Cipolotti, 2000).

Conventional transgressions are considered to be bad because of their disruption of the social order (Nucci et al., 1983; Turiel et al., 1987; Smetana, 1993). Societal rules concerning conventional transgressions function to allow higher status individuals to constrain the behavior of lower status individuals. They may also, by their operation, serve to reduce within species hierarchy conflict.

Principal neural systems implicated in SRR include dorsal anterior cingulate cortex/dorsomedial frontal cortex and ventrolateral prefrontal cortex (Brodmann's area 47). These regions are activated more generally when behavior needs to be altered (Botvinick, Cohen, & Carter, 2004; Budhani, Marsh, Pine, & Blair, 2007). Cues to alter behavior include the angry expression of others and expectations of another's anger. Angry expressions, and other expressions, activate dorsomedial frontal cortex and ventrolateral prefrontal cortex (Sprengelmeyer et al., 1998; Blair, Morris, Frith, Perrett, & Dolan,, 1999). In addition, this region shows increased activity when the individual becomes angry (Dougherty et al., 1999). Importantly, this region shows activity when participants consider conventional transgressions (Berthoz, Armony, Blair, & Dolan, 2002).

Damage to frontal regions is a risk factor for engaging in inappropriate behaviors (Damasio, Tranel, & Damasio, 1990; Damasio, 1994; Blair & Cipolotti, 2000;

Lough et al., 2006). In addition, patients with such lesions show difficulties processing conventional transgressions as indexed by performance on Dewey's (1991) social contexts task (Blair & Cipolotti, 2000; Lough et al., 2006). Interestingly, a developmental form of SRR dysfunction may be represented by childhood bipolar disorder (Leibenluft, Blair, Charney, & Pine, 2003). Children with this disorder show indications of inappropriate processing of hierarchy information and impairment in social conventional rule knowledge (McClure et al., 2005).

It is worthwhile considering work on the social emotion of embarrassment here. A variety of authors have suggested that embarrassment serves an important social function by signaling appeasement to others (Leary, Landel, & Patton, 1996; Gilbert, 1997; Keltner & Anderson, 2000). When a person's untoward behavior threatens his/her standing in an important social group, visible signs of embarrassment function as a non-verbal acknowledgement of shared social standards. The basic idea is that embarrassment serves to aid the restoration of relationships following social transgressions (Keltner & Buswell, 1997). Considerable data supports this "appeasement" or remedial function of embarrassment in humans and non-human primates reviews (Leary & Meadows, 1991; Gilbert, 1997; Keltner & Buswell, 1997; Keltner & Anderson, 2000). Patients with lesions to orbital and inferior frontal cortex show difficulty in responding to the embarrassment of others (Beer et al., 2003). Moreover, recent work has shown that embarrassing scenes lead to activation in those regions necessary for social response reversal; i.e., dorsomedial and inferior frontal cortex (Finger, Marsh, Kamel, Mitchell, & Blair, 2006).

Theory of Mind and Morality

Although it has long been understood that information on the perpetrator's intent has an important impact when assigning moral blame or praise (Piaget, 1932), there has been a recent resurgence in understanding the relationship between Theory of Mind and moral reasoning (Zelazo, Helwig, & Lau, 1996; Knobe, 2003; Leslie et al., 2006; Young, Cushman, Hauser, & Saxe, 2007). Theory of Mind refers to the ability to represent the mental states of others, i.e., their thoughts, desires, beliefs, intentions and knowledge (Premack & Woodruff, 1978; Leslie, 1987; Frith, 1989). Individuals with autism have been consistently shown to be impaired in Theory of Mind (Frith & Happe, 2005; Frith & Frith, 2006).

Early views suggested that the representation of the mental states of others is necessary for empathy (Batson, Fultz, & Schoenrade, 1987; Feshbach, 1987). However, if empathy is necessary for care-based morality, as was argued above, then if these positions were correct, individuals with autism, lacking the ability to represent the mental states of others, should show pronounced impaired in moral development. However, they do not (Blair, 1996; Steele et al., 2003; Leslie et al., submitted). Individuals with autism make the moral/conventional distinction (Blair, 1996; Steele et al., 2003; Leslie et al., submitted). They even, like typically developing children, fail to consider it wrong to commit an action that resulted in another's distress if the other individual's distress was unreasonable (e.g., they were crying because, although they had already eaten a cookie, they were not being allowed to eat the protagonist's cookie) (Leslie et al., 2006).

I argued above that the basis of care-based morality relied on the capacity to respond to the distress cues of another and the ability to perform stimulus-reinforcement learning. There are data indicating that individuals with autism show some impairment in processing the expressions of others (Hobson, 1986; Bormann-Kischkel, Vilsmeier, & Baude, 1995; Howard et al., 2000; Humphreys, Minshew, Leonard, & Behrmann, 2007). However, this may simply relate to more general problems in processing face stimuli (Boucher & Lewis, 1992; Klin et al., 1999; Blair, Frith, Smith, Abell, & Cipolotti, 2002; Klin, Jones, Schultz, Volkmar, & Cohen, 2002). Certainly, the impairment does not appear to be the relatively selective dysfunction for fearful and sad expressions seen in psychopathy (Blair et al., 2001; Montagne et al., 2005; Dadds et al., 2006; Dolan & Fullam, 2006). With respect to stimulus-reinforcement learning, the current data indicates that this is intact in individuals with autism (Bernier, Dawson, Panagiotides, & Webb,, 2005; Gaigg & Bowler, 2007); though there are problems with over-generalization of the learning (Gaigg & Bowler, 2007), similar to those seen in patients with anxiety disorders (Lissek et al., 2005). In short, the capacities thought to be necessary for the development of care-based morality are intact in individuals with autism and the basics of care-based morality appear intact also (Blair, 1996; Steele et al., 2003; Leslie et al., 2006).

However, as noted above, information on the perpetrator's intent is important when assigning moral blame or praise. The individual who intentionally swings a baseball bat into another individual's face has behaved far more "wrongly" than the individual who unintentionally swings a baseball bat into another individual's face (cf. Zelazo et al., 1996). In short, information on the mental state of an individual can qualify the information provided by the basic emotion systems described above, reducing their impact on decision making if the action is regarded as unintentional. Individuals with Theory of Mind impairment (e.g., due to autism for example) show less use of mental state information in the qualification of their moral judgments based on intent information (Steele et al., 2003).

Theory of Mind and Social Convention

The activity of the SRR system is thought to be modulated by information on hierarchy and mental state (the latter provided by systems involved in Theory of Mind) (Blair & Cipolotti, 2000; Berthoz et al., 2002). The form of modulation will be dependent on whether the individual is the perpetrator (or considering being the perpetrator) of, or is the witness to, the conventional transgression.

If embarrassment does serve an important social function by signaling appeasement, the individual's perceived intention is likely to be crucial in determining whether they are expected to show embarrassment. If an individual intends to socially transgress, we might suspect that he/ she will not display appeasement (i.e., embarrassment) afterwards. If the transgression is intentional, the transgressor is unlikely to be interested in the social relationship that has been broken. In contrast, if the violation of the social convention was unintended then we might expect clear displays of embarrassment; the individual will have realized that they have transgressed and wish to restore the social relationship. Some work suggests that this is indeed the case (Berthoz et al., 2002).

Summary

In conclusion, six main claims are made in this chapter. First, there is no single neuro-cognitive architecture that processes all moral rules. Second, there are at least four different emotion-based learning mechanisms that allow the acquisition of social rules and provide them with emotive force. These are: care-based, reciprocity, social convention and disgust based. Any individual's concept of morality may include social rules learnt as a result of at least one of these emotion-based learning mechanisms (though not necessarily all; cf. Haidt & Graham, 2007). Third, care-based morality relates to those rules which proscribe actions that have been associated with the distress (fear and sadness) of others. The basis of care-based morality involves the recruitment of neural systems that process these expressions, allow stimulus-reinforcement learning and allow the use of this information in decision making. These systems include the amygdala and ventromedial frontal cortex. Fourth, disgust-based morality relates to those rules which proscribe actions that have been associated with the disgust of others. The basis of disgust-based morality involves the recruitment of neural systems that allow taste aversion learning. These systems include the insula. Fifth, social convention relates to those rules which proscribe actions that have been associated with the anger of others. The basis of social convention involves the recruitment of neural systems that allow response reversal to social stimuli. These systems include dorsomedial and ventrolateral frontal cortex. Sixth, Theory of Mind may not be necessary for the development of any of these bases of morality. However, representations of the mental states of others, their intentions, clearly modify moral reasoning. Intentional transgressions, whether care or disgust based or social conventional, are regarded as significantly worse than unintentional transgressions.

Acknowledgments This research was supported by the Intramural Research Program of the National Institutes of Health: National Institute of Mental Health.

References

Aniskiewicz, A. S. (1979). Autonomic components of vicarious conditioning and psychopathy. *Journal of Clinical Psychology, 35*, 60–67.

Asendorpf, J. B., & Nunner-Winkler, G. (1992). Children's moral motive strength and temperamental inhibition reduce their immoral behaviour in real moral conflicts. *Child Development, 63*, 1223–1235.

Batson, C. D., Fultz, J. & Schoenrade, P. A. (1987). Adults' emotional reactions to the distress of others. In N. Eisenberg & J. Strayer (Eds.), *Empathy and its development* (pp. 163–185). Cambridge: Cambridge University Press.

Bernier, R., Dawson, G., Panagiotides, H., & Webb, S. (2005). Individuals with autism spectrum disorder show normal responses to a fear potential startle paradigm. *Journal of Autism and Developmental Disorders, 35*(5), 1–9.

Beer, J. S., Heerey, E. A., Keltner, D., Scabini, D., & Knight, R. T. (2003). The regulatory function of self-conscious emotion: Insights from patients with orbitofrontal damage. *Journal of Personality and Social Psychology, 85*, 594–604.

Berthoz, S., Armony, J., Blair, R. J. R., & Dolan, R. (2002). Neural correlates of violation of social norms and embarrassment. *Brain, 125*, 1696–1708.

Birbaumer, N., Veit, R., Lotze, M., Erb, M., Hermann, C., Grodd, W., et al. (2005). Deficient fear conditioning in psychopathy: A functional magnetic resonance imaging study. *Archives of General Psychiatry, 62*, 799–805.

Blair, K. S., Marsh, A. A., Morton, J., Vythilingham, M., Jones, M., Mondillo, K., et al. (2006). Choosing the lesser of two evils, the better of two goods: Specifying the roles of ventromedial prefrontal cortex and dorsal anterior cingulate cortex in object choice. *Journal of Neuroscience, 26*, 11379–11386.

Blair, K. S., Smith, B. W., Mitchell, D. G., Morton, J., Vythilingam, M., Pessoa, L., et al. (2007). Modulation of emotion by cognition and cognition by emotion. *NeuroImage, 35*(1), 430–440.

Blair, R. J., Frith, U., Smith, N., Abell, F. & Cipolotti, L. (2002). Fractionation of visual memory: agency detection and its impairment in autism. *Neuropsychologia, 40*, 108–118.

Blair, R. J. R. (1995). A cognitive developmental approach to morality: Investigating the psychopath. *Cognition, 57*, 1–29.

Blair, R. J. R. (1996). Brief report: Morality in the autistic child. *Journal of Autism and Developmental Disorders, 26*, 571–579.

Blair, R. J. R. (1997). Moral reasoning in the child with psychopathic tendencies. *Personality and Individual Differences, 22*, 731–739.

Blair, R. J. R. (2003). A neurocognitive model of the psychopathic individual. In M. A. Ron & T. W. Robbins (Eds.), *Disorders of brain and mind 2* (pp. 400–420). Cambridge: Cambridge University Press.

Blair, R. J. R. (2004). The roles of orbital frontal cortex in the modulation of antisocial behavior. *Brain and Cognition, 55*, 198–208.

Blair, R. J. R., & Cipolotti, L. (2000). Impaired social response reversal: A case of "acquired sociopathy". *Brain, 123*, 1122–1141.

Blair, R. J. R., Colledge, E., Murray, L., & Mitchell, D. G. (2001). A selective impairment in the processing of sad and fearful expressions in children with psychopathic tendencies. *Journal of Abnormal Child Psychology, 29*, 491–498.

Blair, R. J. R., Jones, L., Clark, F., & Smith, M. (1997). The psychopathic individual: A lack of responsiveness to distress cues? *Psychophysiology, 34*, 192–198.

Blair, R. J. R., Marsh, A. A., Finger, E., Blair, K. S., & Luo, Q. (2006). Neuro-cognitive systems involved in morality. *Philosophical Explorations, 9*, 13–27.

Blair, R. J. R., Morris, J. S., Frith, C. D., Perrett, D. I., & Dolan, R. (1999). Dissociable neural responses to facial expressions of sadness and anger. *Brain, 122*, 883–893.

Borg, J. S., Hynes, C., Van Horn, J., Grafton, S., & Sinnott-Armstrong, W. (2006). Consequences, action, and intention as factors in moral judgments: an FMRI investigation. *Journal of Cognitive Neuroscience, 18*, 803–817.

Bormann-Kischkel, C., Vilsmeier, M., & Baude, B. (1995). The development of emotional concepts in autism. *Journal of Child Psychology and Psychiatry, 36*, 1243–1259.

Botvinick, M. M., Cohen, J. D., & Carter, C. S. (2004). Conflict monitoring and anterior cingulate cortex: An update. *Trends in Cognitive Science, 8*, 539–546.

Boucher, J., & Lewis, V. (1992). Unfamiliar face recognition in relatively able autistic children. *Journal of Child Psychology and Psychiatry and Allied Disciplines, 33*, 843–859.

Brody, G. H., & Shaffer, D. R. (1982). Contributions of parents and peers to children's moral socialisation. *Developmental Review, 2*, 31–75.

Budhani, S., Marsh, A. A., Pine, D. S., & Blair, R. J. (2007). Neural correlates of response reversal: Considering acquisition. *NeuroImage, 34*, 1754–1765.

Church, R. M. (1959). Emotional reactions of rats to the pain of others. *Journal of Comparative & Physiological Psychology, 52*, 132–134.

Colby, A., & Kohlberg, L. (1987). *The measurement of moral judgement*. New York: Cambridge University Press.

Cubero, I., Thiele, T. E., & Bernstein, I. L. (1999). Insular cortex lesions and taste aversion learning: Effects of conditioning method and timing of lesion. *Brain Research, 839*(2), 323–330.

Dadds, M. R., Perry, Y., Hawes, D. J., Merz, S., Riddell, A. C., Haines, D. J., et al. (2006). Attention to the eyes and fear-recognition deficits in child psychopathy. *British Journal of Psychiatry, 189*, 280–281.

Damasio, A. R. (1994). *Descartes' error: Emotion, rationality and the human brain*. New York: Putnam (Grosset Books) .

Damasio, A. R., Tranel, D., & Damasio, H. (1990). Individuals with sociopathic behaviour caused by frontal damage fail to respond autonomically to social stimuli. *Behavioural Brain Research 41*, 81–94.

Desimone, R., & Duncan, J. (1995). Neural mechanisms of selective visual attention. *Annual Review of Neuroscience, 18*, 193–222.

Dewey, M. (1991), Living with Asperger's syndrome. In U. Frith (Ed.), *Autism and Asperger's syndrome* (pp. 184–206). Cambridge: Cambridge University Press.

Dolan, M., & Fullam, R. (2006). Face affect recognition deficits in personality-disordered offenders: association with psychopathy. *Psychological Medicine, 36*, 1563–1569.

Dolan, R. J., & Vuilleumier, P. (2003). Amygdala automaticity in emotional processing. *Annals of the New York Academy of Sciences, 985*, 348–355.

Dougherty, D. D., Shin, L. M., Alpert, N. M., Pitman, R. K., Orr, S. P., Lasko, M., et al. (1999). Anger in healthy men: A PET study using script-driven imagery. *Biological Psychiatry, 46*, 466–472.

Everitt, B. J., Cardinal, R. N., Parkinson, J. A., & Robbins, T. W. (2003). Appetitive behavior: Impact of amygdala-dependent mechanisms of emotional learning. *Annual New York Academy of Sciences, 985*, 233–250.

Feigenson, L., Dehaene, S., & Spelke, E. (2004). Core systems of number. *Trends in Cognitive Sciences, 8*, 307–314.

Feshbach, N. D. (1987). Parental empathy and child adjustment/ maladjustment. In N. Eisenberg & J. Strayer (Eds.), *Empathy and its development*. New York: Cambridge University Press.

Finger, E. C., Marsh, A. A., Kamel, N., Mitchell, D. G., & Blair, J. R. (2006). Caught in the act: The impact of audience on the neural response to morally and socially inappropriate behavior. *NeuroImage, 33*, 414–421.

Frith, C. D., & Frith, U. (2006). The neural basis of mentalizing. *Neuron, 50*, 531–534.

Frith, U. (1989). *Autism: Explaining the enigma*. Oxford: Blackwell.

Frith, U., & Happe, F. (2005). Autism spectrum disorder. *Current Biology, 15*, R786–R790.

Gaigg, S. B., & Bowler, D. M. (2007). Differential fear conditioning in Asperger's syndrome: Implications for an amygdala theory of autism. *Neuropsychologia, 45*, 2125–2134.

Gilbert, P. (1997). The evolution of social attractiveness and its role in shame, humiliation, guilt and therapy. *British Journal of Medical Psychology, 70*, 113–147.

Gottfried, J. A., & Dolan, R. J. (2004). Human orbitofrontal cortex mediates extinction learning while accessing conditioned representations of value. *Nature Neuroscience, 7*, 1144–1152.

Greene, J. D., Nystrom, L. E., Engell, A. D., Darley, J. M., & Cohen, J. D. (2004). The neural bases of cognitive conflict and control in moral judgment. *Neuron, 44*, 389–400.

Greene, J. D., Sommerville, R. B., Nystrom, L. E., Darley, J. M., & Cohen, J. D. (2001). An fMRI investigation of emotional engagement in moral judgment. *Science, 293*, 1971–1972.

Grusec, J. E., & Goodnow, J. J. (1994). Impact of parental discipline methods on the child's internalization of values: A reconceptualization of current points of view. *Developmental Psychology, 30*, 4–19.

Haidt, J. (2001). The emotional dog and its rational tail: A social intuitionist approach to moral judgment. *Psychological Review, 108*, 814–834.

Haidt, J. (2007). The new synthesis in moral psychology. *Science, 31*, 998–1002.

Haidt, J., & Graham, J. (2007). When morality opposes justice: Conservatives have moral intuitions that liberals may not recognize. *Social Justice Research, 20*(1), 98–116.

Haidt, J., & Joseph, C. (2004). Intuitive ethics: How innately prepared intuitions generate culturally variable virtues. *Daedalus, Issue on Human Nature, 135*(4), 55–66.

Harenski, C. L., & Hamann, S. (2006). Neural correlates of regulating negative emotions related to moral violations. *NeuroImage, 30*, 313–324.

Heekeren, H. R., Wartenburger, I., Schmidt, H., Prehn, K., Schwintowski, H. P., & Villringer, A. (2005). Influence of bodily harm on neural correlates of semantic and moral decision-making. *NeuroImage, 24*, 887–897.

Heekeren, H. R., Wartenburger, I., Schmidt, H., Schwintowski, H. P., & Villringer, A. (2003). An fMRI study of simple ethical decision-making. *Neuroreport, 14*, 1215–1219.

Hobson, P. (1986). The autistic child's appraisal of expressions of emotion. *Journal of Child Psychology and Psychiatry, 27*, 321–342.

Hoffman, M. L. (1970). Conscience, personality and socialiszation techniques. *Human Development, 13*, 90–126.

Howard, M. A., Cowell, P. E., Boucher, J., Broks, P., Mayes, A., Farrant, A., et al. (2000). Convergent neuroanatomical and behavioural evidence of an amygdala hypothesis of autism. *Neuroreport, 11*, 1931–1935.

Humphreys, K., Minshew, N., Leonard, G. L., & Behrmann, M. (2007). A fine-grained analysis of facial expression processing in high-functioning adults with autism. *Neuropsychologia, 45*, 685–695.

Kagan, J., & Lamb, S. (1987). *The emergence of morality in young children* University of Chicago Press: Chicago.

Keltner, D., & Anderson, C. (2000). Saving face for Darwin: The functions and uses of embarrassment. *Current Directions in Psychological Science, 9*, 187–192.

Keltner, D., & Buswell, B. N. (1997). Embarrassment: its distinct form and appeasement functions. *Psychological Bulletin, 122*, 250–270.

Klin, A., Jones, W., Schultz, R., Volkmar, F., & Cohen, D. (2002). Visual fixation patterns during viewing of naturalistic social situations as predictors of social competence in individuals with autism. *Archives of General Psychiatry, 59*, 809–816.

Klin, A., Sparrow, S. S., de Bildt, A., Cicchetti, D. V., Cohen, D. J., & Volkmar, F. R. (1999). A normed study of face recognition in autism and related disorders. *Journal of Autism and Developmental Disabilities, 29*, 499–508.

Knobe, J. (2003). Intentional action and side effects in ordinary language. *Analysis, 63*, 190–193.

Knutson, B., Fong, G. W., Adams, C. M., Varner, J. L., & Hommer, D. (2001). Dissociation of reward anticipation and outcome with event-related fMRI. *Neuroreport, 12*, 3683–3687.

Kochanska, G. (1997). Multiple pathways to conscience for children with different temperaments: From toddlerhood to age 5. *Developmental Psychology, 33*, 228–240.

Kochanska, G., De Vet, K., Goldman, M., Murray, K., & Putman, P. (1994). Maternal reports of conscience development and temperament in young children. *Child Development, 65*, 852–868.

Leary, M. R., Landel, J. L., & Patton, K. M. (1996). The motivated expression of embarrassment following a self-presentational predicament. *Journal of Personality, 64*, 619–637.

Leary, M. R., & Meadows, S. (1991). Predictors, elicitors, and concomitants of social blushing. *Journal of Personality and Social Psychology, 60*, 254–262.

LeDoux, J. (1998). *The Emotional Brain*. New York: Weidenfeld & Nicolson.

Leibenluft, E., Blair, R. J., Charney, D. S., & Pine, D. S. (2003). Irritability in pediatric mania and other childhood psychopathology. *Annual New York Academy of Sciences, 1008*, 201–218.

Leslie, A. M. (1987). Pretense and representation: The origins of "theory of mind". *Psychological Review, 94*, 412–426.
Leslie, A. M., Knobe, J., & Cohen, A. (2006). Acting intentionally and the side-effect effect. *Psychological Science, 17*, 421–427.
Leslie, A. M., Mallon, R., & DiCorcia, J. A. (2006). Transgressors, victims and cry babies: Is basic moral judgment spared in autism? *Social Neuroscience, 1*, 270–283.
Lissek, S., Powers, A. S., McClure, E. B., Phelps, E. A., Woldehawariat, G., Grillon, C., et al. (2005). Classical fear conditioning in the anxiety disorders: a meta-analysis. *Behaviour Research and Therapy, 43*, 1391–1424.
Lough, S., Kipps, C. M., Treise, C., Watson, P., Blair, J. R. & Hodges, J. R. (2006). Social reasoning, emotion and empathy in frontotemporal dementia. *Neuropsychologia, 44*, 950–958.
Luo, Q., Nakic, M., Wheatley, T., Richell, R., Martin, A., & Blair, R. J. (2006). The neural basis of implicit moral attitude–an IAT study using event-related fMRI. *NeuroImage, 30*, 1449–1457.
Masserman, J. H., Wechkin, S. & Terris, W. (1964). "Altruistic" behavior in rhesus monkeys. *American Journal of Psychiatry, 121*.
McClure, E. B., Treland, J. E., Snow, J., Schmajuk, M., Dickstein, D. P., Towbin, K. E., et al. (2005). Deficits in social cognition and response flexibility in pediatric bipolar disorder. *American Journal of Psychiatry, 162*, 1644–1651.
Mineka, S., & Zinbarg, R. (2006). A contemporary learning theory perspective on the etiology of anxiety disorders: it's not what you thought it was. *American Psychologist, 61*, 10–26.
Mitchell, D. G., Nakic, M., Fridberg, D., Kamel, N., Pine, D. S., & Blair, R. J. (2007). The impact of processing load on emotion. *NeuroImage, 34*, 1299–1309.
Moll, J., Oliveira-Souza, R., Bramati, I. E., & Grafman, J. (2002). Functional networks in emotional moral and nonmoral social judgments. *NeuroImage, 16*, 696–703.
Moll, J., Oliveira-Souza, R., Eslinger, P. J., Bramati, I. E., Mourao-Miranda, J., Andreiuolo, P. A., et al. (2002). The neural correlates of moral sensitivity: a functional magnetic resonance imaging investigation of basic and moral emotions. *Journal of Neuroscience, 22*, 2730–2736.
Moll, J., Oliveira-Souza, R., Moll, F. T., Ignacio, F. A., Bramati, I. E., Caparelli-Daquer, E. M., et al. (2005). The moral affiliations of disgust: A functional MRI study. *Cognitive and Behavioral Neurology, 18*, 68–78.
Moll, J., Zahn, R., Oliveira-Souza, R., Krueger, F., & Grafman, J. (2005). Opinion: The neural basis of human moral cognition. *Nature Reviews Neuroscience, 6*, 799–809.
Montagne, B., van Honk, J., Kessels, R. P. C., Frigerio, E., Burt, M., van Zandvoort, M. J. E., et al. (2005). Reduced efficiency in recognising fear in subjects scoring high on psychopathic personality characteristics. *Personality and Individual Differences, 38*, 5–11.
Nichols, S. (2002). Norms with feeling: Towards a psychological account of moral judgment. *Cognition, 84*, 221–236.
Nichols, S. (2004). *Sentimental rules: On the natural foundations of moral judgment*. New York: Oxford University Press.
Nucci, L., Turiel, E., & Encarnacion-Gawrych, G. E. (1983). Social interactions and social concepts: Analysis of morality and convention in the Virgin Islands. *Journal of Cross Cultural Psychology, 14*, 469–487.
Nucci, L. P. (1982). Conceptual development in the moral and conventional domains: Implications for values education. *Review of Educational Research, 52*, 93–122.
Nucci, L. P., & Nucci, M. (1982). Children's social interactions in the context of moral and conventional transgressions. *Child Development, 53*, 403–412.
Nucci, L. P., & Turiel, E. (1978). Social interactions and the development of social concepts in preschollo children. *Child Development, 49*, 400–407.
Ochsner, K. N., & Gross, J. J. (2005). The cognitive control of emotion. *Trends in Cognitive Sciences, 9*, 242–249.

Oxford, M., Cavell, T. A., & Hughes, J. N. (2003). Callous-unemotional traits moderate the relation between ineffective parenting and child externalizing problems: A partial replication and extension. *Journal of Clinical Child and Adolescent Psychology, 32*, 577–585.

Paulus, M. P., Hozack, N., Frank, L., & Brown, G. G. (2002). Error rate and outcome predictability affect neural activation in prefrontal cortex and anterior cingulate during decision-making. *NeuroImage, 15*, 836.

Pessoa, L., McKenna, M., Gutierrez, E., & Ungerleider, L. G. (2002). Neural processing of emotional faces requires attention. *Proceedings of the National Academy of Sciences of the United States America, 99*, 11458–11463.

Pessoa, L., Padmala, S,. & Morland, T. (2005). Fate of unattended fearful faces in the amygdala is determined by both attentional resources and cognitive modulation. *NeuroImage, 28*, 249–255.

Phillips, M. L., Young, A. W., Scott, S. K., Calder, A. J., Andrew, C., Giampietro, V., et al. (1998). Neural responses to facial and vocal expressions of fear and disgust. *Proceedings of the Royal Society of London. Series B, 265*, 1809–1817.

Phillips, M. L., Young, A. W., Senior, C., Brammer, M., Andrews, C., Calder, A. J., et al. (1997). A specified neural substrate for perceiving facial expressions of disgust. *Nature, 389*, 495–498.

Piaget, J. (1932). *The moral development of the child*. London: Routledge and Kegan Paul.

Premack, D., & Woodruff, G. (1978). Does the chimpanzee have a theory of mind? *Behavioural and Brain Sciences, 1*, 515–526.

Prinz, J. (2007). *The emotional construction of morals*. New York: Oxford University Press.

Raine, A., & Yang, Y. (2006). Neural foundations to moral reasoning and antisocial behavior. *Social Cognitive and Affective Neuroscience, 1*, 203–213.

Rice, G. E. (1965). Aiding responses in rats: Not in guinea pigs. *Proceedings of the Annual Convention of the American Psychological Association, 73*, 105–106.

Rice, G. E., & Gainer, P. (1962). "Altruism" in the albino rat. *Journal of Comparative & Physiological Psychology, 55*, 123–125.

Robertson, D., Snarey, J., Ousley, O., Harenski, K., DuBois Bowman, F., Gilkey, R., et al. (2007). The neural processing of moral sensitivity to issues of justice and care. *Neuropsychologia, 45*, 755–766.

Rogers, J., Viding, E., James Blair, R., Frith, U., & Happe, F. (2006). Autism spectrum disorder and psychopathy: Shared cognitive underpinnings or double hit? *Psychological Medicine*, 1–10.

Rogers, R. D., Ramnani, N., Mackay, C., Wilson, J. L., Jezzard, P., Carter, C. S., et al. (2004). Distinct portions of anterior cingulate cortex and medial prefrontal cortex are activated by reward processing in separable phases of decision-making cognition. *Biological Psychiatry, 55*, 594.

Rothbart, M., Ahadi, S., & Hershey, K. L. (1994). Temperament an social behaviour in children. *Merrill-Palmer Quarterly, 40*, 21–39.

Rozin, P., Haidt, J., & McCauley, C. R. (1993). Disgust. In M. Lewis & J. M. Haviland (Eds.), *Handbook of emotions* (pp. 575–594). New York: The Guilford Press.

Shweder, R. A., Mahapatra, M., & Miller, J. G. (1987). Culture and moral development. In (J. Kagan & S. Lamb (Eds.), *The emergence of morality in young children* (pp. 1–83).Chicago: University of Chicago Press.

Smetana, J. G. (1981). Preschool children's conceptions of moral and social rules. *Child Development, 52*, 1333–1336.

Smetana, J. G. (1993). Understanding of social rules. In M. Bennett (Ed.), *The child as psychologist: An introduction to the development of social cognition* (pp. 111–141). New York: Harvester Wheatsheaf.

Song, M., Smetana, J. G., & Kim, S. Y. (1987). Korean children's conceptions of moral and conventional transgressions. *Developmental Psychology, 23*, 577–582.

Sprengelmeyer, R., Rausch, M., Eysel, U. T., & Przuntek, H. (1998). Neural structures associated with the recognition of facial basic emotions. *Proceedings of the Royal Society of London.B, 265*, 1927–1931.

Steele, S., Joseph, R. M., & Tager-Flusberg, H. (2003). Brief report: Developmental change in theory of mind abilities in children with autism. *Journal of Autism and Developmental Disorders, 33*, 461–467.
Trasler, G. (1978). Relations between psychopathy and persistent criminality – methodological and theoretical issues. In R. D. Hare & D. Schalling (Eds.), *Psychopathic behaviour: Approaches to research*.Chichester, England: Wiley.
Turiel, E., Killen, M., & Helwig, C. C. (1987). Morality: Its structure, functions, and vagaries. In J. Kagan & S. Lamb (Eds.), *The emergence of morality in young children* (pp. 155–245). Chicago: University of Chicago Press.
Valentin, V. V., Dickinson, A., & O'Doherty, J. P. (2007). Determining the neural substrates of goal-directed learning in the human brain. *Journal of Neuroscience, 27*, 4019–4026.
Vuilleumier, P., Armony, J. L., Driver, J., & Dolan, R. J. (2001). Effects of attention and emotion on face processing in the human brain: An event-related fMRI study. *Neuron, 30*, 829–841.
Wheatley, T., & Haidt, J. (2005). Hypnotic disgust makes moral judgments more severe. *Psychological Science, 16*, 780–784.
Wootton, J. M., Frick, P. J., Shelton, K. K., & Silverthorn, P. (1997). Ineffective parenting and childhood conduct problems: The moderating role of callous-unemotional traits. *Journal of Consulting and Clinical Psychology, 65*, 292–300.
Young, L., Cushman, F., Hauser, M., & Saxe, R. (2007). The neural basis of the interaction between theory of mind and moral judgment. *Proceedings of the National Academy of Sciences of the United States of America, 104*, 8235–8240.
Zelazo, P. D., Helwig, C. C., & Lau, A. (1996). Intention, act, and outcome in behavioural prediction and moral judgment. *Child Development, 67*, 2478–2492.

Jorge Moll's Response on James Blair's Paper

In his provocative chapter, Blair puts forth a "multiple moralities" approach, a view he shares with Jon Haidt and others, and contrasts it to what he interpreted from our work to be "a unitary view of morality". His criticisms, although well articulated, simply do not do apply to our research, less so to our theoretical premises, as we will attempt to briefly show below.

Firstly, we have never claimed that that we viewed morality as "unitary", either phenomenologically or in terms of brain mechanisms, less so that this was in opposition to any purported multiple model of morality. Our early reference to Colby, Kohlberg, Speicher, Hewer, Candee, and Gibss's work (Moll, Eslinger, & Oliveira-Souza, 2001) was a matter of borrowing an *operational* definition of moral judgment to start with, and cannot be taken as an argument to classify our views as "rationalistic" – even if that was the only experimental work ever carried out by our group. As often happens with flourishing lines of research, one regularly finds himself straddling on uncharted grounds, and must draw first on very general hypotheses. As far as we're aware, we were among the first to study the functional correlates of moral judgment with fMRI (Oliveira-Souza & Moll, 2000), and at that time we necessarily *had* to depart from general ideas. In our view, considering this methodological strategy to be a solid theoretical system of premises is an undue overinterpretation of our work.

Although we would refrain from using the expression "multiple moralities" (for reasons which fall beyond the scope of this short reply), we see no contradiction at all between Blair's views of multiple processes underlying morality and our own perspective on morality. While exposing his arguments, Blair contrasts his "multiple moralities" view to a "unitary morality" and refers to the following definition employed in our theoretical article (Moll, Zahn, Oliveira-Souza, Krueger, & Grafman, 2005), which was taken from MacIntyre's *After Virtue* (1984): "Here, morality is considered as the sets of customs and values that are embraced by a cultural group to guide social conduct, a view that does not assume the existence of absolute moral values". Once more, the fact that we have used an operational definition to delimit what we consider to be the moral domain sphere in no way should be taken to imply that we have adopted a "unitary" stance regarding either the phenomenology or the neural underpinnings of morality. On the contrary, in our model paper we explicitly describe how different types of impairment of moral conduct do emerge from dysfunction in discrete brain regions, and provide clear hypotheses on how damage to subcomponents of the various cortical and limbic regions that have been linked to moral cognition and behavior would lead to distinct types of moral dysfunction. Assuming that we believe in a "unitary morality" because we have pointed to the involvement of a broadly defined but stable neural architecture in diverse aspects of moral cognition and behavior is therefore a misinterpretation of our theoretical and experimental work. This would be equivalent to claim that if one believes that visual function, broadly speaking, relies on a well defined set of brain regions (the visual system), one would necessarily believe that all distinctive visual experiences (e.g., motion, color, space) rely on *exactly* the same neural substrates.

Secondly, Blair's claim that we favor rationalistic views in moral psychology is also surprising. We were among the first to recognize and experimentally explore the pivotal role of emotions both in moral judgment and in moral sensitivity – and this was in fact largely influenced by the work of Jon Haidt on moral and basic emotions. Our views on how emotions are critically linked to moral appraisals were articulated in two papers in which we devoted particular attention to the influence of distinct moral emotions in moral judgment (Moll, Oliveira-Souza, Bramati, & Grafman, 2002; Moll, Oliveira-Souza, Eslinger, et al., 2002), a fact that is overlooked in Blair's chapter. Furthermore, it is also interesting that Blair dedicates considerable space in his chapter discussing disgust as evidence for his multiple moralities view, and yet he fails to acknowledge our fMRI study on moral disgust, which was actually the first to directly probe the contribution of basic disgust and moral disgust to brain activation (Moll, Oliveira-Souza, et al., 2005).

Finally, James Blair unfortunately did not have the opportunity to assess our latest views on the contribution of attachment to morality, another piece of work which clearly reflects our view on how distinct neurobiological components and functions support diverse aspects of human morality.

In summary, Blair's qualification of "unitary moral system" is not appropriately applicable to our work. As such, our work is not actually a good framework against which to build criticisms in support of multiple moralities. If anything, were our work useful for the purpose of contrasting unitary *versus* multiple moralities, it would probably fit much more naturally as evidence *in support of*, not against, the latter view.

References

Daquer, E. M., et al. (2005). The moral affiliations of disgust – A functional MRI study. *Cognitive and Behavioral Neurology, 18*(1), 68–78.

MacIntyre, A. (1984). *After virtue: A study in moral theory* (2nd ed.). Notre Dame, IN: University of Notre Dame Press.

Moll, J., Oliveira-Souza, R., Bramati, I. E., & Grafman, J. (2002). Functional networks in emotional moral and nonmoral social judgments. *NeuroImage, 16*(3), 696–703.

Moll, J., Oliveira-Souza, R., Eslinger, P. J., Bramati, I. E., Mourao-Miranda, J., Andreiuolo, P. A., et al. (2002). The neural correlates of moral sensitivity: A functional magnetic resonance imaging investigation of basic and moral emotions. *Journal of Neuroscience, 22*(7), 2730–2736.

Moll, J., Oliveira-Souza, R., Moll, F. T., Ignacio, F. A., Bramati, I. E., Caparelli-Daquer, E. M., et al. (2005). The moral affiliations of disgust: A functional MRI study. *Cognitive and Behavioral Neurology, 18*(1), 68–78.

Moll, J., Eslinger, P. J., & Oliveira-Souza, R. (2001). Frontopolar and anterior temporal cortex activation in a moral judgment task – Preliminary functional MRI results in normal subjects. *Arquivos De Neuro-Psiquiatria, 59*(3B), 657–664.

Moll, J., Zahn, R., Oliveira-Souza, R., Krueger, F., & Grafman, J. (2005). The neural basis of human moral cognition. *Nature Reviews Neuroscience, 6*(10), 799–809.

Oliveira-Souza, R., & Moll, J. (2000). The moral brain: A functional MRI study of moral judgment. *Neurology, 54*(7), A104–A104.

Empathy and Morality: Integrating Social and Neuroscience Approaches

Jean Decety and C. Daniel Batson

Philosophers and psychologists have long debated the nature of empathy (e.g., Batson, 1991; Eisenberg, 2000; Ickes, 2003; Thompson, 2001), and whether the capacity to share and understand other people's emotions sets humans apart from other species (e.g., de Waal, 2005). Here, we consider empathy as a construct accounting for a sense of similarity in feelings experienced by the self and the other, without confusion between the two individuals (Decety & Jackson, 2004; Decety & Lamm, 2006). The experience of empathy can lead to empathic concern (concern for another based on the apprehension or comprehension of the other's emotional state or condition) or to personal distress (i.e., an aversive, self-focused emotional reaction to the apprehension or comprehension of another's emotional state or condition). Understanding empathic processes is essential for understanding human social and moral development (Eisenberg et al., 1994). Furthermore, various psychopathologies are marked by empathy deficits, and a wide array of psychotherapeutic approaches stress the importance of clinical empathy as a fundamental component of treatment (Farrow & Woodruff, 2007).

In recent years, there has been a great upsurge in neuroimaging investigations of empathy. Most of these studies reflect the new approach of social neuroscience, which combines research designs and behavioral measures used in social psychology with neuroscience markers (Decety & Batson, 2007). Such an approach plays an important role in disambiguating competing theories in social psychology in general and in empathy-related research in particular (Decety & Hodges, 2006). For instance, two critical questions debated among social psychologists are whether perspective-taking instructions induce empathic concern and/or personal distress, and to what extent prosocial motivation springs from self-other overlap.

In this chapter, we focus on recent social neuroscience research exploring how people respond behaviorally and neurally to the pain of others. The perception of others in painful situations constitutes an ecologically valid way to investigate the mechanisms underpinning the experience of empathy. Findings from these studies

J. Decety (✉)
Departments of Psychology and Psychiatry, and Center for Cognitive and Social Neuroscience, The University of Chicago, 5848 S. University Avenue, Chicago, IL 60637, USA
e-mail: decety@uchicago.edu

J. Verplaetse et al. (eds.), *The Moral Brain*, DOI 10.1007/978-1-4020-6287-2_5,

Table 1 Major neurological components involved in the experience of empathy

In social neuroscience, empathy refers to a psychological construct that involves representations (i.e., memories that are localized in distributed neural networks that encode information and, when temporarily activated, enable access to this stored information, e.g., shared affective representations) and processes (i.e., computational procedures that are neurally localized and are independent of the nature or modality of the stimulus that is being processed – e.g., decoupling mechanism between self and other).

Current neuroscience research suggests that empathy can be influenced by both bottom-up (rapid and unconscious) and top-down (conscious) information processing and can be broken down into a number of interacting macro-components (Decety & Jackson, 2004; Decety & Meyer, 2008):

1. Motor and physiological resonance mediated by the perception-action direct coupling mechanism and the autonomic nervous system that regulates bodily states, emotion and reactivity. This aspect primarily draws on motor, premotor, and somatosensory cortices, limbic system, and insula.
2. Meta-cognitive abilities to infer or imagine another person's thoughts or feelings, to infer or imagine one's own thoughts or feelings in another's situation, including the capacity to distinguish between one's own thoughts and those of others, which is a key component of interpersonal interactions. The medial prefrontal cortex, dorsal anterior cingulate cortex and temporo-parietal junction play a critical role in these processes.
3. Emotion regulation modulates negative arousal and adds flexibility, allowing the individual to be less dependent on external cues. The lateral prefrontal cortex and the anterior cingulate with their reciprocal connections to the amygdala orbitofrontal cortex are part of a circuit that regulates emotion and cognition.

It is worth noting that there are bidirectional anatomical and functional links between the (widely distributed) areas in which representation of emotions are temporarily activated (including autonomic and somatic responses) during empathic experience and the areas involved in emotion regulation and meta-cognition. Each region has unique patterns of intracerebral connections, which determine its function, and differences in neural activity during the experience of empathy are produced by distributed subsystems of brain regions. Even though there is massive parallel processing, the dynamic interaction of these regions is also an important aspect to be investigated further.

demonstrate that the mere perception of another individual in pain results, in the observer, in the activation of part of the neural network involved in the processing of first-hand experience of pain. This intimate overlap between the neural circuits responsible for our ability to perceive the pain of others and those underpinning our own self-experience of pain supports the shared-representation theory of social cognition. This theory posits that perceiving someone else's emotion and having an emotional response or subjective feeling state both fundamentally draw on the same computational processes and rely on somatosensory and motor representations (see Table 1). However, we argue that a complete self-other overlap can lead to personal distress and possibly be detrimental to empathic concern (Decety & Lamm, 2009). Personal distress may result in egoistic motivation to reduce it, by withdrawing from the stressor, for example, thereby decreasing the likelihood of prosocial behavior (Tice, Bratslavsky, & Baumeister, 2001).

We first consider the neuroanatomy of empathy from an evolutionary perspective. Then we present recent functional neuroimaging studies showing the involvement of shared neural circuits during the observation of pain in others and during the experience of pain in the self. Third, we discuss how perspective taking and the ability to differentiate the self from the other affect this sharing mechanism. Fourth, we examine how some interpersonal variables modulate empathic concern and personal distress. Finally, we suggest that empathic concern may be a source of immoral as well as moral behavior.

Evolutionary Origins of Empathy

Natural selection has fine-tuned the mechanisms that serve the specific demands of each species' ecology, and social behaviors are best understood in the context of evolution. MacLean (1985) has proposed that empathy emerged in relation with the evolution of mammals (180 million year ago). In the evolutionary transition from reptiles to mammals, three key developments were (1) nursing, in conjunction with maternal care; (2) audiovocal communication for maintaining maternal-offspring contact; and (3) play. The development of this behavioral triad may have depended on the evolution of the thalamocingulate division of the limbic system, a derivative from early mammals. The thalamocingulate division (which has no distinctive counterpart in the reptilian brain) was, in turn, followed by development of the prefrontal neocortex that, in human beings, may be inferred to play a key role in familial acculturation. When mammals developed parenting behavior the stage was set for increased exposure and responsiveness to emotional signals of others including signals of pain, separation and distress. Indeed, parenting involves the protection and transfer of energy, information, and social skills to offspring. African hominoids, including chimpanzees, gorillas, and humans, share a number of parenting mechanisms with other placental mammals, including internal gestation, lactation, and attachment mechanisms involving neuropeptides such as oxytocin (Geary & Flinn, 2001).

The phylogenic origin of behaviors associated with social engagement has been linked to the evolution of the autonomic nervous system and how it relates to emotion. According to Porges (2007), social approach or withdrawal stem from the implicit computation of feelings of safety, discomfort, or potential danger. Basic to survival is the capacity to react to challenges or stressors and maintain visceral homeostatic states necessary for vital processes such as oxygenation of tissues and supplying nutrients to the body. He proposed that evolution of the autonomic nervous system (sympathetic and parasympathetic systems) provides a means to understand the adaptative significance of mammalian affective processes including empathy and the establishment of lasting social bonds. Thus empathy can be viewed in terms of adaptive neuroendocrine and autonomic processes, including changes in neuromodulatory systems that regulate bodily states, emotions, and reactivity (Carter, Harris, & Porges, 2009). These basic evaluative systems are associated with motor responses that aid the adaptive responding of the organism.

At this primitive level, appetitive and aversive behavioral responses are modulated by specific neural circuits in the brain that share common neuroarchitectures among mammals (Parr & Waller, 2007). These brain systems are genetically hard-wired to enable animals to respond unconditionally to threatening, or appetitive, stimuli using specific response patterns that are most adaptive to the particular species and environmental condition. The limbic system, which includes the hypothalamus, the parahippocampal cortex, the amygdala, and several interconnected areas (septum, basal ganglia, nucleus accumbens, insula, retrospenial cingulate cortex and prefrontal cortex) is primarily responsible for emotion processing. What unites these regions are their roles in motivation and emotion, mediated by connections with the autonomic system. The limbic system also projects to the cingulate and orbitofrontal cortices, which are involved with the regulation of emotion.

There is evidence for a lateralization of emotion processing in humans and primates which has been marshaled under two distinct theories. One theory states that the right hemisphere is primarily responsible for emotional processing (Cacioppo & Gardner, 1999), while another one suggests that the right hemisphere regulates negative emotion and the left hemisphere regulates positive emotion (Davidson, 1992). This asymmetry is anatomically based on an asymmetrical representation of homeostatic activity that originates from asymmetries in the peripheral autonomic nervous system, and fits well with the homeostatic model of emotional awareness, which posits that emotions are organized according to the fundamental principle of autonomic opponency for the management of physical and mental energy (Craig, 2005). Supporting evidence for the lateralization of emotion comes from neuroimaging studies and neuropsychological observations with brain damaged patients, but also studies in non-human primates.

In one study, tympanic membrane temperature (Tty) was used to assess asymmetries in the perception of emotional stimuli in chimpanzees (Parr & Hopkins, 2000). The tympanic membrane is an indirect, but reliable, site from which to measure brain temperature, and is strongly influenced by autonomic and behavioral activity. In that study, chimpanzees were shown positive, neutral, and negative emotional videos depicting scenes of play, scenery, and severe aggression, respectively. During the negative emotion condition, right Tty was significantly higher than the baseline temperature. This effect was relatively stable, long lasting, and consistent across individuals. Temperatures did not change significantly from baseline in the neutral or positive emotion condition, although a significant number of measurements showed increased left Tty during the neutral emotion condition. These data suggest that viewing emotional stimuli results in asymmetrical changes in brain temperature, in particular increased right Tty during the negative emotion condition, evidence of emotional arousal in chimpanzees, and providing support for right hemispheric asymmetry in our closest living relative.

At the behavioral level it is evident from the descriptions of comparative psychologists and ethologists that behaviors homologous to empathy can be observed in other mammalian species. Notably, a variety of reports on ape empathic reactions suggest that, apart from emotional connectedness, apes may have an appreciation of the other's situation (de Waal, 1996). A good example is consolation, defined as reassurance behavior by an uninvolved bystander towards one of the combatants in a previous aggressive incident (de Waal & van Roosmalen, 1979). De Waal (1996) has argued that empathy is not an all-or-nothing phenomenon, and many forms of empathy exist between the extremes of mere agitation at the distress of another and full understanding of their predicament. Many other comparative psychologists however view empathy as a kind of induction process by which emotions, both positive and negative, are shared, and by which the probabilities of similar behavior are increased in the participants. In the view developed in this paper, this may occur but is not a sufficient mechanism to account for the full-blown ability of human empathy. Indeed, some aspects of empathy may be present in other species, such as motor mimicry and emotion contagion (see de Waal & Thompson, 2005). For instance, Parr (2001) conducted an experiment in which peripheral skin temperature (indicating greater negative arousal) was measured in chimpanzees while they were exposed to emotionally negative video scenes. Results demonstrate that skin temperature decreased when subjects viewed videos of conspecifics injected with needles or videos of needles themselves, but not videos of a conspecific chasing the veterinarian. Thus, when chimpanzee are exposed to meaningful emotional stimuli, they are subject to physiological changes similar to those observed during fear in humans, which some see as evidence of emotional contagion (Hatfield, 2009).

In humans, the construct of empathy accounts for a more complex psychological state than the one associated with the automatic sharing of emotions. Like in other species, emotions and feelings may be shared between individuals, but humans are also able to intentionally "feel for" and act on behalf of other people whose experiences may differ greatly from their own (Batson et al., 1991; Decety & Hodges, 2006). This phenomenon, called empathic concern or sympathy, is often associated with prosocial behaviors such as helping, and has been considered as a chief enabling process for altruism. According to Wilson (1988), empathic helping behavior has evolved because of its contribution to genetic fitness. In humans and other mammals, an impulse to care for offspring is almost certainly genetically hard-wired. It is far less clear that an impulse to care for siblings, more remote kin, and similar non-kin is genetically hard-wired (Batson, 2006). The emergence of altruism, of empathizing with and caring for those who are not kin is thus not easily explained within the framework of neo-Darwinian theories of natural selection such as kin selection or reciprocal altruism. It seems more plausibly explained by cognitive extension of parental nurturance to non-offspring.

One of the most striking aspects of human empathy is that it can be felt for virtually any target – even targets of a different species (Batson, Lishner, Cook, & Sawyer, 2005). In addition, as emphasized by Harris (2000), humans, unlike other primates, can put their emotions into words, allowing them not only to express emotion but also to report on current as well as past emotions. These reports provide an opportunity to share, explain, and regulate emotional experience with others that is not found in other species. Conversation helps to develop empathy, for it is often here that one learns of shared experiences and feelings. Moreover, this self-reflexive capability (which includes emotion regulation) may be an important difference between humans and other animals (Povinelli, 2001).

Interestingly two key regions, the anterior insula and anterior cingulate cortex (ACC), involved in affective processing in general and empathy for pain in particular have singularly evolved in apes and humans. Cytoarchitectonic work by Allman, Watson, Tetreault, and Hakeem (2005) indicates that a population of large spindle neurons is uniquely found in the anterior insula and anterior cingulate of humanoid primates. Most notably, they reported a trenchant phylogenetic correlation, in that spindle cells are most numerous in aged humans, but progressively less numerous in children, gorillas, bonobos and chimpanzees, and nonexistent in macaque monkeys. It was recently suggested that these spindle neurons interconnect the most advanced portions of limbic sensory (anterior insula) and limbic motor (ACC) cortices, both ipsilaterally and contralaterally, which, in sharp contrast to the tightly interconnected and contiguous sensorimotor cortices, are situated physically far apart as a consequence of the pattern of evolutionary development of limbic cortices (Craig, 2007). Thus, the spindle neurons could enable fast, complex and highly integrated emotional behaviors.

In support of this view, convergent functional imaging findings reveal that the anterior insula and the anterior cingulate cortices are conjointly activated during all human emotions. This, according to Craig (2002), indicates that the limbic representation of subjective "feelings" (in the anterior insula) and the limbic representation of volitional agency (in the anterior cingulate) together form the fundamental neuroanatomical basis for all human emotions, consistent with the definition of an emotion in humans as both a feeling and a motivation with concomitant autonomic sequelae (Rolls, 1999).

Overall, this evolutionary conceptual view is compatible with the hypothesis that advanced levels of social cognition may have arisen as an emergent property of powerful executive functioning assisted by the representational properties of language (Barrett, Henzi, & Dunbar, 2003). However, these higher levels operate on previous levels of organization, and should not be seen as independent of, or conflicting with one another. Evolution has constructed layers of increasing complexity, from non-representational (e.g., emotion contagion) to representational and meta-representational mechanisms (e.g., sympathy), which need to be taken into account for a full understanding of human empathy.

Shared Neural Circuits Between Self and Other

It has long been suggested that empathy involves resonating with another person's unconscious affect. For instance, Basch (1983) speculated that, because their respective autonomic nervous systems are genetically programmed to respond in a similar fashion, a given affective expression by a member of a particular species can trigger similar responses in other members of that species. The view that unconscious automatic mimicry of a target generates in the observer the autonomic response associated with that bodily state and facial expression subsequently received empirical support from a variety of behavioral and physiological studies marshaled under the perception-action coupling mechanism (see Preston & de Waal, 2002). The core assumption of the perception-action model of empathy is that perceiving a target's state automatically activates the corresponding representations of that state in the observer which in turn activates somatic and autonomic responses. The discovery of sensory-motor neurons (called mirror neurons) in the premotor and posterior parietal cortex that discharge during both the production of a given action and during the perception of the same action performed by another individual, provides one possible physiological mechanism for this direct link between perception and action (Rizzolatti & Craighero, 2004).

Behavioral studies demonstrate that viewing facial expressions triggers similar expressions on one's own face, even in the absence of conscious recognition of the stimulus. One functional magnetic resonance imaging (fMRI) experiment confirmed these results by showing that when participants are required to observe or to imitate facial expressions of various emotions, increased neurodynamic activity is detected in the brain regions implicated in the facial expressions of these emotions, including the superior temporal sulcus, the anterior insula, and the amygdala, as well as specific areas of the premotor cortex (Carr, Iacoboni, Dubeau, Mazziotta, & Lenzi, 2003).

Accumulating evidence suggests that a "mirroring" or resonance mechanism is also at play both when one experiences sensory and affective feelings in the self and perceives them in others. Even at the level of the somatosensory cortex, seeing another's neck or face being touched elicits appropriately organized somatotopic activations in the mind of the observer (Blakemore & Frith, 2005). Robust support for the involvement of shared neural circuits in the perception of affective states comes from recent neuroimaging and transcranial magnetic stimulation (TMS). For instance, the first-hand experience of disgust and the sight of disgusted facial expressions in others both activate the anterior insula (Wicker et al., 2003). Similarly, the observation of hand and face actions performed with an emotion engages regions that are also involved in the perception and experience of emotion and/or communication (Grosbras & Paus, 2006).

A number of neuroimaging studies recently demonstrated that the observation of pain in others recruits brain areas chiefly involved in the affective and motivational processing of direct pain perception.

In one study, participants in the scanner received painful stimuli in some trials while in other trials simply observed a signal indicating that their partner, who was present in the same room, would receive the painful stimuli (Singer, Seymour, O'Doherty, Kaube, et al., 2004). During both types of trials, the medial and anterior cingulate cortex (MCC and ACC) and the anterior insula were activated (see also Morrison, Lloyd, di Pellegrino, & Roberts, 2004). These regions contribute to the affective and motivational processing of noxious stimuli, i.e., aspects of pain that pertain to desires, urges, or impulses to avoid or terminate a painful experience. Similar results were reported in a study by Jackson, Meltzoff, and Decety (2005) in which participants were shown pictures of people's hands or feet in painful or neutral everyday-life situations. Significant activation in regions involved in the affective aspects of pain processing (MCC/ACC and anterior insula) was detected but, as in Singer, Seymour, O'Doherty, Kaube, et al.'s study (2004), no signal change was found in the somatosensory cortex. However, a recent TMS study did report changes in corticospinal motor representations of hand muscles in individuals observing needles penetrating hands or feet of a human model (Avenanti, Bueti, Galati, & Aglioti, 2005), indicating that the observation of pain can also involve sensorimotor representations. It is worth mentioning that the activation of these regions (i.e., ACC, AI, PAG, SMA) is not specific to the processing of noxious stimuli. The same neural network, which includes the amygdala, responds to any unpleasant and salient stimuli (e.g., disgusting), and its involvement in empathy for pain may thus reflect a general aversive response (Benuzzi et al., 2008; Yamada & Decety, 2009).

Summing up, current neuroscientific evidence suggests that merely observing another individual in a painful situation yields responses in the neural network associated with the coding of the motivational-affective dimension of pain in oneself. A recent meta-analysis of neuroimaging studies, however, indicates that this overlap is not complete (Jackson, Rainville, & Decety, 2006b). Both in the insula and the cingulate cortex, the perception of pain in others results in more rostral activations than the first-hand experience of pain. Also, vicariously instigated activations in the pain matrix are not necessarily specific to the emotional experience of pain, but to other processes such as somatic monitoring, negative stimulus evaluation, and the selection of appropriate skeletomuscular movements of aversion. Thus, the shared neural representations in the affective-motivational part of the pain matrix might not be specific to the sensory qualities of pain, but instead be associated with more general survival mechanisms such as aversion and withdrawal.

The discovery that the observation of pain in others activates brain structures involved in negative emotional experiences has important implications for the question of whether observing another's plight will result in empathic concern or personal distress. Appraisal theory views emotions as resulting from the assessed personal relevance of external or internal stimuli (Scherer, Schorr, & Johnstone, 2001). Perceiving the emotions of others is a powerful instigator of physiological responses, leading to distinct changes in both the central and the autonomic nervous system. Interestingly, closer linkage in psychophysiological indicators such as heart rate and electrodermal activity between observer and target predicts better understanding of the target's emotional state (Levenson & Ruef, 1992). Note also that parts of the insula and the MCC active during the observation of pain in others contribute to the monitoring of bodily changes, such as visceral and somatic responses. Hence, it is plausible that depending upon whether these responses are attributed to the self or to the other, they might result in more or fewer other- vs. self-oriented emotions.

Perspective-Taking, Self-Other Awareness, and Empathy

There is general consensus among theorists that the ability to adopt and entertain the psychological perspective of others has a number of important consequences. Well-developed perspective-taking abilities allow us to overcome our usual egocentrism and tailor our behaviors to others' expectations (Davis, Conklin, Smith, & Luce,

1996). Further, successful perspective taking has been linked to altruistic motivation (Batson et al., 1991). Using mental imagery to take the perspective of another is a powerful way to place oneself in the situation or emotional state of that person. Mental imagery not only enables us to see the world of our conspecifics through their eyes or in their shoes, but may also result in similar sensations as the other person's (Decety & Grèzes, 2006).

Social psychology has for a long time been interested in the distinction between imagining the other and imagining oneself, and in particular in the emotional and motivational consequences of these two perspectives. A number of these studies show that focusing on another's feelings (imagine other) may evoke stronger empathic concern, while explicitly putting oneself into the shoes of the target (imagine self) induces both empathic concern and personal distress. In one such study, Batson, Early, and Salvarini (1997) investigated the affective consequences of different perspective-taking instructions when participants listened to a story about Katie Banks, a young college student struggling with her life after the death of her parents. This study demonstrated that different instructions had distinct effects on how participants perceived the target's situation. Notably, participants imagining themselves to be in Katie's place (imagine self) showed stronger signs of discomfort and personal distress than participants focusing on the target's responses and feelings (imagine other), or than participants instructed to take on an objective, detached point of view. Also, both perspective-taking instructions differed from the detached perspective by resulting in higher empathic concern. This observation may help explain why observing a need situation does not always yield to prosocial behavior: if perceiving another person in an emotionally or physically painful circumstance elicits personal distress or a detached, objective perspective, then the observer may tend not to fully attend to the other's experience and as a result lack sympathetic behaviors.

Interestingly, cognitive neuroscience research demonstrates that when individuals adopt the perspective of others, neural circuits common to the ones underlying first-person experiences are activated as well. However, taking the perspective of the other produces further activation in specific parts of the frontal cortex that are implicated in executive functions, particularly inhibitory control (e.g., Ruby & Decety, 2003, 2004). In line with these findings, the frontal lobes may functionally serve to separate perspectives, or resist interference from one's own perspective when adopting the subjective perspective of others (Decety & Jackson, 2004). This ability is of particular importance when observing another's distress, since a complete merging with the target would lead to confusion as to who is experiencing the negative emotions and therefore to different motivations as to who should be the target of supportive behavior.

In two successive functional MRI studies, we recently investigated the neural mechanisms subserving the effects of perspective-taking during the perception of pain in others. In the first study, participants were shown pictures of hands and feet in painful situations and asked to either imagine themselves or to imagine another individual experiencing these situations and rate the level of pain they would induce (Jackson et al., 2006a). Both the self-perspective and the other-perspective were associated with activation in the neural network involved in pain processing.

This finding is consistent with the shared neural representations account of social perception discussed above. However, the self-perspective yielded higher pain ratings and quicker response times, and it involved the pain matrix more extensively in the secondary somatosensory cortex, a sub-area of the MCC, and the insula.

In a second neuroimaging study, the distinction between empathic concern and personal distress was investigated in more detail by using a number of additional behavioral measures and a set of ecological and extensively validated dynamic stimuli (Lamm, Batson, & Decety, 2007). Participants watched a series of video-clips featuring patients undergoing painful medical treatment. They were asked to either put themselves explicitly in the shoes of the patient (imagine self), or to focus on the patient's feelings and affective expressions (imagine other). The behavioral data confirmed that explicitly projecting oneself into an aversive situation leads to higher personal distress – while focusing on the emotional and behavioral reactions of another in distress is accompanied by higher empathic concern and lower personal distress. The neuroimaging data are consistent with this finding and provide some insights into the neural correlates of these distinct behavioral responses. The self-perspective evoked stronger hemodynamic responses in brain regions involved in coding the motivational-affective dimensions of pain, including bilateral insular cortices, anterior MCC, the amygdala, and various structures involved in action control. The amygdala plays a critical role in fear-related behaviors, such as the evaluation of actual or potential threats. Imagining oneself to be in a painful and potentially dangerous situation might therefore have triggered a stronger fearful and/or aversive response than imagining someone else to be in the same situation.

Corresponding with Jackson and colleagues (2006a), the insular activation was also located in a more posterior, mid-dorsal sub-section of this area. The mid-dorsal part of the insula plays a role in coding the sensory-motor aspects of painful stimulation, and it has strong connections with the basal ganglia where activity was also higher during the self-perspective. Taken together, it appears that the insular activity during the self-perspective reflects simulation of sensory aspects of the painful experience. Such a simulation might server to mobilize motor areas for the preparation of defensive or withdrawal behaviors, as well as instigate the interoceptive monitoring associated with autonomic changes evoked by this simulation process (Critchley, Wiens, Rotshtein, Öhman, & Dolan, 2005). Such an interpretation also accounts for the activation difference present in the somatosensory cortex. Finally, the higher activation in premotor structures might connect with a stronger mobilization of motor representations by the more stressful and discomforting first-person perspective. Further support for this interpretation is provided by a positron emission tomography study investigating the relationship between situational empathic accuracy and brain activity. It also found higher activation in medial premotor structures, partially extending into MCC, when participants witnessed the distress of others (Shamay-Tsoory et al., 2005). This study also pointed to the importance of prefrontal areas in the understanding of distress.

Overall, the Lamm, Batson, and Decety results fit well with the pioneering research of Stotland (1969) on the effects of perspective taking on empathy and distress. Participants observed an individual experiencing a painful diathermy while adopting either an imagine self or an imagine other perspective. Stotland found higher vasoconstriction for the other perspective, and more palmar sweat and higher tension and nervousness for the self-perspective. This finding was interpreted as indicating that a focus on the target's feelings (imagine other) produced more other-oriented feeling for the target when focusing on his affective expressions and motor responses, while the first person perspective (imagine self) led to more self-oriented responding that was less a response to the presumed feelings of the target.

Altogether, the available empirical findings reveal important differences between the neural systems involved in first- and third-person perspective-taking, and contradicts the hypothesis/notion that the self and other completely merge in empathy. The specific activation differences in both the affective and sensorimotor aspects of the pain matrix, along with the higher pain and distress ratings, reflects the self-perspective's more direct and personal involvement. One key region that might facilitate self vs. other distinctions is the right temporo-parietal junction (TPJ). The TPJ is activated in most neuroimaging studies on empathy (Decety & Lamm, 2007), and seems to play a decisive role in self-awareness and the sense of agency. Agency (i.e., the awareness of oneself as an agent who is the initiator of actions, desires, thoughts and feelings) is crucial/essential for successful navigation of shared representations between self and other (Decety, 2005).

Thus, self-awareness and a sense of agency both play pivotal roles in empathy and significantly contribute to social interaction. These important aspects are likely to be involved in distinguishing emotional contagion, which relies heavily on the automatic link between perceiving the emotions of another and one's own experience of the same emotion, from empathic responses which call for a more detached relation. The neural responses identified in these studies as non-overlapping between self and other may take advantage of available processing capacities to plan appropriate future actions concerning the other.

Awareness of one's own feelings, furthermore, and the ability to (consciously and automatically) regulate one's own emotions may allow us to disconnect empathic responses to others from our own personal distress, with only the former leading to altruistic motivation.

Modulation of Empathic Responding

Despite the fact that the mere perception of the behavior of others can, at times, activate similar circuits in the self, and in the case of empathy for pain can activate part of the neural circuit involved in the first-hand experience of pain, there is also evidence that this unconscious level can be modulated by various situational and dispositional variables. Research in social psychology has identified a number of these factors, such as the relationship between target and empathizer, the empathizer's dispositions, and the context in which the social interaction takes place. Therefore, whether observing the distress of a close friend results in empathic concern and helping behavior or withdrawal from the situation is influenced by the complex interaction between these factors.

Emotion regulation seems to have a particularly important role in social interaction, and it has a clear adaptive function for both the individual and the species (Ochsner & Gross, 2005). Importantly, it has been demonstrated that individuals who can regulate their emotions are more likely to experience empathy, and also to interact in morally desirable ways with others (Eisenberg et al., 1994). In contrast, people who experience their emotions intensely, especially negative emotions, are more prone to personal distress, an aversive emotional reaction, such as anxiety or discomfort based on the recognition of another's emotional state or condition.

In the case of perception of others in pain, the ability to down-regulate one's emotions is particularly valuable when the distress of the target becomes overwhelming. For example, a mother alarmed by her baby's cries at night has to cope with her own discomfort in order to provide appropriate care for her distressed offspring. One strategy to regulate emotions is based on cognitive re-appraisal. This involves reinterpreting a stimulus in order to change the way in which we respond to it. It can either be intentionally achieved, or result from additional information provided about the emotion-eliciting stimulus. By providing different context information about the consequences of the observed pain, we investigated the effects of cognitive appraisal on the experience of empathy in the above-mentioned fMRI study by Lamm and colleagues (2007). The observed target patients belonged to two different groups. In one group, health and quality of life improved after the painful therapy, while members of the other group did not benefit from the treatment. Thus, stimuli of identically arousing and negatively valenced emotional content were watched with different possibilities to appraise the patients' pain. The results confirmed our hypotheses and demonstrated that appraisal of an aversive event can considerably alter one's responses to it. Patients undergoing non-effective treatment were judged to experience higher levels of pain, and personal distress in the observers was more pronounced when watching videos of these patients. Brain activation was modulated in two sub-regions of the orbitofrontal cortex (OFC) and the rostral part of aMCC. The OFC is known to play an important role in the evaluation of positive and negative reinforcements and is also involved in emotion reappraisal. Activity in the OFC may thus reflect the evaluation of presented stimuli's valence. Interestingly, watching effectively versus non-effectively treated patients did not modulate the hemodynamic activity in either the visual-sensory areas, or the insula. This suggests that both patient groups triggered an emotional reaction, and that top-down appraisal did not alter stimulus processing at an early perceptual stage.

Another intrapersonal factor affecting the empathic response is the emotional background state of the observer (Niedenthal, Halberstadt, Margolin, & Innes-Ker, 2000). For example, a depressive mood can affect the way in which we perceive the expression of emotions by others. In a recent developmental neuroscience study, limbic structures such as the amygdala and the nucleus accumbens became hyperactive when participants with pediatric bipolar disorder attended to the facial expression of emotion (Pavuluri, O'Connor, Harral, & Sweeney, 2006). Similarly, patients with generalized social phobia show increased amygdala activation when exposed to angry or contemptuous faces (Stein, Goldin, Sareen, Zorrilla, & Brown, 2002).

Whether individual differences in dispositional empathy and personal distress modulate the occurrence and intensity of self- vs. other-centered responding is currently a matter of debate. Several recent neuroimaging studies demonstrate specific relationships between questionnaire measures of empathy and brain activity. For example, both Singer, Seymour, O'Doherty, Kaube, colleagues (2004) and Lamm and colleagues (2007) detected significantly increased activation in insular and cingulate cortices in participants with higher self-reported empathy during perception. This shows modulation of neural activity in the very brain regions that are involved in coding the affective response to the other's distress. Note however, that no such correlations were found in a similar study (Jackson et al., 2005). Also, no correlations with self-report data on personal distress were observed by Lamm et al. (2007) or Jackson et al. (2006a). However, Lawrence et al. (2006) did report such correlations in cingulate and prefrontal regions of participants labeling a target's mental and affective state. Part of this discrepancy between neuroscience research and dispositional measures may be related to the low validity of self-report measures in predicting actual empathic behavior (Davis & Kraus, 1997). It is our conviction that brain-behavior correlations should be treated with caution, and care must be taken to formulate specific hypotheses both about the neural correlates of the dispositional measures as well as what the questionnaire actually measures. For example, the personal distress subscale of the Interpersonal Reactivity Index (Davis, 1994) showed correlations close to zero with the experimentally derived distress measures, and no significant correlations with brain activation. This indicates that the subscale is probably not an appropriate measure of situative discomfort evoked by the observation of another's distress.

The effects of *interpersonal* factors – such as the similarity or closeness of empathizer and target – have been investigated at the behavioral, psychophysiological, and neural levels. For instance, Cialdini, Brown, Lewis, Luce, and Neuberg (1997) have documented that for some types of need, relationship closeness is an important predictor of helping behavior and correlates strongly with empathic concern. Lanzetta and Englis (1989) made interesting observations concerning the effects of attitudes on social interaction. Their studies show that, in a competitive relationship, observation of joy can result in distress, while pain in the competitor leads to positive emotions. These findings reflect an important and often ignored aspect of perspective taking, namely that this ability can also be used in a malevolent way – as when knowledge about the emotional or cognitive state of competitors is used to harm them. A recent study by Singer, Seymour, O'Doherty, Stephan, (2006) revealed the neural correlates of such counter-empathic responding. In that study, participants were first engaged in a sequential Prisoner's Dilemma game with confederate targets who were playing the game either fairly or unfairly. Following this behavioral manipulation, fMRI measures were taken during the observation of fair and unfair players receiving painful stimulation. Compared to the observation of fair players, activation in brain areas coding the affective components of pain was significantly reduced when participants observed unfair players in pain. This effect, however, was detected in male participants only, who also exhibited a concurrent increase of activation in reward-related areas.

In sum, there is strong behavioral evidence demonstrating that the experience of empathy and personal distress can be modulated by a number of social-cognitive factors. In addition, a few recent neuroscience studies indicate that such a modulation leads to activity changes in the neural systems that process social information. Further studies are required to increase our knowledge about the various factors, processes and (neural and behavioral) effects involved in and resulting from the modulation of empathic responses. This knowledge will inform us how empathy can be promoted to ultimately increase humankind's ability to act in more prosocial and altruistic ways.

Empathy and Morality

We come, finally, to the relation of empathy to morality. Support for the empathy-altruism hypothesis suggests that empathic concern produces altruistic motivation (Batson, 1991). One of the more surprising implications of the empathy-altruism hypothesis is that empathy-induced altruism does not necessarily produce moral behavior; it can produce immoral behavior as well. This implication is surprising because many people equate altruism with morality. The empathy-altruism hypothesis does not. In the empathy-altruism hypothesis, altruism refers to a motivational state with the ultimate goal of increasing another's welfare. What is morality? The dictionary gives the following as the first two definitions: (1) "Of or concerned with principles of right conduct." (2) "Being in accord with such principles." Typically, moral principles are universal and impartial – for example, principles of fairness or justice. Given these definitions of altruism and morality, altruism stands in the same relation to morality as does egoism. An egoistic desire to benefit myself may lead me to unfairly put my needs and interests in front of the parallel needs and interests of others. An altruistic desire to benefit another may lead me to unfairly put that person's needs and interests in front of the parallel needs and interests of others. Each action violates the moral principle of fairness. Egoism, altruism, and morality are three independent motives, each of which may conflict with another.

To test this derivation from the empathy-altruism hypothesis, Batson, Klein, Highberger, and Shaw (1995) conducted two experiments. Results of each experiment supported the proposal that empathy-induced altruism can lead one to act in a way that violates the moral principle of fairness. In each, participants were asked to make an allocation decision that affected the welfare of other individuals. Participants who were not induced to feel empathic concern for one of the other individuals tended to adhere strictly to a principle of fairness. Participants who were induced to adopt an imagine-other perspective and, thereby, feel empathic concern were significantly more likely to violate this principle, allocating resources preferentially to the individual for whom empathy was felt. This was true even though high-empathy participants who showed partiality agreed with other participants that acting with partiality in this situation was less fair and less moral than acting impartially. Overall, results suggested that empathy-induced altruism and the desire to uphold a moral principle of fairness are independent motives that can at times conflict.

The power of empathy-induced altruism to override moral motivation is, however, only half the story. A more positive implication of recognizing the independence of these two prosocial motives is that one can think about using them in concert. Consider the moral motive to uphold a principle of justice. This is a powerful motive but vulnerable to rationalization; it is easily co-opted (Lerner, 1980; Solomon, 1990). Empathy-induced altruism also is a powerful motive but limited in scope; it produces partiality. Perhaps if empathic concern can be evoked for the victims of injustice, then these two motives can be made to work together rather than at odds. Desire for justice may provide perspective and reason; empathy-induced altruism may provide emotional fire and a push toward seeing the victims' suffering end, preventing rationalization and derogation.

Something of this sort occurred, we believe, among the rescuers of Jews in Nazi Europe. A careful look at data collected by Oliner and Oliner (1988) and their colleagues suggests that involvement in rescue activity frequently began with concern for a specific individual or individuals for whom compassion was felt – often individuals known previously. This initial involvement subsequently led to further contacts and rescue activity, and to a concern for justice, that extended well beyond the bounds of the initial empathic concern. Clearly, we still have a long way to go before we fully understand how empathy-induced altruism and moral motives can compete and cooperate. To recognize that empathic concern can, at times, lead to immoral action is a first step. Social neuroscience work will certainly enrich our understanding of the biological and computational mechanisms that subserve morality (see Decety, Michalska, & Akitsuki, 2008; Greene, Sommerville, Nystrom, Darley, & Cohen, 2001; Moll et al., 2007).

Conclusion

The combined results of functional neuroimaging studies demonstrate that when individuals perceive others in pain or distress, some of the same neural mechanisms are activated as when they are in painful situations themselves. Such a shared neural mechanism offers an interesting foundation for intersubjectivity because it provides a functional bridge between first-person information and third person information, grounded in self-other equivalence (Decety & Sommerville, 2003; Sommerville & Decety, 2006), which allows analogical reasoning, and offers a possible route to understanding others. Yet a minimal distinction between self and other is essential social interaction in general and for empathy in particular, and new work in social neuroscience has demonstrated that the self and other are distinguished at both the behavioral and neural levels. Recent cognitive neuroscience research indicates that the neural response to others in pain can be modulated by various situational and dispositional variables. Finally, it is important to recognize that empathic concern does not necessarily produce moral action.

All together, these data support the view that empathy operates by way of conscious and automatic processes which, far from functioning independently, really represent different aspects of a common mechanism. These accounts of empathy are in harmony with theories of embodied cognition, which contend that cognitive representations and operations are fundamentally grounded in bodily states and in the brain's modality-specific systems (Niedenthal, Barsalou, Ric, & Krauth-Gruber, 2005).

Acknowledgments The writing of this chapter was supported by grant from the National Science Foundation to Jean Decety (BCS-0718480).

References

Allman, J. M., Watson, K. K., Tetreault, N. A., & Hakeem, A. Y. (2005). Intuition and autism: A possible role for Von Economo neurons. *Trends in Cognitive Sciences, 9*, 367–373.

Avenanti, A., Bueti, D., Galati, G., & Aglioti, S. M. (2005). Transcranial magnetic stimulation highlights the sensorimotor side of empathy for pain. *Nature Neuroscience 8*, 955–960.

Barrett, L., Henzi, P., & Dunbar, R. I. M. (2003). Primate cognition: From what now to what if. *Trends Cognitive Sciences, 7*, 494–497.

Basch, M. F. (1983). Empathic understanding: A review of the concept and some theoretical considerations. *Journal of the American Psychoanalytic Association, 31*, 101–126.

Batson, C. D. (1991). *The altruism question: Toward a social-psychological answer*. Hillsdale, NJ: Erlbaum Associates.

Batson, C. D. (2006). Folly bridges. In P. A. M. van Lange (Ed.), *Bridging social psychology* (pp. 59–64). Mahwah, NJ: Erlbaum.

Batson, C. D., Batson, J. G., Singlsby, J. K., Harrell, K. L., Peekna, H. M., & Todd, R. M. (1991). Empathic joy and the empathy-altruism hypothesis. *Journal of Personality and Social Psychology, 61*, 413–426.

Batson, C. D., Early, S., & Salvarini, G. (1997). Perspective taking: Imagining how another feels versus imagining how you would feel. *Personality & Social Personality Bulletin, 23*, 751–758.

Batson, C. D., Klein, T. R., Highberger, L., & Shaw, L. L. (1995). Immorality from empathy-induced altruism: When compassion and justice conflict. *Journal of Personality and Social Psychology, 68*, 1042–1054.

Batson, C. D., Lishner, D. A., Cook, J., & Sawyer, S. (2005). Similarity and nurturance: Two possible sources of empathy for strangers. *Basic and Applied Social Psychology, 27*, 15–25.

Blakemore, S.-J., & Frith, C. D. (2005). The role of motor contagion in the prediction of action. *Neuropsychologia, 43*, 260–267.

Benuzzi, F., Lui, F., Duzzi, D., Nichelli, P. F., & Porro, C. A. (2008). Does it look painful or disgusting? Ask your parietal and cingulated cortex. *Journal of Neuroscience, 28*, 923–931.

Cacioppo, J. T., & Gardner, W. L. (1999). Emotion. *Annual Review of Psychology, 50*, 191–214.

Carr, L., Iacoboni, M., Dubeau, M. C., Mazziotta, J. C., & Lenzi, G. L. (2003). Neural mechanisms of empathy in humans: A relay from neural systems for imitation to limbic areas. *Proceedings of the National Academy of Science United States of America, 100*, 5497–5502.

Carter, C. S., Harris, J., & Porges, S. W. (2009). Neural and evolutionary perspective on empathy. In: J. Decety & W. Ickes (Eds.), *The social neuroscience of empathy* (pp. 169–182). Cambridge: MIT Press.

Cialdini, R. B., Brown, S. L., Lewis, B. P., Luce, C., & Neuberg, S. L. (1997). Reinterpreting the empathy-altruism relationship: When one into one equals oneness. *Journal of Personality and Social Psychology, 73*, 481–494.

Craig, A. D. (2002). How do you feel? Interoception: The sense of the physiological condition of the body. *Nature Reviews in Neuroscience, 3*, 655–666.

Craig, A. D. (2005). Forebrain emotional asymmetry: A neuroanatomical basis? *Trends in Cognitive Sciences, 9*, 566–571.

Craig, A. D. (2007). Interoception and emotion: A neuroanatomical perspective. In: M. Lewis J. M. Haviland-Jones, & L. F. Barrett (Eds.), *Handbook of emotion* (3rd ed., pp. 272–288). New York: Guilford Press.

Critchley, H. D., Wiens, S., Rotshtein, P., Öhman, A., & Dolan, R. D. (2005). Neural systems supporting interoceptive awareness. *Nature Neuroscience, 7*, 189–195.

Davidson, R. J. (1992). Anterior cerebral asymmetry and the nature of emotion. *Brain and Cognition, 20*, 125–151.

Davis, M. H. (1994). *Empathy: A social psychological approach.* Madison, WI: Westview Press.

Davis, M. H., Conklin, L., Smith, A., & Luce, C. (1996). Effect of perspective taking on the cognitive representation of persons: A merging of self and other. *Journal of Personality and Social Psychology, 70*, 713–726.

Davis, M. H., & Kraus, L. A. (1997). Personality and empathic accuracy. In W. Ickes (Ed.), *Empathic accuracy* (pp. 144–168). New York: The Guilford Press.

Decety, J. (2005). Perspective taking as the royal avenue to empathy. In B. F. Malle & S. D. Hodges (Eds.), *Other minds: How humans bridge the divide between self and other* (pp. 135–149). New York: Guilford Publications.

Decety, J., & Batson, C. D. (2007). Social neuroscience approaches to interpersonal sensitivity. *Social Neuroscience, 2*(3–4), 151–157.

Decety, J., & Grèzes, J. (2006). The power of simulation: Imagining one's own and other's behavior. *Brain Research, 1079*, 4–14.

Decety, J., & Hodges, S. D. (2006). A social cognitive neuroscience model of human empathy. In P. A. M. van Lange (Ed.), *Bridging social psychology: Benefits of transdisciplinary approaches* (pp. 103–109). Mahwah, NJ: Lawrence Erlbaum Associates.

Decety, J., & Jackson, P. L. (2004). The functional architecture of human empathy. *Behavioral and Cognitive Neuroscience Reviews, 3*, 71–100.

Decety, J., & Lamm, C. (2006). Human empathy through the lens of social neuroscience. *The Scientific World Journal, 6*, 1146–1163.

Decety, J., & Lamm, C. (2007). The role of the right temporoparietal junction in social interaction: How low-level computational processes contribute to meta-cognition. *The Neuroscientist, 13*(6), 580–593.

Decety, J., & Lamm, C. (2009). Empathy versus personal distress – Recent evidence from social neuroscience. In J. Decety & W. Ickes (Eds.), *The Social neuroscience of empathy* (pp. 199–213). Cambridge, MA: MIT press.

Decety, J., Michalska, K. J., & Akitsuki, Y. (2008). Who caused the pain? A functional MRI investigation of empathy and intentionality in children. *Neuropsychologia, 46*, 2607–2614.

Decety, J., & Sommerville, J. A. (2003). Shared representations between self and others: A social cognitive neuroscience view. *Trends in Cognitive Sciences, 7*, 527-533.

Decety, J., & Meyer, M. (2008). From emotion resonance to empathic understanding: A social developmental neuroscience account. *Development and Psychopathology, 20*, 1053–1080.

De Waal, F. (2005). Primates, monks and the mind. *Journal of Consciousness Studies, 12*, 1–17.

De Waal, F. B. M. (1996). *Good natured: The origins of right and wrong in humans and other animals*. Harvard: Harvard University Press.

De Waal, F. B. M., & van Roosmalen, A. (1979). Reconciliation and consolation among chimpanzees. *Behavioral Ecology and Sociobiology, 5*, 55–66.

De Waal, F. B. M., &, Thompson, E. (2005). Primates, monks and the mind: The case of empathy. *Journal of Consciousness Studies, 12*, 38–54.

Eisenberg, N. (2000). Emotion, regulation, and moral development. *Annual Review in Psychology, 51*, 665–697.

Eisenberg, N., Fabes, R. A., Murphy, B., Karbon, M., Maszk, P., Smith, M., et al. (1994). The relations of emotionality and regulation to dispositional and situational empathy-related responding. *Journal of Personality and Social Psychology, 66*, 776–797.

Farrow, T., & Woodruff, P. W. (2007). *Empathy in mental illness and health.* Cambridge: Cambridge University Press.

Geary, D. C., & Flinn, M. (2001). Evolution of human parental behavior and the human family. *Parenting: Science and Practice, 1*, 5–61.
Greene, J. D., Sommerville, R. B., Nystrom, L. E., Darley, J. M., & Cohen, J. D. (2001). An fMRI investigation of emotional engagement in moral judgment. *Science, 293*, 2105–2108.
Grosbras, M. H., & Paus, T. (2006). Brain networks involved in viewing angry hands or faces. *Cerebral Cortex, 16*, 1087–1096.
Harris, P. L. (2000). Understanding emotion. In M. Lewis & J. M. Haviland-Jones (Eds.), *Handbook of emotions* (pp. 281–292). New York: The Guilford Press.
Hatfield, E. (2009). Emotional contagion and empathy. In: J. Decety & W. Ickes (Eds.), *The social neuroscience of empathy* (pp. 19–30). Cambridge: MIT Press.
Ickes, W. (2003). *Everyday mind reading: Understanding what other people think and feel.* Amherst, NY: Prometheus Books.
Jackson, P. L., Brunet, E., Meltzoff, A. N., & Decety, J. (2006a) Empathy examined through the neural mechanisms involved in imagining how I feel versus how you feel pain. *Neuropsychologia, 44*, 752–761.
Jackson, P. L., Meltzoff, A. N., & Decety, J. (2005). How do we perceive the pain of others: A window into the neural processes involved in empathy. *NeuroImage, 24*, 771–779.
Jackson, P. L., Rainville, P., & Decety, J. (2006b). From nociception to empathy: The neural mechanism for the representation of pain in self and in others. *Pain, 125*, 5–9.
Lamm, C., Batson, C. D., & Decety, J. (2007). The neural basis of human empathy – effects of perspective-taking and cognitive appraisal. *Journal of Cognitive Neuroscience, 19*, 1–17.
Lanzetta, J. T., & Englis, B. G. (1989). Expectations of cooperation and competition and their effects on observers' vicarious emotional responses. *Journal of Personality and Social Psychology, 56*, 543–554.
Lawrence, E. J., Shaw, P., Giampietro, V. P., Surguladze, S., Brammer, M. J., & David, A. S. (2006). The role of 'shared representations' in social perception and empathy: An fMRI study. *NeuroImage, 29*, 1173–1184.
Lerner, M. J. (1980). *The belief in a just world: A fundamental delusion*. New York: Plenum.
Levenson, R. W., & Ruef, A. M. (1992). Empathy: A physiological substrate. *Journal of Personality and Social Psychology, 63*, 234–246.
MacLean, P. D. (1985). Brain evolution relating to family, play, and the separation call. *Archives of General Psychiatry, 42*, 405–417.
Moll, J., Oliveira-Souza, R., Garrido, G. J., Bramati, I. E., Caparelli-Daquer, E. M.., Paiva, M. M. F., et al. (2007). The self as a moral agent: Linking the neural bases of social agency and moral sensitivity. *Social Neuroscience, 2*(2–3), 336–352.
Morrison, I., Lloyd, D., di Pellegrino, G., & Roberts, N. (2004). Vicarious responses to pain in anterior cingulate cortex: Is empathy a multisensory issue? *Cognitive & Affective Behavioral Neuroscience, 4*, 270–278.
Niedenthal, P. M., Barsalou, L. W., Ric, F., & Krauth-Gruber, S. (2005). Embodiment in the acquisition and use of emotion knowledge. In L. Feldman Barrett, P. M. Niedenthal, & P. Winkielman (Eds.), *Emotions and consciousness* (pp. 21–50). New York: The Guilford Press.
Niedenthal, P. M., Halberstadt, J. B., Margolin, J., & Innes-Ker, A. H. (2000). Emotional state and the detection of change in the facial expression of emotion. *European Journal of Social Psychology, 30*, 211–222.
Ochsner, K. N., & Gross, J. J. (2005). The cognitive control of emotion. *Trends in Cognitive Sciences, 9*, 242–249.
Oliner, S. P., & Oliner, P. M. (1988). *The altruistic personality: Rescuers of Jews in Nazi Europe*. New York: The Free Press.
Parr, L. A. (2001). Cognitive and physiological markers of emotional awareness in chimpanzees (Pan troglodytes). *Animal Cognition, 4*, 223–229.
Parr, L. A., & Hopkins, W. D. (2000). Brain temperature asymmetries and emotional perception in chimpanzees, Pan troglodytes. *Physiology and Behavior, 71*, 363–371.
Parr, L. A., &, Waller, B. (2007). The evolution of human emotion. In J. Kaas (Ed.), *Evolution of the nervous system* (Vol. 5, pp. 447–472). New York: Elsevier.

Pavuluri, M. N., O'Connor, M. M., Harral, E., & Sweeney, J. A. (2006). Affective neural circuitry during facial emotion processing in pediatric bipolar disorder. *Biological Psychiatry*, Epub ahead of print.
Porges, S. W. (2007). The polyvagal perspective. *Biological Psychology, 74*, 116–143.
Povinelli, D. J. (2001). *Folk physics for apes*. New York: Oxford University Press.
Preston, S. D., & de Waal, F. B. M. (2002). Empathy: Its ultimate and proximate bases. *Behavioral Brain Science, 25*, 1–72.
Rizzolatti, G., & Craighero, L. (2004). The mirror-neuron system. *Annual Review in Neuroscience, 27*, 169–92.
Rolls, E. T. (1999). *The brain and emotion*. Oxford: Oxford University Press.
Ruby, P., & Decety, J. (2003). What you believe versus what you think they believe? A neuroimaging study of conceptual perspective taking. *European Journal of Neuroscience, 17*, 2475–2480.
Ruby, P., & Decety, J. (2004). How would you feel versus how do you think she would feel? A neuroimaging study of perspective taking with social emotions. *Journal of Cognitive Neuroscience, 16*, 988–999.
Scherer, K. R., Schorr, A., & Johnstone, T. (2001). *Appraisal processes in emotion*. New York: Oxford University Press.
Shamay-Tsoory, S. G., Lester, H., Chisin, R., Israel, O., Bar-Shalom, R., Peretz, A., et al. (2005). The neural correlates of understanding the other's distress: A positron emission tomography investigation of accurate empathy. *NeuroImage, 15*, 468–472.
Singer, T., Seymour, B., O'Doherty, J., Kaube, H., Dolan, R. J., & Frith, C. D. (2004). Empathy for pain involves the affective but not the sensory components of pain. *Science, 303*, 1157–1161.
Singer, T., Seymour, B., O'Doherty, J. P., Stephan, K. E., Dolan, R. J., & Frith, C. D. (2006). Empathic neural responses are modulated by the perceived fairness of others. *Nature, 439*, 466–469.
Solomon, R. C. (1990). *A passion for justice: Emotions and the origins of the social contract*. Reading, MA: Addison-Wesley.
Sommerville, J. A., & Decety, J. (2006). Weaving the fabric of social interaction: Articulating developmental psychology and cognitive neuroscience in the domain of motor cognition. *Psychonomic Bulletin & Review, 13*(2), 179–200.
Stein, M. B., Goldin, P. R., Sareen, J., Zorrilla, L. T., & Brown, G. G. (2002). Increased amygdala activation to angry and contemptuous faces in generalized social phobia. *Archives of General Psychiatry, 59*, 1027–1034.
Stotland, E. (1969). Exploratory investigations of empathy. In L. Berkowitz (Ed.), *Advances in experimental social psychology* (Vol. 4, pp. 271–313). New York: Academic Press.
Thompson, E. (2001). Empathy and consciousness. *Journal of Consciousness Studies, 8*, 1–32.
Tice, D. M., Bratslavsky, E., & Baumeister, R. F. (2001). Emotional distress regulation takes precedence over impulse control: If you feel bad, do it! *Journal of Personality and Social Psychology, 80*, 53–67.
Wicker, B., Keysers, C., Plailly, J., Royet, J. P., Gallese, V., & Rizzolatti, G. (2003). Both of us disgusted in my insula: The common neural basis of seeing and feeling disgust. *Neuron, 40*, 655–664.
Wilson, E. O. (1988). *On human nature*. Cambridge MA: Harvard University Press.
Yamada, M., & Decety, J. (2009). Unconscious affective processing and empathy: An investigation of subliminal priming on the detection of painful facial expressions. *Pain, 143*, 71–75.

Moral Judgment and the Brain: A Functional Approach to the Question of Emotion and Cognition in Moral Judgment Integrating Psychology, Neuroscience and Evolutionary Biology

Kristin Prehn and Hauke R. Heekeren

Moral judgment can be defined as the evaluation of actions with respect to social norms and values established in a society (such as not stealing or being an honest citizen).[1] Judging whether our own actions or of those of another person are good or bad (respectively harmful for individuals or the society as a whole) is very central to everyday social life because it guides our behavior in a community. Therefore, the question how humans think about right and wrong has recurred over the centuries in many disciplines including philosophy, arts, religion, economics, or law studies (see Goodenough & Prehn, 2004). More recently, the question how moral judgments are made and which processes are involved has triggered much research in psychology and neuroscience. One key issue comprises the question to what extent the processes involved are open to conscious deliberation and whether our moral sense is a product of education (i.e., the acquisition of knowledge on social norms and values) or rather a result of an innate mechanism activated during childhood. In particular, the issue whether moral judgments are caused by emotional or rather cognitive[2] processes and whether emotional responses make moral judgments better or worse has caused much controversies and debates.

K. Prehn (✉)
Neuroscience Research Center, Berlin NeuroImaging Center and Department of Neurology, Charité University Medicine, Berlin, Germany; Department of Psychology, Humboldt University, Berlin, Germany; Max-Planck-Institute for Human Development, Berlin, Germany
e-mail: kristin.prehn@fu-berlin.de

[1] For our purposes, we use the term "moral judgment" as an inclusive description of all judgments and decision making processes about those things that ought to be done and those that ought not to be done, particularly in the social context of interactions with other people and do not distinguish between moral and socio-conventional judgments. In the literature (Turiel, 1983; Smetana, 1993; Blair, 1995; Nucci, 2001) this distinction is used to differentiate cases where harm is caused to a person (= moral transgressions) from cases where only socio-conventional norms are violated (= conventional transgressions) without necessarily causing harm (e.g., spitting in a glass of wine at a dinner party).

[2] Here we follow the convention that the term "cognitive" suggests processes on the reasoning-rational-conscious end of the spectrum which are opposed to emotion-linked "affective" processes. Note that sometimes the term "cognitive" is used in a more extended sense representing a wider range of mental processes and information processing in general (e.g., in the term "cognitive neuroscience").

J. Verplaetse et al. (eds.), *The Moral Brain*, DOI 10.1007/978-1-4020-6287-2_6,

Two Competing Psychological Models on Moral Judgment

Moral Reasoning from a Cognitive-Developmental Perspective

Psychological research on moral judgment has long been dominated by a cognitive-developmental approach investigating the maturation of moral orientations and principles. This line of research emphasizes the role of conscious and rational reasoning processes (Piaget, 1965; Kohlberg, 1969).

In his empirical studies, Kohlberg investigated moral reasoning and the underlying moral orientations and principles by presenting child and adolescent participants with moral dilemmas. In these dilemmas a protagonist is caught in an awkward position; whatever he or she decides to do will conflict with some rule of conduct. In one of Kohlberg's best known dilemmas, for instance, a man must decide whether he should break into a druggist's shop to steal a medicine that would save the life of his dying wife (Kohlberg, 1969). After presenting children and adolescents with such dilemmas, Kohlberg asked the participants to argue why it would be justified to choose a certain action (i.e., breaking into the druggists shop or not).

Based on how children and adolescents argued, Kohlberg discerned a widely cited *six-stage model* of the development of moral reasoning, through which, he argued, humans progress as their cognitive abilities mature and come to a more sophisticated understanding of social relationships. These stages are characterized, for instance, by the growing ability for perspective taking: In higher stages of moral reasoning, it is assumed that people come to see situations not only from their own perspective but also from the perspectives of all other people involved in the conflict.

At stage 1 (= obedience and punishment orientation), children think a behavior is right when an authority says it is. Doing the right thing means obeying an authority and avoiding punishment. At stage 2 (= self interest and exchange orientation), children see that there can be different sides to an issue and each person is free to pursue his or her own interests. Additionally, children understand that it is often useful to do someone else a favor (pre-conventional level). Later on at the conventional level, young people think of themselves as members of the society with its values, norms, and expectations. At stage 3 (= interpersonal accord and conformity orientation), they aim to be a "good boy or girl", which basically means being helpful toward other people who are close to them. At stage 4 (= authority and social order maintaining orientation), the concern shifts toward obeying the laws to maintain society as a whole. At the post-conventional level, people start to think about the principles and values that constitute a good society. At stage 5 (= social contract orientation), laws are regarded as social contracts rather than rigid dictums. Those laws that do not promote the general welfare should be changed when necessary to meet the greatest good for the greatest number of people (e.g., by a democratic majority decision). Finally at stage 6 (= universal ethical principles), moral reasoning is based on abstract reasoning using the universal ethical principles of justice and of the reciprocity and equality of human rights with respect for the dignity of human beings as individuals (see Table 1; Kohlberg, 1969).

Kohlberg's theory of the development of moral reasoning and the underlying moral orientations and principles has strongly influenced the discourse about

Table 1 Kohlberg's six stages of moral development

Level 1: Pre-conventional morality	Stage 1: Obedience and punishment orientation Stage 2: Self interest and exchange orientation
Level 2: Conventional morality	Stage 3: Interpersonal accord and conformity orientation Stage 4: Authority and social order maintaining orientation
Level 3: Post-conventional morality	Stage 5: Social contract orientation Stage 6: Universal ethical principles

morality and the subsequent research on moral education. However, there also exists much criticism. For instance, it has been criticized that Kohlberg investigated only post-hoc justifications on moral judgments that have already occurred rather than actual reasoning processes leading to moral judgments by using this interview technique (see below, Haidt, 2001). Others criticized the stage concept in general and, in particular, the assumption that there exists a universal and invariant sequence of developmental stages (Snarey, 1985). Another major point of criticism on Kohlberg's theory is that it emphasizes justice to the exclusion of other values. Carol Gilligan, for instance, has argued that Kohlberg's theory was mainly based on empirical research on male participants, and thus, does not adequately describe the concerns of women. Therefore, she developed an alternative theory of moral reasoning that is not based on justice, but on the ethics of caring (Gilligan, 1977; Gilligan & Attanucci, 1988).

At least for our purposes to explore which processes are involved in moral judgment, the relevance of this cognitive-developmental theory can be seen in the idea that morality does not only rely on the acquisition of social knowledge and moral values and virtues, but also on the way in which an individual understands and thinks about social situations. This way – how an individual thinks about social situations – qualitatively changes as a result of an active interaction of the individual and his or her social environment. Following this approach, moral judgment is based on a person's representations of social interactions and relationships, and thus, on cognitive functions rather than on emotions of fear, anxiety, shame, or guilt.[3]

The Role of Emotions and Intuitive Feelings in Moral Judgment

While psychological research on moral judgment has long been dominated by the Kohlbergian approach, more recent theories and models question the assumption that moral judgment is primarily reached by formal reasoning and emphasize the

[3]Please note that the cognitive-developmental theory is not affect-free. While emphasizing the role of cognition, Kohlberg also explicitly proposed that emotions of sympathy for other people, as well as altruism or the spontaneous interest in helping others also contribute to an individual's moral development (especially during the process of taking the perspective of others).

role of intuition as well as automatic, subconscious, and emotional processes (e.g., Haidt, 2001, 2003, 2007; Hauser, 2006; Mikhail, 2007).

The social intuitionist model by Haidt (2001), for instance, posits that fast and automatic intuitions (like gut feelings or aesthetic judgments) are the primary source of moral judgments, whereas conscious deliberations are only used to construct post hoc justifications for judgments that have already occurred. This would mean that moral reasoning is less relevant to moral judgment and behavior than Kohlberg's theory suggests and implies that people often make moral judgments without weighing concerns such as fairness, law, human rights, and abstract ethical values. Haidt (2001) views moral judgments as affective "evaluations (good versus bad) of the actions or a character of a person that are made with respect to a set of virtues held to be obligatory by a culture or subculture" (Haidt, 2001, p. 817). He describes the minor role of moral reasoning in moral judgment provocatively as the "rational tail of the emotional dog" and provided some striking examples of "moral dumbfounding" in which participants were unable to generate adequate reasons for an intuitively given moral judgment. For instance, if people speak about consensual sex between adult siblings, almost everyone insists that it is wrong, even though they cannot articulate reasons for their view.

The universal moral grammar theory (by using concepts and models in analogy to those used in the study of language; e.g., Hauser, 2006; Mikhail, 2007), very similar to the social intuitionist model, proposes that the human mind is endowed with an innate moral grammar consisting of a domain specific and complex set of rules, concepts, and principles that guide our social behavior in a community. For instance, there is evidence that people consistently judge harm caused by action as morally worse than the same harm caused by omission (action principle). Harm intended as the means to a goal is also judged as morally worse than the same harm foreseen as the side effect of a goal (intention principle). Using physical contact to cause harm to a victim, moreover, is seen as morally worse than causing equivalent harm to a victim without using physical contact (contact principle, Hauser, 2006; Mikhail, 2007). According to the universal grammar theory, at least some of these mechanisms and principles are supposed to be innate, which means that their ontogenetic development is pre-determined by the inherent structure of the mind. The development of these principles, however, is assumed to be triggered and shaped by environmental experiences.

Further highlighting the role of emotion, Haidt (2003) moreover suggests some useful distinctions, sorting "moral emotions" (i.e., emotions in response to moral violations) into other-condemning emotions (contempt, anger, and disgust), self-conscious emotions (shame, embarrassment, and guilt), the other-suffering family (sympathy and compassion) and the other-praising family (gratitude, awe, and elevation).

Following the social intuitionist model and the universal grammar theory, our moral judgments rely on intuitive feelings. Furthermore, it is stated that we often have no conscious understanding of why we feel what we feel. The love we feel toward our children and the anger at those who cheat us, thus, can be considered as an innate moral sense and an adaptive mechanism of selective advantage that has been shaped over the course of evolution. While there is some evidence supporting this view (e.g., Haidt, Koller, & Dias, 1993; Cushman, Young, & Hauser, 2006; Koenigs et al., 2007; Hauser, Cushman, Young, Jin, & Mikhail, 2007), others argue that immediate intuitions can also be informed by conscious deliberation (e.g., Pizarro & Bloom, 2003; Takezawa, Gummerum, & Keller, 2006) and that some moral principles are available to conscious reason while others are not (e.g., the above mentioned intention principle with its distinction between intended and foreseen consequences appears to be inaccessible to conscious reflection, see Cushman et al., 2006).

The Neuroscientific Study of Moral Judgment

Lesion Studies Provide First Evidence for a Neurobiological Basis of Morality

A first hint that morality and moral behavior might have a neurobiological basis came from the classic case of Phineas Gage, a railroad worker who's decision making in real life was impaired after his ventromedial prefrontal cortex was damaged in an accidental explosion (Harlow, 1848; Damasio, Grabowski, Frank, Galaburda, & Damasio, 1994). After his surprising recovery from this injury, he showed preserved basic cognitive abilities and social knowledge (indexed by IQ-tests and other measures) but an irresponsible and inappropriate social behavior, impaired moral decision making in everyday life, and a limited ability to experience emotions.

More recent case reports also indicate that damage to the prefrontal cortex (especially its ventromedial and orbitofrontal portions) leads to deficits in social and moral behavior (e.g., Saver & Damasio, 1991; Dimitrov, Phipps, Zahn, & Grafman, 1999). Patients with lesions in the orbitofrontal cortex, for instance, show a defective ability for anticipating negative consequences of their choices during a gambling task and do not experience regret (Camille et al., 2004; for an illustration of the location of these regions, namely the ventromedial and the orbitofrontal cortex see Fig. 1).

Based on such observations on patients with lesions of the "orbitomedial" prefrontal cortex, Damasio et al. (1994; Damasio, 1996) deduced the "somatic marker hypothesis". This hypothesis states that emotional responses that involve bodily changes and feelings (labeled as "somatic markers" such as an increase in heart rate or skin conductance) become associated with reinforcing stimuli (or punishment, respectively) and bias real life decision making in the future, especially in very complex situations with a high degree of uncertainty and ambiguity. After the repeated experience of certain bodily changes as a response to a certain outcome of a certain action in a certain situation (for instance, a bad feeling when caught red-handed), the brain areas that monitor these bodily changes begin to respond whenever a similar situation with similar behavioral options arises. The mere thought of a particular action, thus, becomes sufficient to trigger an "as if"

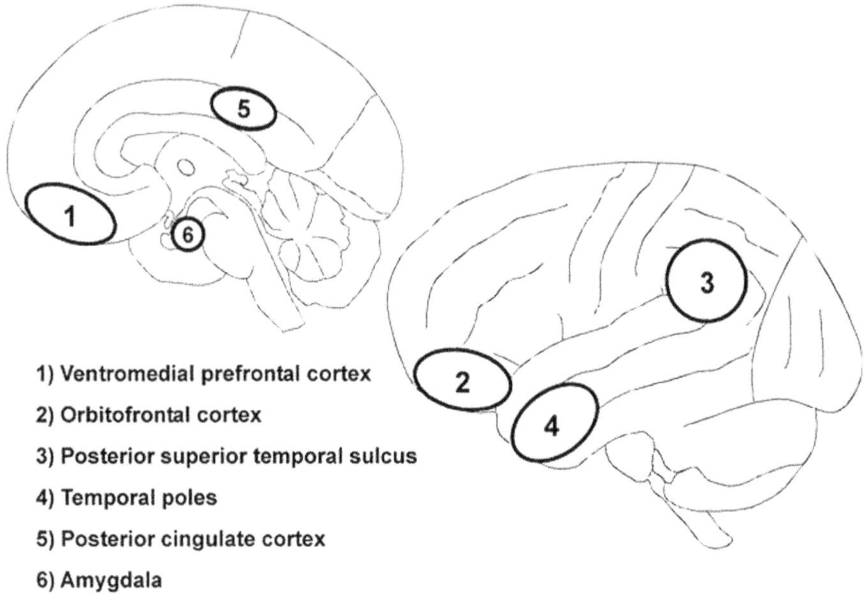

Fig. 1 A neural network involved in moral cognition

response in the brain and the person experiences the same bodily feelings he or she has experienced after performing this action in the past. The orbitomedial prefrontal cortex is assumed to integrate these feelings with other knowledge and planning functions.

To further investigate the role of emotion in moral judgment, Koenigs and colleagues (2007), tested a group of patients with damage to the ventromedial prefrontal cortex which showed reduced responses to emotionally charged pictures and, according to their spouses, reduced feelings of empathy and guilt. When confronted with moral dilemmas, the patients with lesions in the ventromedial prefrontal cortex were more likely to choose the utilitarian option (e.g., to sacrifice one person's life to save a number of other lives that normally would elicit a strong emotional response) than were control participants and patients with lesions in other brain regions (Koenigs et al., 2007; Koenigs & Tranel, 2007).

Notably, the age at which the brain injury occurred also has an effect on the degree and nature of the deficits. Anderson, Bechara, Damasio, Tranel, & Damasio (1999) showed that lesions in the ventromedial and orbitofrontal cortex (very similar to those mentioned above) acquired in early childhood not only lead to impaired social and moral behavior but also seem to prevent the acquisition of factual knowledge about the accepted standards of moral behavior in general (see Eslinger & Biddle, 2000).

In summary, lesion studies displayed evidence that at least some of the processes and components involved in moral judgment are dissociable (e.g., the distinction mentioned above between acquisition versus application of social rules, the identification of subcomponents such as the ability to anticipate punishment or to experience moral emotions). Regarding the question of emotion and cognition in moral judgment, the Somatic Marker Theory claims that emotions are important to adapt behavior to environmental demands, while other studies showed that emotional responses should be suppressed or regulated when utilitarian moral judgments are required. Following lesion studies, emotions clearly have a role in moral judgment. However, it is still unclear whether emotions improve or impair the decision making.

Some Methodological Considerations on Imaging Brain Activity

It is important to keep in mind that lesion studies rely only on very few cases with mostly very large and heterogeneous lesions. They give important hints but they cannot really tell us how the process of behaving appropriately or making moral judgments and ethical choices is organized in the intact human brain. In recent years, cognitive neuroscientists have taken great advantage of neuroimaging methods like functional magnetic resonance imaging (fMRI) that make it possible to measure brain activity during a specific sensory, motor, or cognitive task (e.g., judging a behavior in terms of being good or bad) in healthy participants. For a non-specialist faced with imaging data, however, it is important to understand what it does – and does not – mean. In particular, it is important to know that the colorful pictures of brains "lighting up" and showing a map of brain regions activated during a specific task are actually artifacts of extensive analysis and selective presentation.

Many fMRI experiments as well as those reviewed in this chapter are using subtraction logic in their experimental designs. This logic was pioneered by the Dutch physiologist Franciscus Cornelius Donders in reaction time experiments (see Donders, 1969, translation of: Die Schnelligkeit psychischer Prozesse, first published in 1868, *Arch Anat Physiol, 8*, 657–681) and relies on the a priori assumption of "pure insertion", which means that one (cognitive) process can be added to a preexisting set of (cognitive) processes without affecting them and asserts that there are no interactions among the different components of a task. Although this assumption has not been validated in any physiological sense (see Friston et al., 1996), it is applied in almost all fMRI studies due to the fact that during the performance of a complex task (e.g., judging whether a behavior is violating a social norm) many if not all parts of the brain are activated to some degree. A way to identify brain regions which are specifically related to the moral judgment process is to compare neural activity during a moral judgment task with neural activity elicited by another judgment task, which shares all sub-processes with the moral judgment task but the moral component.

In our own work, for instance, we compared neural activity during a moral judgment task with activity during a grammatical judgment task (for an example of

Table 2 Examples of sentence material used in an fMRI study

First sentence (Intro)		Moral judgment	Grammatical judgment
A uses public transportation [A fährt mit der S-Bahn]	Non-violation	He looks out of the window [Er sieht aus dem Fenster]	He looks out of the window [Er sieht aus dem Fenster]
	Violation	He smashes the window [Er wirft das Fenster ein]	He look out of the window [Er sehen aus dem Fenster]

During both tasks (moral and grammatical judgment), the first sentence of a trial introduced the participants to a specific situation. Half of the second sentences contained a violation of a social norm or grammatical rule. After the appearance of the second sentence, participants were instructed to decide whether the action described in the second sentence was a social norm violation or not, or whether the sentence was grammatically correct or incorrect.

such a task and material see Table 2). During both the moral and the grammatical judgment task, participants have to read sentences on a screen, to judge either whether the actions described are morally appropriate or not, or whether the sentences are grammatically correct or not, respectively, and then to respond as quickly and correctly with a button press. Following the subtraction logic in this case, the grammatical judgment task controls for visual input, language processing, decision making, and motor output.

Results of an fMRI experiment typically show a colorful projection of those brain regions onto a model brain, in which some statistically significant level of increase or decrease in BOLD signal occurred during a task (e.g., a moral judgment task) relative to a control state (e.g., a grammatical judgment task), mostly on some kind of accumulated basis over a sample of several subjects.

Because the results of fMRI studies only show differences of activations between compared tasks, it is clear that the data on neural correlates of mental phenomena can only be as good as the underlying tasks and experimental paradigms. Therefore, experimenters have to carefully design the experimental tasks that specifically activate the cognitive functions of interest and rule out influences of possible confounds. To be able to interpret a certain pattern of brain activity as a response to a specific task, one also needs very clear hypotheses about the involved mental processes derived from psychological theories and hypotheses about the underlying neuronal mechanisms, for instance, derived from lesion data or electrophysiology in monkeys (Henson, 2005, 2006; Poldrack, 2006).

Moreover, it is necessary to use a combination of methods and data such as behavioral parameters (e.g., response times and error rates to indicate information processing speed and accuracy) and psychophysiological parameters (e.g., skin conductance responses to indicate emotional arousal or pupillary responses to indicate the amount of mental resource consumption).

A Distributed Functional Network of Brain Regions Activated During Moral Judgment

As far as we know to date, we cannot expect that any complex representation such as morality is located in a special and distinct brain area (i.e., in “a moral centre”). Our current brain model is the interconnected networking model of information processing. Even when compared with a control task, a complex task such as judging whether a presented behavior is wrong in regard to social norms or conventions comprises a number of cognitive and emotional processes that are represented by a distributed network of brain regions.

Additionally, different complex tasks often show highly overlapping neural networks. For instance, in contrast to common belief even cognition and emotion are not subserved by separate and independent circuits (Davidson, 2000, 2003; Dolan, 2002). As Davidson (2000) stated “it turned out that the duality between reason and emotion that has been perpetuated through the ages is a distinction that is not honored by the architecture of the brain” (p. 91). Subcortical structures which are part of the limbic system that is commonly assumed as the seat of emotion are also crucial for certain cognitive processes (e.g., the hippocampus for memory) while cortical regions once thought to be the exclusive seat of cognition and complex thought are now known to be intimately involved in emotion as well (e.g., the orbitomedial prefrontal cortex).

Using functional MRI, recent neuroimaging studies have helped to discover which brain regions contribute to moral cognition in healthy subjects. These studies used very different tasks ranging from simple moral decisions (e.g., Moll, Eslinger, & Oliveira-Souza, 2001; Moll, Oliveira-Souza, Bramati, & Grafman, 2002; Moll, Oliveira-Souza, Bramati et al. 2002; Heekeren, Wartenburger, Schmidt, Schwintowski, & Villringer, 2003; Heekeren et al., 2005; Luo et al., 2006) to complex dilemmatic moral reasoning (e.g., Greene, Sommerville, Nystrom, Darley, & Cohen, 2001; Greene, Nystrom, Engell, Darley, & Cohen, 2004; Borg, Hynes, Van, Grafton, & Sinnott-Armstrong, 2006).

For instance, Greene et al. (2001) investigated moral judgment by presenting participants moral dilemmas. In this experiment two types of dilemmas were contrasted, represented by the “trolley dilemma” and the “footbridge dilemma”. In the trolley dilemma the participant is asked to consider the following situation: A runaway trolley is quickly approaching a fork in the tracks. On the tracks extending to the left is a group of five railway workmen. On the tracks extending to the right is a single railway workman. If one does nothing, the trolley will proceed to the left, causing the deaths of the five workmen. The only way to avoid the deaths of these workmen is to hit a switch on your dashboard that will cause the trolley to proceed to the right, causing the death of the single workman. After presenting this story, the participant in the experiment is asked to respond whether it is appropriate to hit the switch to avoid the deaths of the five workmen. In the footbridge dilemma the situation is slightly different. Again, a runaway trolley is heading down the tracks toward five workmen who will be killed if the trolley proceeds on its present course. The participant now has to imagine being on a footbridge over the tracks and in between the approaching trolley and the five workmen. Next on the footbridge stands a large stranger. The only way to save the lives of the five workmen is to push the stranger off the bridge and onto the tracks below where his large body will stop the trolley. The stranger will die as a result of this action, but the five workmen will be saved. After presenting this situation, the participants again were asked to respond whether it is appropriate to push the stranger onto the tracks to save the five workmen. By comparing neural activity during reasoning about these different types of dilemmas, Greene et al. (2001) found that reasoning about dilemmas that are emotionally engaging like the footbridge dilemma (i.e., personal dilemmas or dilemmas in which physical harm is caused to another person directly by the agent) as compared to dilemmas that are less emotionally engaging like the trolley dilemma (i.e., impersonal dilemmas or dilemmas in which physical harm is caused to another person only indirectly) activate the medial prefrontal cortex, the posterior cingulate cortex and the posterior superior temporal sulcus.

Moll, Oliveira-Souza, Bramati et al. (2002) used passive viewing of pictures portraying emotionally charged, unpleasant social scenes representing moral violations and reported that the orbital and the medial prefrontal cortex as well as the posterior superior temporal sulcus are recruited during passively viewing scenes evocative of moral emotions (defined as emotions that are intrinsically linked to the interests or welfare either of the society as a whole or of persons other than the agent).

In our own studies, we used short sentences and compared neural activity during moral judgments with neural activity during semantic or grammatical judgments (Heekeren et al., 2003, 2005; Prehn et al., 2008; see Table 2 for examples of this material).

During both tasks (e.g., moral and grammatical judgment), the first sentence of a trial introduced the participants to a specific situation. Half of the second sentences contained a violation of a social norm or grammatical rule. After the appearance of the second sentence, participants were instructed to decide whether the action described in the second sentence was a social norm violation or not, or whether the sentence was grammatically correct or incorrect.

Using this short sentence material, we found that the ventromedial prefrontal cortex and the posterior superior temporal sulcus are a common neural substrate of moral judgment (Heekeren et al., 2003).

Some authors, additionally, have variously focused on moral emotions like guilt, shame, regret, or gratitude (Harenski & Hamann, 2006), the evaluation of one's own actions and whether actions are intentionally or accidentally (Berthoz, Armony, Blair, & Dolan, 2002; Borg et al., 2006; Berthoz, Grezes, Armony, Passingham, & Dolan, 2006), the influence of bodily harm on neural correlates of moral decision making (Heekeren et al., 2005, see below), the role of cognitive control and conflict processing (Greene et al., 2004, see below), or the impact of audience on moral judgments (Finger, Marsh, Kamel, Mitchell, & Blair, 2006).

In summary, the fMRI studies on moral judgment revealed a functional network of brain regions including the ventromedial prefrontal cortex, the orbitofrontal cortex, the temporal poles, the amygdala, the posterior cingulate cortex, and the

Table 3 Overview of the possible functions of the different brain regions involved in moral judgment

Brain region	Functions
Ventromedial prefrontal cortex	– Understanding other people's behavior in terms of their intentions or mental states (Theory of mind, e.g., Berthoz et al., 2002; Frith & Frith, 2006) – Generation of social emotions such as compassion, shame, and guilt, that are closely related with moral values when confronted with social norm violations (e.g., Koenigs et al., 2007; Koenigs & Tranel, 2007)
Orbitofrontal cortex	– Representation of the expected value of possible outcomes of a behavior in regards to rewards and punishments (e.g., Walton, Devlin, & Rushworth, 2004; Camille et al., 2004; Amodio & Frith, 2006)
Posterior superior temporal sulcus	– Representation of socially significant information from different modalities [e.g., multisensory integration (e.g., Beauchamp, Lee, Argall, & Martin, 2004), processing of biological motion cues (e.g., Beauchamp, Lee, Haxby, & Martin, 2002; Beauchamp, Lee, Haxby, & Martin, 2003; Schultz, Friston, O'Doherty, Wolpert, & Frith, 2005), detection and analysis of goals and intentions of another person's behavior (e.g., Schultz, Imamizu, Kawato, & Frith, 2004; Young, Cushman, Hauser, & Saxe, 2007)]
Temporal poles	– Episodic and autobiographical memory (e.g., Fink et al., 1996; Dolan, Lane, Chua, & Fletcher, 2000)
Posterior cingulate cortex	– Processing the emotional significance of words and objects (e.g., Maddock, 1999), imagery (e.g., Mantani, Okamoto, Shirao, Okada, & Yamawaki, 2005)
Amygdala	– Emotional processing especially negative stimuli and threat cues like fearful or angry faces (e.g., Adolphs, Russell, & Tranel, 1999; Adolphs, 1999)

Greene & Haidt (2002), Moll et al. (2003), Casebeer (2003), Casebeer & Churchland (2003), Goodenough & Prehn (2004), Moll et al. (2005).

posterior superior temporal sulcus and support a theory of moral judgment according to which both emotional and cognitive components play an important role (Greene et al., 2004; see Fig. 1 and Table 3 for an overview of the different involved brain regions and their functions; for reviews see also: Greene & Haidt, 2002; Moll, Oliveira-Souza, & Eslinger, 2003; Casebeer, 2003; Casebeer & Churchland, 2003; Goodenough & Prehn, 2004; Moll, Zahn, Oliveira-Souza, Krueger, & Grafman, 2005).

Because our current brain model is the interconnected networking model of information processing, we additionally will need to investigate how different brain regions interact during moral judgment tasks, for instance, by using measures related to connectivity. To answer the question how moral judgment is influenced by emotional responses, it might be necessary to clarify how a change of neural activity in brain regions associated with emotion processing (e.g., the amygdala) modulates neural activity in other brain areas engaged in the moral judgment process (e.g., the temporal poles or the posterior superior temporal sulcus, see below).

Neuroimaging Studies with a Focus on Emotion and Cognition in Moral Judgment

As presented above, the results of the neuroimaging studies on moral judgment revealed a distributed functional network of brain regions associated with emotional as well as cognitive processes.

However, the precise role of the different brain regions within the network still remains unclear. In detail, all the identified regions are active during a number of tasks, for instance, control of behavior, processing of socially relevant cues, mind and intention reading, semantic memory retrieval, and processing of emotional stimuli (see Greene & Haidt, 2002). Thus, rather than searching for a "moral centre" of the brain, one of the challenges will be to better understand how the different brain regions act together to perform such a complicated task building on emotional and cognitive processes. Therefore, careful experimental manipulation of the different processes contributing to moral judgment will help us to better understand how moral judgments are made and which processes are involved. In the following section, we will review results from three recent studies that have addressed the question of emotion and cognition in moral judgment.

Evidence for Competing Emotional and Cognitive Subsystems During Dilemmatic Moral Judgments

As described above, in their first study on moral judgment, Greene et al. (2001) distinguished between "personal" (e.g., the "footbridge dilemma") and "impersonal" moral dilemmas (e.g., the "trolley dilemma"). In a personal moral dilemma like the footbridge dilemma, an actor directly causes serious bodily harm to a victim who is represented as a person and this harm "must not result from the deflection of an existing threat onto a different party" (e.g., throwing a man off the bridge onto the tracks where he will die; Greene et al., 2004, p. 389). Moral judgments regarding personal moral dilemmas are supposed to be driven largely by emotional responses while judgments regarding impersonal moral dilemmas are supposed to be driven more by cognitive processes.

In another study, Greene et al. (2004) investigated how people solve particularly difficult personal moral dilemmas. An example for such a very difficult personal dilemma is the "crying baby dilemma" that was used especially to bring cognitive and emotional processes into tension.

In this dilemma the participant has to decide whether it is appropriate to smother his own crying baby to save his life and the lives of other refugees which are hiding from enemy soldiers in the cellar. This dilemma has been proven to be very difficult. That is, participants answer usually very slow when presented with such a dilemma and do not reach a consensus on this issue. It has been argued that the difficulty is due to a competition between the negative emotional response associated with the thought of killing one's own child with the more abstract and utilitarian understanding that many lives could be saved by carrying out this simple but horrific act.

Using such complicated personal moral dilemmas, Greene et al. (2004) conducted an fMRI study and compared neuronal activity during utilitarian and non-utilitarian decisions (i.e., neural activity when participants decided that smothering the crying baby to save the more lives is appropriate vs. neural activity when participants decided that smothering the crying baby is not appropriate). The authors found increased activity in brain regions associated with abstract reasoning, conflict processing, and cognitive control such as the anterior cingulate cortex and the dorsolateral prefrontal cortex when participants made a utilitarian decision on this issue compared to trials when participants made a non-utilitarian decision.

One interpretation of these results is that a conflict associated with such a difficult moral question is detected by the anterior cingulate cortex which then recruits control mechanisms and rational reasoning processes associated with neuronal activity in the dorsolateral prefrontal cortex. These control processes that are reflected in activity in the dorsolateral prefrontal cortex help to resolve the conflict and to override the prepotent emotional response to make a utilitarian decision (Greene et al., 2004).

The Influence of Bodily Harm on Neural Correlates of Moral Decision Making

In most fMRI studies on moral judgment, the stimulus material contained bodily harm or violence (like in the moral dilemmas used by Greene et al., 2001, 2004; or the emotionally charged pictures used by Moll, Oliveira-Souza, Bramati et al., 2002).

To investigate how the neural and behavioral correlates of moral judgment are modulated by the presence of bodily harm or violence in the stimulus material, we conducted an fMRI study and measured neural activity and response times while participants made moral and semantic decisions about sentences describing actions of agents that either contained bodily harm or not (Heekeren et al., 2005). In the moral decision-making task participants were required to decide whether the sentences represent a moral violation or not, whereas in the semantic decision-making task the participants had to decide whether the sentences were semantically correct or incorrect. The sentence material was similar to the material used in our previous studies, except that half of the sentences in each task contained descriptions of bodily harm or violence.

In line with the previous studies, a network of brain regions including the ventromedial prefrontal cortex, the posterior superior temporal sulcus, the posterior cingulate cortex, and the temporal poles was more activated during moral decision making compared to semantic decision making (Heekeren et al., 2005).

Regarding the impact of illustrations of bodily harm, at the behavioral level we, moreover, found that during both the moral and the semantic decision-making task the presence of bodily harm or violence resulted in shorter response times. This result is in line with the literature on emotion processing showing that "threat cues" such as illustrations of bodily harm or violence signify great importance to the organism, and therefore, these cues are usually processed faster (Ochsner & Feldman Barrett, 2001; Compton, 2003; Phelps, 2006).

At the neural level, however, we found no increase in neural activity in brain regions associated with emotion processing such as the amygdala, the ventromedial prefrontal cortex and the posterior cingulate cortex in response to trials containing bodily harm compared to trials not containing bodily harm during both tasks. Rather we found decreased neural activity in the anterior temporal poles during trials containing bodily harm relative to trials not containing bodily harm. This effect was also not specific to the moral judgment task (Heekeren et al., 2005).

Moral decision making (i.e., deciding whether a behavior is appropriate or not in respect to norms and values held to be virtuous in society) as well as semantic decision making (i.e., deciding whether a sentence is semantically correct or not) might require the recall of episodic and autobiographic knowledge stored in long-term memory that have been acquired during socialization. The temporal poles have been implicated in episodic and autobiographic memory retrieval (e.g., Fink et al., 1996; Dolan et al., 2000). To enable a fast processing and appropriate responses (flight or fight reaction), memory retrieval is reduced in the presence of threat cues such as bodily harm. In other words, cognitive processing depth is reduced in favor of more intuitive and automated processing. In our study this was reflected in weaker activity in the temporal poles and faster response times (Heekeren et al., 2005).

Based on the results of this study, we argue that moral judgment does not entirely rely on emotional processes. In contrast, it seems that strong emotional cues or features are distracting and detrimental for cognitive or rational processing during the moral decision-making process.

The Influence of Individual Differences in Moral Judgment Competence on Neural Correlates of Moral Judgment

So far, there is evidence from clinical and neuroimaging studies that cognitive processes, namely reason, *as well as* emotion play an important role during moral judgment and sometimes both come into conflict with each other. We conclude that it might depend on the context and the particular social situation whether moral judgments are dominated by cognitive or by emotional processes and whether emotion makes moral judgments better or worse. As Talmi and Frith (2007, p. 866) put it "the challenge, then, is for decision makers to cultivate an intelligent use of their emotional responses by integrating them with a reflective reasoning process, sensitive to the context and goals of the moral dilemmas they face. If decision makers meet this challenge, they may be better able to decide when to rely upon their emotions, and when to regulate them." Along these lines of reasoning, one might postulate that decision makers *individually differ* in their ability to integrate the emotional and cognitive processes into moral judgments, decisions, and behavior.

All studies on the neural correlates of moral judgment conducted so far relied on group analyses and individual differences in information processing were treated as "noise" (see methodological considerations on imaging brain activity above). However, the results of imaging studies may crucially depend on the specific sample and their characteristics in emotional and cognitive information processing (see Thompson-Schill, Braver, & Jonides, 2005; Meriau et al., 2006).

A current approach (Lind, 2008), in particular, points out the role of individual differences within the moral domain. Here, morality is defined as consisting of two inseparable, yet distinguishable aspects: (a) a person's *moral orientations and principles* and (b) a person's *competence* to act accordingly. According to this theory, moral judgment competence is defined as the ability to apply moral orientations and principles in a consistent and differentiated manner in varying social situations. Thus, social norms and values represented as affectively laden moral orientations are linked by means of moral judgment competence with everyday behavior and decision making. While most people commonly agree upon moral orientations and principles that are considered to be virtuous in their society, it is evident that people differ considerably with respect to their moral judgment competence (Lind & Wakenhut, 1985; Lind, 2008).

In a recent fMRI study (Prehn et al., 2008), we therefore investigated how individual differences in moral judgment competence are reflected in changes in brain activity during a simple moral judgment task (socio-normative judgments) very similar to the task used in our previous studies (see Table 2). We measured neural activity while participants made either moral or grammatical judgments (i.e., participants were required to decide whether sentences were morally or grammatically correct or not) and correlated neural activity during these tasks with individual scores in moral judgment competence. Individual moral judgment competence was measured using the Moral Judgment Test (MJT; Lind, 1998, 2008; www.uni-konstanz.de/ag-moral/mut/mjt-intro.htm).

The MJT confronts a participant with two complex moral dilemmas. In one dilemma (the doctor dilemma), for instance, a woman had cancer with no hope for being cured. She suffered terrible pain and begged the doctor to aid her in committing medically assisted suicide. She said she could no longer endure the pain and would be dead in a few weeks anyway. The doctor complied with her wish. After presentation of this short story, the participant has to indicate to which degree he or she agrees or disagrees with the solution chosen by the protagonist. After that, the participant is presented with six arguments supporting (pro-arguments) and six arguments rejecting (counter-arguments) the protagonist's solution of mercy-killing which the participant has to rate with regard to its acceptability on a nine point rating scale ranging from –4 (highly unacceptable) to +4 (highly acceptable). Each pro- and counter- argument represents a certain moral orientation (according to the six Kohlbergian stages; Kohlberg, 1969). An example for a lower level argument against the doctors solution would be: "The doctor acted wrongly because he could get himself into much trouble. They have already punished others for doing the same thing.", whereas the argument: "The doctor acted wrongly because the protection of life is everyone's highest moral obligation. We have no clear moral criteria for distinguishing between mercy-killing and murder." represents a more elaborated argument against the given solution of mercy-killing. A person's moral orientation can be assessed by calculating the median acceptability ratings for all arguments which refer to a certain moral orientation (i.e., one pro- and one counter-argument for each of the two presented moral dilemmas).

In general, adult participants in contrast to children or adolescents prefer the more elaborated arguments (i.e., adults rate more elaborated arguments as more acceptable than low level arguments) due to their advanced moral orientation or their higher developmental stage of moral judgment. However, adult participants differ greatly in their ability of applying this high moral orientation consistently in different situations especially when confronted with counter-arguments (i.e., arguments which are against their own opinion). Some participants rate the more elaborated arguments as more acceptable than the lower level arguments only when these arguments represent their own opinion, but do not judge arguments differentially when these arguments are against their own opinion (e.g., reject all counter-arguments regardless whether they are elaborated or not).

The moral judgment competence score (C-score, the MJT's main score) reflects this ability how consistently or, in Lind's terms how competently a person applies a certain moral orientation in the decision-making process independently of whether arguments are in line with the personal opinion on a particular issue or not, and is calculated as an individual's total response variation concerning the underlying moral orientations of the given arguments. A highly competent person (indicated by a high C-score close to 100) will consistently appreciate all arguments referring to a certain socio-moral perspective, irrespective of whether this argument is a pro- or counter-argument. In contrast, a person with low moral judgment competence will appreciate only arguments which support their own solution of the dilemma (only pro- or counter-arguments, respectively).

The concept of moral judgment competence is based on Kohlberg who introduced the term moral judgment competence as "the capacity to make decisions and judgments which are moral (i.e., based on internal principles) and to act in accordance with such judgments" (Kohlberg, 1964, p. 425). However, by defining moral judgment competence more precisely, Lind's approach clearly goes beyond what we may ordinarily call "moral competence" as well as the Kohlbergian approach which focused merely on moral orientations and the level of reasoning. The MJT to measure this competence has proved to be a valid and reliable psychometric test. For instance, moral judgment competence has been associated with responsible and democratic behavior (see Heidbrink, 1985; Sprinthall, Sprinthall, & Oja, 1994; Gross, 1997). Translated in many languages, it also has been successfully used in scientific research (i.e., testing theoretical assumptions on moral development) and in evaluation of educational programs (see Vernon, 1983; Lind, 2006; Lerkiatbundit, Utaipan, Laohawiriyanon, & Teo, 2006; Lind, 2008).

Contrasting activity during socio-normative judgments with grammatical judgments revealed activation in the left ventromedial prefrontal cortex, the left orbitofrontal cortex, the temporal poles, and the left posterior superior temporal sulcus. By this, we replicated previous findings on a cerebral network involved in moral cognition (Prehn et al., 2008; see Fig. 1 and Table 3).

Our data, moreover, showed that neuronal activity in the functional network of brain regions involved in moral judgment is modulated by individual differences in moral judgment competence. Participants with lower moral judgment competence recruited the left ventromedial prefrontal cortex and the left posterior superior temporal sulcus more than participants with greater competence in this domain when identifying social norm violations. Activity in both regions has been associated with social-cognitive and emotional processes (see Table 3) and was found in almost all studies investigating moral judgment.

In the literature, greater neural activity in participants with lower competence in a certain task has been associated with compensation (e.g., Kosslyn, Thompson, Kim, Rauch, & Alpert, 1996; Rypma et al., 2006). In detail, efficiency theories suggest that individuals differ in the efficiency with which fundamental cognitive operations are performed. According to these theories, efficient individuals are able to perform fundamental cognitive operations faster than inefficient individuals and/or with minimized resource allocation (Vernon, 1983). Reduced neuronal activation in individuals with higher competence, thus, may be due to the minimized recruitment of socio-cognitive and emotional processes.

Additionally, we found that moral judgment competence scores were inversely correlated with activity in the right dorsolateral prefrontal cortex during moral relative to grammatical judgments (Prehn et al., 2008). That is, participants with comparably lower moral judgment competence recruited the dorsolateral prefrontal cortex more during moral judgment than participants with more competence in this domain. As described in detail above, moral judgment competence represents the ability to apply a moral orientation in a consistent and differentiated manner in varying social situations (Lind, 2008). Because the dorsolateral prefrontal cortex seems to be eminent in the implementation of control processes, task monitoring, and inhibitory control during rule-based response selection (e.g., Miller, 2000; Bunge, 2004), increased activity in right dorsolateral prefrontal cortex in participants with comparably low moral judgment competence might be interpreted as higher processing demand due to a more controlled application of rule-based behavioral knowledge during the decision-making process (i.e., deciding whether a behavior is appropriate in regard to the norms and values held to be virtuous in a society).

More evidence for a role of the right dorsolateral prefrontal cortex in moral judgment and the implementation of morally appropriate behavior comes from a study using repetitive transcranial magnetic stimulation (rTMS). Here, a disruption also of the right, but not the left dorsolateral prefrontal cortex reduced the subject's willingness to reject their partner's intentionally unfair monetary offers. Importantly, subjects were still able to judge the unfair offers as unfair, which indicates that the right dorsolateral prefrontal cortex plays a key role especially in the *implementation* of fairness-related behaviors (Knoch, Pascual-Leone, Meyer, Treyer, & Fehr, 2006). That rTMS study, thus, provides complementary evidence to our study in showing that this region is also crucial for the execution of morally appropriate behavior.

An additional analysis of effective connectivity (psychophysiological interaction analysis; Friston et al., 1997) between the right dorsolateral prefrontal cortex and other brain regions revealed an increased coupling of this cortical region during the moral judgment task with the limbic system, namely, the left posterior cingulate cortex and the precuneus. Both regions are also part of the previously identified neural network contributing to moral judgment and are associated with emotion processing, for instance, the processing the emotional significance of words and objects (see Maddock, 1999). Thus, the coupling of the right dorsolateral prefrontal cortex with these structures also possibly reflects a controlled recruitment of emotional information (e.g., the affectively laden moral orientations) in the moral judgment process.

In summary, individual differences in moral judgment competence are reflected in two brain regions identified in other neuroimaging studies and commonly associated with emotional and social-cognitive processes (namely, the ventromedial prefrontal cortex and the posterior superior temporal sulcus) as well as in the dorsolateral prefrontal cortex associated with the controlled application of social rules. That is, activity in these regions that reflect emotional as well as cognitive processes during the moral judgment task is modulated by individual differences in moral judgment competence. Moreover, the coupling of cortical regions (dorsolateral prefrontal cortex) with limbic brain structures (e.g., the posterior cingulate cortex) provides neuroimaging evidence for an interaction between cognitive and emotional processes during moral judgment and might be also modulated by the individual moral judgment competence.

Because moral judgment competence represents a more cognitive aspect of morality it would be interesting to also investigate individual characteristics that refer to emotional information processing like the above mentioned sensitivity to issues of care (see Gilligan, 1977; Robertson et al., 2007) or individual differences in affective style or empathy (Davidson, 2004).

A Functional Approach to Moral Judgment Integrating Psychological Models, Neuroscientific Results, and Evolutionary Biology

As presented in this chapter, there are competing and unconnected psychological theories on moral judgment emphasizing either intuitive feelings or rational reasoning processes in moral judgment as well as isolated neuroscientific findings that to some extent support the one theoretical model and to some extent the other. The current work on the question of how moral judgments are made has largely rejected the Kohlbergian conception of morality as properly seated in the realm of affect-free, rational, and conscious thought. However, the emerging consensus about the role of emotions raises its own concerns. While we agree strongly with the importance of giving proper value to emotion and intuition in many forms of moral judgment, we are concerned that the pendulum may swing too far, and that cognitive processes at the reasoning end of the spectrum will be undervalued. The examples of current research on moral judgment discussed above also show that it is quite a challenge to disentangle the contributions of different cognitive and emotional processes to moral judgment.

As already mentioned in the introductory remarks at the beginning of this chapter, judging whether our own or another person's actions are good or bad (respectively harmful for individuals or the society as a whole) is very central to everyday social life and it has probably been central since the beginning of mankind.

From an evolutionary point of view, moral judgment (i.e., the ability to judge whether our own or another person's actions are good or bad) might be considered as a *psychological function* that evolved together with other cognitive and emotional functions, for instance, language, the fear of snakes and spiders, memory, empathy, or inductive reasoning. All psychological functions might have been shaped by natural selection over vast periods of time to solve the recurrent information processing problems our ancestors were faced with. Information processing problems of survival and reproduction which our ancestors recurrently had to deal with comprised such things as choosing which foods to eat, negotiating social hierarchies, dividing investment among offspring, the detection of cheaters, and selecting mates (Buss, 1995; Tooby & Cosmides, 2005). The ability to judge whether our own or another's actions are good or bad, in particular, might also be regarded as a function that evolved in the ancestral social environment to solve special problems like reducing uncertainty in social interactions by providing a system of norms and values, and thus, a common social knowledge how to behave in a community. In this sense moral judgment prevents anti-social behavior and protects the individual as well as the social community. Living in a social group is, in general, regarded as the primary survival strategy of humans (Buss, 1995), and therefore, would have selected for adaptations such as cooperativeness, loyalty, and the fear of social exclusion. Individuals who have been uncooperative, deviant from group norms, or disloyal presumably would have had more trouble with surviving, finding mates, and raising their offspring than individuals which have shown the opposite behavior.

Using this evolutionary framework, we here suggest a *functional* account of moral judgment, integrating psychological models and neuroscientific results with an evolutionary point of view (see Mundale & Bechtel, 1996). From this point of view the key questions are not debates whether moral judgment is "innate versus learned", "nature versus nurture" or "emotionally driven versus cognitive". The key questions from an evolutionary point of view, in contrast, are: Why do people judge and behave morally? What specific adaptive information processing problems are moral judgment and behavior meant to solve? By means of which internal (emotional and cognitive) information processing mechanisms could these specific adaptive problems in the human ancestral environment have been solved? The field of evolutionary psychology then focuses on identifying these information processing problems, developing models of the functions that may have evolved to solve these problems and models of how these functions are implemented in our brains and minds, and testing these models in research (Buss, 1995).

The mechanisms that are built to solve the adaptive problems vary along many dimensions. Some mechanisms are more domain-general, others are more domain-specific, some are more cognitively penetrable, others are more difficult to be overridden by other mechanisms. The mechanisms are very different because they have to solve very different adaptive problems. The adaptive problems, in addition, can be more or less complex. The problem of preventing anti-social behavior and protecting the community, of course, is more complex than the problem of avoiding snake-bites. From an evolutionary point of view the problem how to avoid snake-bites could have easily been solved by the fear of snakes that leads to a flight reaction when an individual sees a snake. In contrast, the problem of preventing anti-social behavior certainly involves many different cognitive and emotional mechanisms.

On the basis of the evidence reviewed above, we argue that moral judgment has both cognitive and emotional mechanisms that were shaped by natural selection over the course of evolution. We propose the following working model of moral judgment (see Fig. 2):

Judging a behavior as good or bad, in general, comprises the internal representation of the social situation as well as the accepted standard of social and moral behavior (norms and values considered to be virtuous in a society, i.e., affectively laden internal moral orientations or principles), and the evaluation of the behavior with respect to these internal representations.

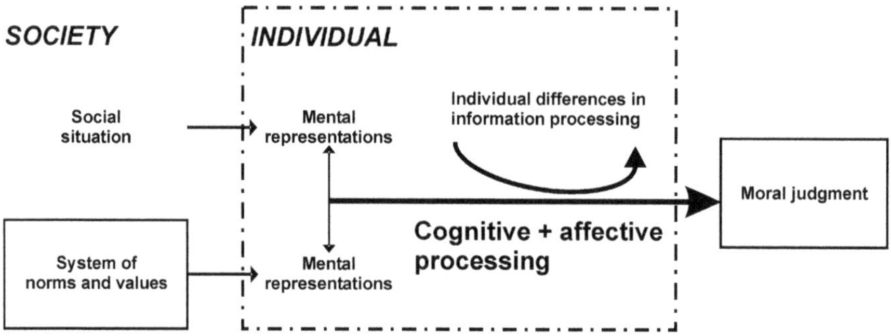

Fig. 2 A working model of moral judgment. During the moral judgment process mental representations of the social situation (i.e., a given behavior in a certain context) and mental representations of moral orientations and principles held to be virtuous in a society are supposed to be compared. The comparison process includes cognitive as well as emotional processes which can be more rational and conscious or more intuitive and automated. Moreover, information processing can be modulated by individual differences, for instance individual differences in moral judgment competence or affective information processing (a certain affective style or individual differences in empathy)

In detail, cognitive mechanisms include retrieving factual knowledge stored in long-term memory, understanding the social situation, inferring intentions of another person (e.g., recognizing that the person did what he or she did on purpose) as well as rational reasoning processes (e.g., the controlled application of these norms and values). The emotional mechanisms include experiencing of emotions such as guilt, sympathy, shame, or anger when social or moral norms are violated as well as automated and intuition-based processes (see Fig. 2).

Depending on the circumstances (the social situation, the emotional content, personal involvement, necessity of a utilitarian decision, etc.), either the cognitive or the emotional processing dominates the decision-making process and the judgment. In situations that contain issues of life and death, murder, bodily harm, or incest (i.e., that immediately affect the survival or the successful reproduction of the individual) mechanisms have probably evolved that may proceed automatically and effortlessly without conscious reflection as a result of biological adaptation over the course of evolution, like the mechanisms proposed in the psychological models of moral judgment developed by Haidt (2001) and Hauser (2006). Therefore, some studies on patients with brain lesions (mainly of the ventromedial prefrontal cortex) and imaging studies using moral judgment tasks that included decisions about issues of life and death provided strong evidence for the role of emotions during moral judgment (e.g., Damasio et al., 1994; Dimitrov et al., 1999; Greene et al., 2001, 2004). Other studies that used complicated and dilemmatic moral judgment tasks showed that emotion should better *not* guide our decisions, and emotional responses should be suppressed or regulated when more abstract and rational utilitarian judgments are required (Koenigs et al., 2007; Koenigs & Tranel, 2007). In situations which are more abstract, other cognitive processes are involved. Coming to the conclusion, for instance, that illegally downloading music files from the internet is harmful for society for various reasons (see Goodenough & Prehn, 2004) requires more abstract cognitive processes as proposed by the Kohlbergian model emphasizing rational reasoning processes in moral judgment (Kohlberg, 1964).

Because several psychological mechanisms are involved in moral judgment, the question is also how well a decision maker is able to integrate these different mechanisms (e.g., intuitive emotional responses with rational reasoning processes) sensitive to the context of the particular social situation he or she faces (see Talmi & Frith, 2007). Thus, in addition, this view points out the role of individual differences, for instance, in moral judgment competence, which is defined as the ability to apply a certain moral orientation in a consistent and differentiated manner in varying social conditions (Lind, 2008). Other individual differences in information processing that are likely to modulate the moral judgment process and that should be taken into account are differences in emotional processing (e.g., the affective style or empathy, see Fig. 2).

The mechanisms of moral judgment might have evolved over the course of evolution and are implemented in our brains and minds. However, the mechanisms are neither static and unchangeable, nor are we not responsible for our moral judgment and behavior. In contrast, there is much evidence that the mechanisms change during ontogenetic development (see the Kohlbergian theory of moral development).

There is also evidence that moral judgment competence can be improved as a result of an active interaction between the individual and his or her social environment (e.g., the Konstanz method of dilemma discussion; Lerkiatbundit et al., 2006).

Furthermore, moral judgments that have been made, the mental representations of the social situation, and the representations of accepted standards of social and moral behavior have in turn an impact on the society (namely, the system of norms and values), subsequent social interactions, and a person's mental representations (see Fig. 2).

Acknowledgments The authors wish to thank the Gruter Institute for Law and Behavioral Research, and in particular Oliver Goodenough for encouragement and financial support for the work reflected in this essay. The work was also financially supported by grants from the Graduate Program Berlin (Scholarship Nachwuchsfoerderung), the BMBF (Berlin NeuroImaging Center, BNIC), and the DFG [Emmy-Noether-Program (HE 3347/2–1)].

References

Adolphs, R. (1999). The human amygdala and emotion. *Neuroscientist, 5*, 125–137.

Adolphs, R., Russell, J. A., & Tranel, D. (1999). A role for the human amygdala in recognizing emotional arousal from unpleasant stimuli. *Psychological Science, 10*, 167–171.

Amodio, D. M., & Frith, C. D. (2006). Meeting of minds: The medial frontal cortex and social cognition. *Nature Reviews Neuroscience, 7*, 268–277.

Anderson, S. W., Bechara, A., Damasio, H., Tranel, D., & Damasio, A. R. (1999). Impairment of social and moral behavior related to early damage in human prefrontal cortex. *Nature Neuroscience, 2*, 1032–1037.

Beauchamp, M. S., Lee, K. E., Argall, B. D., & Martin, A. (2004). Integration of auditory and visual information about objects in superior temporal sulcus. *Neuron, 41*, 809–823.

Beauchamp, M. S., Lee, K. E., Haxby, J. V., & Martin, A. (2002). Parallel visual motion processing streams for manipulable objects and human movements. *Neuron, 34*, 149–159.

Beauchamp, M. S., Lee, K. E., Haxby, J. V., & Martin, A. (2003). FMRI responses to video and point-light displays of moving humans and manipulable objects. *Journal of Cognitive Neuroscience, 15*, 991–1001.

Berthoz, S., Armony, J. L., Blair, R. J., & Dolan, R. J. (2002). An fMRI study of intentional and unintentional (embarrassing) violations of social norms. *Brain, 125*, 1696–1708.

Berthoz, S., Grezes, J., Armony, J. L., Passingham, R. E., & Dolan, R. J. (2006). Affective response to one's own moral violations. *NeuroImage, 31*, 945–950.

Blair, R. J. R. (1995). A cognitive developmental-approach to morality – investigating the psychopath. *Cognition, 57*, 1–29.

Borg, J. S., Hynes, C., Van, H. J., Grafton, S., & Sinnott-Armstrong, W. (2006). Consequences, action, and intention as factors in moral judgments: An FMRI investigation. *Journal of Cognitive Neuroscience, 18*, 803–817.

Bunge, S. A. (2004). How we use rules to select actions: A review of evidence from cognitive neuroscience. *Cognitive, Affective and Behavioral Neuroscience, 4*, 564–579.

Buss, D. M. (1995). Evolutionary psychology – a new paradigm for psychological science. *Psychological Inquiry, 6*, 1–30.

Camille, N., Coricelli, G., Sallet, J., Pradat-Diehl, P., Duhamel, J. R., & Sirigu, A. (2004). The involvement of the orbitofrontal cortex in the experience of regret. *Science, 304*, 1167–1170.

Casebeer, W. D. (2003). Moral cognition and its neural constituents. *Nature Reviews Neuroscience, 4*, 840–846.

Casebeer, W. D., & Churchland, P. S. (2003). The neural mechanisms of moral cognition: A multiple-aspect approach to moral judgment and decision-making. *Biology and Philosophy, 18*, 169–194.
Compton, R. J. (2003). The interface between emotion and attention: A review of evidence from psychology and neuroscience. *Behavioral and Cognitive Neuroscience Reviews, 2*, 115–129.
Cushman, F., Young, L., & Hauser, M. (2006). The role of conscious reasoning and intuition in moral judgment: Testing three principles of harm. *Psychological Science, 17*, 1082–1089.
Damasio, A. R. (1996). The somatic marker hypothesis and the possible functions of the pre-frontal cortex. *Philosophical Transactions of the Royal Society of London: Series B, Biological Sciences, 351*, 1413–1420.
Damasio, H., Grabowski, T., Frank, R., Galaburda, A. M., & Damasio, A. R. (1994). The return of gage, phineas – clues about the brain from the skull of a famous patient. *Science, 264*, 1102–1105.
Davidson, R. J. (2000). Cognitive neuroscience needs affective neuroscience (and vice verse). *Brain and Cognition, 42*, 89–92.
Davidson, R. J. (2003). Seven sins in the study of emotion: Correctives from affective neuroscience. *Brain and Cognition, 52*, 129–132.
Davidson, R. J. (2004). Well-being and affective style: Neural substrates and biobehavioural correlates. *Philosophical Transactions of the Royal Society of London: Series B, Biological Sciences, 359*, 1395–1411.
Dimitrov, M., Phipps, M., Zahn, T. P., & Grafman, J. (1999). A thoroughly modern gage. *Neurocase, 5*, 345–354.
Dolan, R. J. (2002). Emotion, cognition, and behavior. *Science, 298*, 1191–1194.
Dolan, R. S., Lane, R., Chua, P., & Fletcher, P. (2000). Dissociable temporal lobe activations during emotional episodic memory retrieval. *NeuroImage, 11*, 203–209.
Donders, F. C. (1969). On speed of mental processes. *Acta Psychologica (AMST), 30*, 412–431.
Eslinger, P. J., & Biddle, K. R. (2000). Adolescent neuropsychological development after early right prefrontal cortex damage. *Developmental Neuropsychology, 18*, 297–329.
Finger, E. C., Marsh, A. A., Kamel, N., Mitchell, D. G., & Blair, J. R. (2006). Caught in the act: The impact of audience on the neural response to morally and socially inappropriate behavior. *NeuroImage, 33*, 414–421.
Fink, G. R., Markowitsch, H. J., Reinkemeier, M., Bruckbauer, T., Kessler, J., & Heiss, W. D. (1996). Cerebral representation of one's own past: Neural networks involved in autobiographical memory. *Journal of Neuroscience, 16*, 4275–4282.
Friston, K. J., Buechel, C., Fink, G. R., Morris, J., Rolls, E., & Dolan, R. J. (1997). Psychophysiological and modulatory interactions in neuroimaging. *NeuroImage, 6*, 218–229.
Friston, K. J., Price, C. J., Fletcher, P., Moore, C., Frackowiak, R. S. J., & Dolan, R. J. (1996). The trouble with cognitive subtraction. *NeuroImage, 4*, 97–104.
Frith, C. D., & Frith, U. (2006). The neural basis of mentalizing. *Neuron, 50*, 531–534.
Gilligan, C. (1977). In a different voice – womens conceptions of self and of morality. *Harvard Educational Review, 47*, 481–517.
Gilligan, C., & Attanucci, J. (1988). Two moral orientations – gender differences and similarities. *Merrill-Palmer Quarterly of Behavior and Development, 34*, 223–237.
Goodenough, O. R., & Prehn, K. (2004). A neuroscientific approach to normative judgment in law and justice. *Philosophical Transactions of the Royal Society of London: Series B, Biological Sciences, 359*, 1709–1726.
Greene, J., & Haidt, J. (2002). How (and where) does moral judgment work? *Trends in Cognitive Sciences, 6*, 517–523.
Greene, J. D., Nystrom, L. E., Engell, A. D., Darley, J. M., & Cohen, J. D. (2004). The neural bases of cognitive conflict and control in moral judgment. *Neuron, 44*, 389–400.
Greene, J. D., Sommerville, R. B., Nystrom, L. E., Darley, J. M., & Cohen, J. D. (2001). An fMRI investigation of emotional engagement in moral judgment. *Science, 293*, 2105–2108.

Gross, M. L. (1997). *Ethics and activism: The theory and practice of political morality*. Cambridge, MA: Cambridge University Press.

Haidt, J. (2001). The emotional dog and its rational tail: A social intuitionist approach to moral judgment. *Psychological Review, 108*, 814–834.

Haidt, J. (2003). The moral emotions. In R. J. Davidson, K. R. Scherer, & H. H. Goldsmith (Eds.), *Handbook of affective sciences* (pp. 852–870). Oxford: Oxford University Press.

Haidt, J. (2007). The new synthesis in moral psychology. *Science, 316*, 998–1002.

Haidt, J., Koller, S. H., & Dias, M. G. (1993). Affect, culture, and morality, or is it wrong to eat your dog. *Journal of Personality and Social Psychology, 65*, 613–628.

Harenski, C. L., & Hamann, S. (2006). Neural correlates of regulating negative emotions related to moral violations. *NeuroImage, 30*, 313–324.

Harlow, J. M. (1848). Passage of an iron rod through the head. *Boston Medical and Surgical Journal, 39*, 389–393.

Hauser, M., Cushman, F., Young, L., Jin, R. K. X., & Mikhail, J. (2007). A dissociation between moral judgments and justications. *Mind and Language, 22*, 1–21.

Hauser, M. D. (2006). The liver and the moral organ. *Social Cognitive and Affective Neuroscience, 1*, 214–220.

Heekeren, H. R., Wartenburger, I., Schmidt, H., Prehn, K., Schwintowski, H. P., & Villringer, A. (2005). Influence of bodily harm on neural correlates of semantic and moral decision-making. *NeuroImage, 24*, 887–897.

Heekeren, H. R., Wartenburger, I., Schmidt, H., Schwintowski, H. P., & Villringer, A. (2003). An fMRI study of simple ethical decision-making. *Neuroreport, 14*, 1215–1219.

Heidbrink, H. (1985). Moral judgment competence and political learning. In G. Lind, H. A. Hartmann, & R. H. Wakenhut (Eds.), *Moral development and social environment. Studies in the philosophy and psychology of moral judgment and education* (pp. 259–271). Chicago: Precedent Publishing.

Henson, R. (2005). What can functional neuroimaging tell the experimental psychologist? *Quarterly Journal of Experimental Psychology A, 58*, 193–233.

Henson, R. (2006). Forward inference using functional neuroimaging: Dissociations versus associations. *Trends in Cognitive Sciences, 10*, 64–69.

Knoch, D., Pascual-Leone, A., Meyer, K., Treyer, V., & Fehr, E. (2006). Diminishing reciprocal fairness by disrupting the right prefrontal cortex. *Science, 314*, 829–832.

Koenigs, M., & Tranel, D. (2007). Irrational economic decision-making after ventromedial prefrontal damage: Evidence from the ultimatum game. *Journal of Neuroscience, 27*, 951–956.

Koenigs, M., Young, L., Adolphs, R., Tranel, D., Cushman, F., Hauser, M., et al. (2007). Damage to the prefrontal cortex increases utilitarian moral judgements. *Nature, 446*, 908–911.

Kohlberg, L. (1964). Development of moral character and moral ideology. In M. L. Hoffman & L. W. Hoffman (Eds.), *Review of child development research* (pp. 381–431). New York: Russel Sage Foundation.

Kohlberg, L. (1969). Stage and sequence: The cognitive-developmental approach to socialization. In D. A. Goslin (Ed.), *Handbook of socialization theory and research* (pp. 347–480). Chicago: Ran McNally.

Kosslyn, S. M., Thompson, W. L., Kim, I. J., Rauch, S. L., & Alpert, N. M. (1996). Individual differences in cerebral blood flow in area 17 predict the time to evaluate visualized letters. *Journal of Cognitive Neuroscience, 8*, 78–82.

Lerkiatbundit, S., Utaipan, P., Laohawiriyanon, C., & Teo, A. (2006). Impact of the Konstanz method of dilemma discussion on moral judgment in allied health students: A randomized controlled study. *Journal of Allied Health, 35*, 101–108.

Lind, G. (2006). The moral judgment test: Comments on Villegas de Posada's critique. *Psychological Reports, 98*, 580–584.

Lind, G. (2008). The meaning and measurement of moral judgment competence revisited – A dual-aspect model. In D. Fasko & W. Willis (Eds.), *Contemporary philosophical and psychological perspectives on moral development and education* (pp. 185–220). Cresskill, NJ: Hampton Press.

Lind, G. (1998). *An introduction to the Moral Judgment Test (MJT). Unpublished manuscript.* Konstanz: University of Konstanz. Retrieved from www.uni-konstanz.de/ag-moral/mut/mjt-intro.htm
Lind, G., & Wakenhut, R. H. (1985). Testing for moral judgment competence. In G. Lind, H. A. Hartmann, & R. H. Wakenhut (Eds.), *Moral development and social environment* (pp. 79–105). Chicago: Precedent Publishing.
Luo, Q. A., Nakic, M., Wheatley, T., Richell, R., Martin, A., & Blair, R. J. R. (2006). The neural basis of implicit moral attitude – an IAT study using event-related fMRI. *NeuroImage, 30*, 1449–1457.
Maddock, R. J. (1999). The retrosplenial cortex and emotion: New insights from functional neuroimaging of the human brain. *Trends in Neurosciences, 22*, 310–316.
Mantani, T., Okamoto, Y., Shirao, N., Okada, G., & Yamawaki, S. (2005). Reduced activation of posterior cingulate cortex during imagery in subjects with high degrees of alexithymia: A functional magnetic resonance imaging study. *Biological Psychiatry, 57*, 982–990.
Meriau, K., Wartenburger, I., Kazzer, P., Prehn, K., Lammers, C. H., van der Meer, E., et al. (2006). A neural network reflecting individual differences in cognitive processing of emotions during perceptual decision making. *NeuroImage, 33*, 1016–1027.
Mikhail, J. (2007). Universal moral grammar: Theory, evidence and the future. *Trends in Cognitive Sciences, 11*, 143–152.
Miller, E. K. (2000). The prefrontal cortex and cognitive control. *Nature Reviews Neuroscience, 1*, 59–65.
Moll, J., Eslinger, P. J., & Oliveira-Souza, R. (2001). Frontopolar and anterior temporal cortex activation in a moral judgment task: Preliminary functional MRI results in normal subjects. *Arquivos de neuro-psiquiatria, 59*, 657–664.
Moll, J., Oliveira-Souza, R., Bramati, I. E., & Grafman, J. (2002). Functional networks in emotional moral and nonmoral social judgments. *NeuroImage, 16*, 696–703.
Moll, J., Oliveira-Souza, R., & Eslinger, P. J. (2003). Morals and the human brain: A working model. *Neuroreport, 14*, 299–305.
Moll, J., Oliveira-Souza, R., Eslinger, P. J., Bramati, I. E., Mourao-Miranda, J., Andreiuolo, P. A., et al. (2002). The neural correlates of moral sensitivity: A functional magnetic resonance imaging investigation of basic and moral emotions. *Journal of Neuroscience, 22*, 2730–2736.
Moll, J., Zahn, R., Oliveira-Souza, R., Krueger, F., & Grafman, J. (2005). Opinion: The neural basis of human moral cognition. *Nature Reviews Neuroscience, 6*, 799–809.
Mundale, J., & Bechtel, W. (1996). Integrating neuroscience, psychology, and evolutionary biology through a teleological conception of function. *Minds and Machines, 6*, 481–505.
Nucci, L. (2001). *Education in the moral domain*. Cambridge: Cambridge University Press.
Ochsner, K. N., & Feldman Barrett, L. (2001). A multiprocess perspective on the neuroscience of emotion. In T. Mayne & G. Bonnano (Eds.), *Emotion: Current issues and future directions* (pp. 38–81). New York: The Guilford Press.
Phelps, E. A. (2006). Emotion and cognition: Insights from studies of the human amygdala. *Annual Review of Psychology, 57*, 27–53.
Piaget, J. (1965). *The moral judgment of the child*. New York: Free Press.
Pizarro, D. A., & Bloom, P. (2003). The intelligence of the moral intuitions: Comment on Haidt (2001). *Psychological Review, 110*, 193–196.
Poldrack, R. A. (2006). Can cognitive processes be inferred from neuroimaging data? *Trends in Cognitive Sciences, 10*, 59–63.
Prehn, K., Wartenburger, I., Mériau, K., Scheibe, C., Goodenough, O. R., Villringer, A., et al. (2008). Individual differences in moral judgment competence influence neural correlates of socio-normative judgments. *Social Cognitive and Affective Neuroscience, 3,* 33–46.
Robertson, D., Snarey, J., Ousley, O., Harenski, K., Bowman, E. D., Gilkey, R., et al. (2007). The neural processing of moral sensitivity to issues of justice and care. *Neuropsychologia, 45*, 755–766.

Rypma, B., Berger, J. S., Prabhakaran, V., Bly, B. M., Kimberg, D. Y., Biswal, B. B., et al. (2006). Neural correlates of cognitive efficiency. *NeuroImage, 33*, 969–979.

Saver, J. L., & Damasio, A. R. (1991). Preserved access and processing of social knowledge in a patient with acquired sociopathy due to ventromedial frontal damage. *Neuropsychologia, 29*, 1241–1249.

Schultz, J., Friston, K. J., O'Doherty, J., Wolpert, D. M., & Frith, C. D. (2005). Activation in posterior superior temporal sulcus parallels parameter inducing the percept of animacy. *Neuron, 45*, 625–635.

Schultz, J., Imamizu, H., Kawato, M., & Frith, C. D. (2004). Activation of the human superior temporal gyrus during observation of goal attribution by intentional objects. *Journal of Cognitive Neuroscience, 16*, 1695–1705.

Smetana, J. (1993). Understanding of social rules. In M. Bennett (Ed.), *The development of social cognition: The child as psychologist* (pp. 111–141). New York: Guilford Press.

Snarey, J. R. (1985). Cross-cultural universality of social moral development – a critical review of Kohlbergian research. *Psychological Bulletin, 97*, 202–232.

Sprinthall, N., Sprinthall, R. C., & Oja, S. N. (1994). *Educational psychology. A developmental approach* (6th ed.). New York: McGraw-Hill.

Takezawa, M., Gummerum, M., & Keller, M. (2006). A stage for the rational tail of the emotional dog: Roles of moral reasoning in group decision making. *Journal of Economic Psychology, 27*, 117–139.

Talmi, D., & Frith, C. (2007). Neurobiology – feeling right about doing right. *Nature, 446*, 865–866.

Thompson-Schill, S. L., Braver, T. S., & Jonides, J. (2005). Individual differences. *Cognitive, Affective and Behavioral Neuroscience, 5*, 115–116.

Tooby, J., & Cosmides, L. (2005). Conceptual foundations of evolutionary psychology. In D. M. Buss (Ed.), *The handbook of evolutionary psychology* (pp. 5–67). Hoboken, NJ: Wiley.

Turiel, E. (1983). *The development of social knowledge: Morality and convention*. Cambridge: Cambridge University Press.

Vernon, P. A. (1983). Speed of information-processing and general intelligence. *Intelligence, 7*, 53–70.

Walton, M. E., Devlin, J. T., & Rushworth, M. F. (2004). Interactions between decision making and performance monitoring within prefrontal cortex. *Nature Neuroscience, 7*, 1259–1265.

Young, L., Cushman, F., Hauser, M., & Saxe, R. (2007). The neural basis of the interaction between theory of mind and moral judgment. *Proceedings of the National Academy of Sciences of the USA, 104*, 8235–8240.

Moral Dysfunction: Theoretical Model and Potential Neurosurgical Treatments

Dirk De Ridder, Berthold Langguth, Mark Plazier, and Tomas Menovsky

Introduction

> The profound irony is that our noblest achievement, morality, has evolutionary ties to our basest behavior, warfare (Frans De Waal in *Primates and philosophers*).

Darwin developed the idea that morality might be an evolutionary byproduct of warfare, even though in his typical prudent style he did not state it that clearly. In his *Descent of Men* he wrote that "In order that primeval men, or the ape-like progenitors of man, should become social, they must have acquired the same instinctive feelings, which impel other animals to live in a body... They would have felt uneasy when separated from their comrades, for whom they would have felt some degree of love; they would have warned each other of danger, and have given mutual aid in attack or defence. . . .When two tribes. . . .came into competition, if. . . .the one tribe included a great number of courageous, sympathetic and faithful members, who were always ready to warn each other of danger, to aid and defend each other, this tribe would succeed better and conquer the other."

In these couple of lines Darwin describes what is necessary for neural systems coding for morality to evolve in the brain by natural selection: our progenitors must have developed a kind of theory of mind, which requires self-awareness, so that social interactions become possible. Group survival (of the family clan) and thus individual survival benefited from group cohesion in warfare against other clans or tribes. "It must not be forgotten that although a high standard of morality gives but a slight or no advantage to each individual man and his children over the other men of the same tribe, yet an increase in the number of well-endowed men and an advancement in the standard of morality will certainly give an immense advantage to one tribe over another. A tribe including many members who, from possessing the spirit of patriotism, fidelity, obedience, courage and sympathy, were always ready to aid one another, and to sacrifice themselves for the common good, would be victorious over most other tribes; and this would be natural selection" (Darwin).

D. De Ridder (✉)
BRAIN & Department of Neurosurgery, University Hospital Antwerp, Belgium
e-mail: Dirk.De.Ridder@uza.be

J. Verplaetse et al. (eds.), *The Moral Brain*, DOI 10.1007/978-1-4020-6287-2_7,

In a visionary statement years before Richard Dawkins' *Selfish Gene* he wrote that this group cohesion would ultimately also lead to an increased survival of the genes of the individual: "Even if they left no children, the tribe would still include their blood-relations..." Morality as seen from an evolutionary neurological point of view thus deals with neural mechanisms that evolved to guide social behavior in everyday life increasing chances of individual gene spreading.

Initially human morality evolved at a time when people lived in small foraging societies and often had to make instant life-or-death decisions, with no time for conscious evaluation of different options. This suggests that moral judgments proceed from emotions, as Hume believed and is not solely based on reason, as Kant thought. Human morality however, involves both emotional and cognitive aspects and cognition might fine-tune emotionally driven moral behavior. Therefore a final moral judgment requires an integrated cognitive and emotional response to the social stimulus.

A common feature of the antisocial, rule-breaking behavior that is central to criminal, violent and psychopathic individuals is the failure to follow moral guidelines (Raine & Yang, 2006). Understanding the functional anatomy of moral judgment and the anatomical and functional differences between social and antisocial brains is a prerequisite for developing socially and morally acceptable neurosurgical interventions that treat social or moral dysfunction.

A simple rationale, which has been used successfully for treating phantom perceptions such as pain (De Ridder, De Mulder, et al., 2007a, c; De Ridder, De Mulder, et al., 2007) and tinnitus (De Ridder, De Mulder, et al., 2006; De Ridder, De Mulder, et al., 2007; De Ridder, De Mulder, et al., 2007a, b), can be further developed to modulate moral dysfunction. It consists of four conceptual steps: (1) morality and social behavior are products of specific brain circuit activations/deactivations, (2) these brain activations can be visualized using functional imaging techniques, (3) the cortical activations or deactivations can be transiently modulated by transcranial magnetic stimulation, the functional changes in deep brain areas by selective amytal testing, and (4) implantation of electrodes on or in these areas can subsequently modulate these areas permanently. Amytal is a barbiturate which can temporarily ($\pm 10\,\text{min}$) suppress neuronal activity in specific brain areas by supraselective infusion of the drug in the arteries supplying bloodflow to the brain areas of interest.

Similar to the above mentioned conditions pain or tinnitus, antisocial behavior is not a diagnostic entity, but a symptom. Therefore we have to assume different forms of antisocial behavior which differ not only in their genesis, symptomatology or comorbidity, but also in their underlying neurological changes. Nevertheless these changes have to occur in neuronal systems which are involved in moral decision making. This can be illustrated by comparison with other brain functions. Disorders of the motor system (e.g. tremor) might be very heterogeneous with respect to their etiology, but functional and structural changes occur in brain areas related to motor function. This justifies the proposed stepwise approach which first identifies brain areas and brain networks of interest, by comparing results from studies which investigate decision making in normal controls with lesion studies and imaging studies in patient groups with antisocial behavior (step 1). The next steps identify whether a specific brain area in an individual patient is involved in his antisocial behavior (step 2) and whether modulation of these brain areas result in behavioral changes (steps 3 and 4).

Recent functional neuroimaging research has revealed the regions most commonly activated in moral judgment tasks. As there are similarities between the neural systems underlying moral decision-making in normal individuals, and brain mechanisms thought to be impaired in delinquent, criminal, violent and psychopathic populations, understanding these mechanisms can help in gaining neural insight into the etiology of antisocial behavior (Raine & Yang, 2006), a prerequisite for neuromodulation in an attempt to treat moral dysfunction. When considering antisocial behavior a pathological form of morality, it comes as no surprise that the functional anatomy of antisocial brains include dorsal and ventral regions of the prefrontal cortex (PFC), amygdala, hippocampus, angular gyrus, anterior cingulate and temporal cortex (Raine & Yang, 2006).

The Moral Brain

When external or internally generated sensory stimuli reach the brain they elicit a motor response and autonomic response. The processing of this information most likely occurs in two separate pathways, an emotional and a cognitive pathway, which integrate at different levels of the information processing, culminating in the dorsolateral prefrontal cortex (Gray, Braver, et al., 2002), which then sends feedback to almost every single anatomical structure in the brain (Selemon & Goldman-Rakic, 1988; Petrides & Pandya, 1999). This is modulated by the mesolimbic reward system, which demonstrates a great overlap with the emotional pathways.

The emotional pathways update the incoming stimulus in the amygdala, from where the information is relayed to the anterior cingulate cortex, ventromedial prefrontal cortex and orbitofrontal cortex. The cognitive updating of the sensory stimulus occurs in the hippocampus, from where the information is relayed to the posterior cingulate cortex and subsequently via a parietal and temporal pathway to the dorsomedial prefrontal cortex. The integration of emotion and cortex is subsequently processed in the dorsolateral prefrontal cortex. It has to be mentioned that most if not all of these pathways are reciprocal and actually consist of a double configuration. The detailed functional anatomy of the pathways involved is however beyond the scope of this chapter.

The neural substrates of "self" and "other" related processing, theory of mind network, mirror neuron and default circuits seem to depend on a common network activation, consisting of a medial ACC and BA32 (emotion), PCC and precuneus co-activion (cognition), and a lateral VLPFC-parietal-STS circuit (imitation). Thus it seems that at rest, the brain keeps a network activated that is ready to process stimuli both emotionally and cognitively in reference to oneself, which are internally "replayed" (internally imitated or simulated) in order to comprehend these stimuli.

The Moral Brain Circuit

Since in evolution (pre)human morality developed before cognition, humans possess emotion-based automatic, implicit and non-conscious moral attitudes. In contrast to laborious deductive reasoning, they enable rapid, automatic, and unconscious cognitive appraisals of interpersonal events (Haidt, 2001). This implicit form of morality is associated with increased activation of the right amygdala and the ventromedial orbitofrontal cortex (Luo, Nakic, et al., 2006). As implicit morality evolved from emotions, both basic and moral emotions share a common subcortical emotional circuit consisting of the amygdala, thalamus, and upper midbrain.

Also for explicit or conscious morality there is no specific region in the brain (Greene & Haidt, 2002), but a network encodes our explicit or conscious morality. Every brain region described in functional neuroanatomy studies of morality has also been implicated in non-moral processes (Greene & Haidt, 2002). Regions most commonly activated in moral judgment tasks involve the orbitofrontal (Moll et al., 2002a, 2002b,; Moll, Oliveira-Souza, et al., 2005; Borg, Hynes, et al., 2006; Luo, Nakic, et al., 2006) and medial prefrontal cortex (BA9/10) (Moll et al., 2001; Moll et al., 2002a, 2002b; Heekeren, et al., 2003; Greene, Nystrom, et al., 2004; Moll, Oliveira-Souza, et al., 2005; Moll, Zahn, et al., 2005; Harenski & Hamann, 2006; Raine & Yang, 2006), amygdala (Greene, Nystrom, et al., 2004; Berthoz, Grezes, et al., 2006; Harenski & Hamann, 2006; Luo, Nakic, et al., 2006; Raine & Yang, 2006), angular gyrus (Heekeren et al., 2003, 2005; Greene, Nystrom, et al., 2004; Borg, Hynes, et al., 2006; Harenski & Hamann, 2006; Raine & Yang, 2006) and superior temporal sulcus (BA21) (Moll et al., 2002a, 2002b; Greene, 2003; Moll, Zahn, et al., 2005) and posterior cingulate (BA31) (Greene, 2003; Raine & Yang, 2006; Robertson, Snarey, et al., 2007), as well as the anterior cingulate and dorsolateral prefrontal cortex (BA9/46)(Greene & Haidt, 2002; Greene, 2003; Greene, Nystrom, et al., 2004).

The affective or emotional component of moral transgression is generated in the amygdala bilaterally and in the right dorsolateral prefrontal cortex (Berthoz, Grezes, et al., 2006). In line with what can be regarded theoretically as the emotional component of the moral circuitry. Based on the general functional anatomy of the brain the emotional component of the moral circuitry involves the amygdala, anterior cingulate and ventromedial and orbitofrontal cortex, whereas the rational component can be localized in the posterior cingulate, the medial prefrontal cortex (BA9-10). In the dorsolateral prefrontal cortex emotional and cognitive moral processing could integrate (see Fig. 1). The ACC involvement in morality could represent a corrective for a neurobiological bias towards immediate reward by the amygdala, the VMPFC could represent the valence (Grimm, Schmidt, et al., 2006) of the moral stimulus, and the orbitofrontal cortex the integration of reward and emotion (Kringelbach, O'Doherty, et al., 2003).

Very interestingly, three brain regions involved in the moral brain circuit also coincide with regions identified in a meta-analysis of the resting brain's activity (Gusnard, Raichle, et al., 2001; Greene & Haidt, 2002): the ventromedial prefrontal cortex (BA9/10), the STS (BA21) extending into inferior parietal area (BA7/39/40)

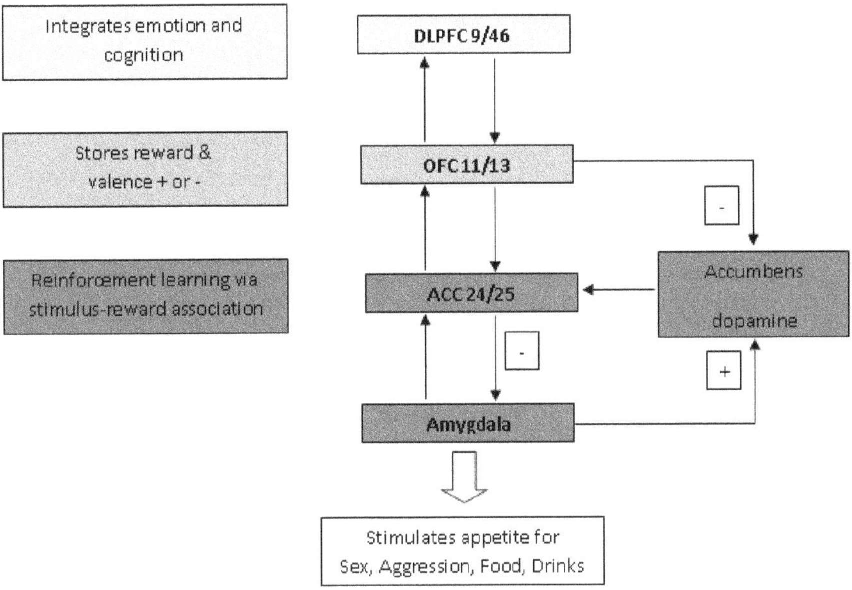

Fig. 1 Simplified model of emotional-reward pathways

and the posterior cingulate-precuneus (BA31/7). These areas could therefore very well represent the self-referenced components or introspection (Greene & Haidt, 2002) in a social setting of the moral brain.

Reward System and Morality

A mesolimbic dopaminergic system has been described that consists of the ventral tegmental area (VTA), nucleus accumbens in the ventral striatum, ventral pallidum, mediodorsal nucleus of the thalamus and orbitofrontal cortex (BA11/13) (Koob, 2006). Apart from the mesolimbic dopaminergic system the dorsolateral prefrontal cortex (DLPFC), the anterior cingulate cortex (ACC) and the posterior cingulate cortex, the frontal eye fields, the parietal cortex and the thalamus have been implicated in reward processing (Walter, Abler, et al., 2005).

From an evolutionary point of view, the capacity to seek rewards as goals is essential for the survival and reproduction of mobile organisms. Generally, rewards can be defined as those stimuli which positively reinforce the frequency or intensity of a behavior pattern. Food, water and sexual stimuli are called primary rewards as they reinforce behavior without being learned (McClure, York, et al., 2004; Walter, Abler, et al., 2005): reward mechanisms pertaining to such stimuli are mostly innate because they are essential for survival and reproduction. Other stimuli, such as cultural goods or money, are called secondary rewards, as they reinforce behavior only after having been learned (McClure, York, et al., 2004; Walter, Abler, et al.,

2005). The reward system is also involved in social behavior, for example in charitable donation (Moll et al., 2006), social cooperation (Rilling, Glenn, et al., 2007), altruism (Rilling, Sanfey, et al., 2004) etc.

In a complicated environment, there is no way to anticipate all possible environmental scenarios, and so hard-coded solutions are doomed to fail at some point (McClure, York, et al., 2004). One solution to this problem is reinforcement learning. Reinforcement learning generates appropriate actions based solely on the intermittent receipt of rewards and punishments, a function controlled by the dopaminergic mesolimbic reward system. Once a brain can predict the amount of reward it will receive, it can calculate the difference between its prediction and the reward it receives on the moment it performs an action (Quartz & Sejnowski, 2002). The ventral striatum responses signal errors in the prediction of reward (Montague, Dayan, et al., 1996). In other words, it tells whether you are doing better or worse than predicted, so that you can remember to select this action pattern for future similar situations. The occurrence of a salient/important reinforcing stimulus may be signaled by the amygdala, which then may trigger a learning signal in the ventral striatum so that the reward may become better predicted in the future via formation of stimulus-reward association. Metaphorically speaking, it behaves as the child's play "hotter/colder". The value of the reward may then be assessed in the OFC, to be used by the prefrontal cortex to decide a course of action consistent with current goals (McClure, York, et al., 2004).

The reward system has three functions (Schultz, 2004): (1) rewards induce learning, as they make an organism come back for more (positive reinforcement). (2) they induce approach behavior for acquiring the reward object and (3) they induce hedonic (positive, pleasure) feelings. By contrast, punishments induce avoidance learning, withdrawal behavior and negative (sad) emotions. The medial OFC is activated by positive reinforcement, approach behavior and positive feelings, whereas the lateral OFC seems to be activated by negative reinforcement, avoidance behavior and negative feelings (Elliott, Dolan, et al., 2000).

Thus the reward system engages you with the world and allows you to learn from it, which explains why the reward system is also involved in social behavior, such as in charitable donation (Moll et al., 2006), social cooperation (Rilling, Glenn, et al., 2007), altruism (Rilling, Sanfey, et al., 2004).

When the ventral tegmental area, from which the dopaminergic projection to the PFC originates, is lesioned after birth, the medial prefrontal cortex does not develop properly, leading to 6% decrease in cortical thickness in VMPFC in rats (Kalsbeek, Buijs, et al., 1987). A full distribution of dopamine fibers in the frontal cortex is not established until adulthood (Verney, Berger, et al., 1982; Kalsbeek, Voorn, et al., 1988). Dopamine, the main neurotransmitter of the reward system, does not directly mediate the hedonic component (liking), which is opiate generated, but rather the motivational component (wanting) in rewarded behavior (Berridge, 2006).

In summary, from a neurobiological perspective, moral decision making could be considered the result of positive or negative reinforcement, and morality could be reduced to an emotional-cognitively integrated approach versus withdrawal reaction to social stimuli as perceived by the self. Morally "good" or "bad" could reflect the hedonic response of the opiate part of the mesolimbic reward system to internally imitated social stimuli, consciously perceived via activation of the medial and lateral OFC respectively. The initial emotional response generated in the amygdala-ACC-BA32 network would be integrated with the cognitive PCC-precuneus influence and rationalized in the DLPFC. The social stimuli processed by the anterior temporal poles are probably internally imitated by the medial and lateral self network and if processed as pleasure, will be consciously perceived as "morally=socially good" and if processed as not pleasurable, it will be interpreted consciously as "morally/socially bad".

Dysfunctional Moral Brain

According to Heraclites everything is in constant movement and this means there is a constant building and tearing down of everything. This leads to the conclusion that war is the basis of everything. Hobbes developed this into his philosophy that the natural state of humans is "a war of all against all", suggesting that our natural state is being inborn killers. This could actually be correct as it has been demonstrated that humans are more physically aggressive between the ages of 24 and 30 months than at any other time in their life. The typical 2 year old engages in 8–9 aggressive acts an hour (Tremblay, 2000). The normal developmental pattern of physical aggression is one of occasional and declining use over time, and most children learn to suppress physical aggression during childhood, but 16% do not (Cote, Vaillancourt, et al., 2006). Less than 6% of children develop chronic antisocial behavior (Lacourse, Cote, et al., 2002).

Antisocial Personality Disorders (APD)/Psychopathy

Psychopathy can be either developmental or acquired (=secondary). There is a fundamental difference between these two forms of psychopathy. Both forms are characterized by an insensitivity to future consequences of decisions, both have defective autonomic responses in the face of punishment and both do not respond to treatment (Quartz & Sejnowski, 2002). But whereas acquired or secondary psychopaths have built up moral knowledge which they cannot apply anymore, developmental psychopaths never acquired moral knowledge at all.

Antisocial populations are neurologically characterized by functionally or structurally impaired brain areas such as the dorsolateral and orbitofrontal regions of the prefrontal cortex, amygdala, hippocampus, angular gyrus, anterior and posterior cingulate, and anterior temporal cortex (Raine & Yang, 2006). The rule-breaking behavior common to antisocial, violent and psychopathic individuals can be partly explained by impairments of these areas, subserving moral cognition and emotion (Raine & Yang, 2006).

Lesion studies have demonstrated that dysfunction in the OFC is associated with decreased inhibitory control, poor emotional decision-making, impulsivity, and disturbed reward/punishment processing, in the sense that patients with OFC dysfunction are unconcerned with the consequences of their behavior (Rolls, Hornak, et al., 1994; Brower & Price, 2001). In antisocial individuals, the additional involvement of DLPFC dysfunction may also cause poor planning/organization (Manes, Sahakian, et al., 2002; Raine & Yang, 2006), resulting in an occupationally and socially dysfunctional lifestyle (Raine & Yang, 2006). But DLPFC may also lead to poor integration of emotion and cognition, and the right DLPFC has been suggested to harbor the capacity for resisting to temptation (Knoch & Fehr, 2007). In addition to the OFC and DLPFC, structure and/or function impairments have also been demonstrated in the anterior cingulate cortex (Kiehl, Smith, et al., 2001; Birbaumer, Veit, et al., 2005), involved in autonomic and emotion regulation (Raine & Yang, 2006), and representing the first integration step of emotion and cognition. Furthermore structural or functional dysfunctions have been demonstrated in the amygdala (emotional drive for food, sex, aggression), hippocampus (cognitive memory updating), temporal pole (social) and middle-superior temporal gyrus (which are separated by the superior temporal sulcus: agency (=the capacity to make choices) and theory of mind), superior and inferior (angular) parietal area.

Antisocial personality disorder is characterized by dysfunctional responses to rewarding and aversive stimuli, but probably not by a hypersensitivity to reward and hyposensitivity to loss (Völlm, Richardson, et al., 2007) as previously thought (Quay, 1965; Newman & Kosson, 1986). It most likely results from disturbed prefrontal responses and reduced activity in the subcortical reward system during positive reinforcement (Völlm, Richardson, et al., 2007). This might be related to the structural abnormalities seen in the prefrontal cortex in APD, characterized by a decrease in gray matter, from 11% (Raine, Lencz, et al., 2000) to 22% (Yang, Raine, et al., 2005), pointing to a significant loss of prefrontal (inhibiting) neurons. Voxel based morphometry specified the gray matter volume losses to the medial and lateral orbitofrontal and frontopolar cortex (Oliveira-Souza, 2008), and histopathological investigations have confirmed this cortical thinning to Brodmann's areas 10, 11, 12, and 32, i.e. the orbitofrontal and prelimbic areas (Narayan, Narr, et al., 2007). It is of interest that heavy smoking (>10 cigarettes/day) during pregnancy is related to an increase in antisocial behavior of the offspring (Huijbregts, Seguin, et al., 2008). This might be due to the fact that adolescents born from women who smoke during pregnancy develop thinner orbitofrontal, middle frontal, and parahippocampal cortices as compared with non-exposed individuals (Toro, Leonard, et al., 2008).

In other words the stimulus-reward association might be deficient, either due to nucleus accumbens dysfunction with secondary orbitofrontal developmental hypotrophy or frontal hypotrophy leading to abnormal estimation of reward value. Acquired psychopathy due to orbitofrontal lesions supports the latter mechanism. This fits with neuropsychological tests, which suggest the predominant functional deficit in psychopathy is located in the orbitofrontal cortex and not the DLPFC or ACC (Blair, Newman, et al., 2006). Secondary psychopathy due to orbitofrontal lesions supports this mechanism as well (Blair & Cipolotti, 2000; Mitchell, Avny, et al., 2006).

But also deficits of cognitive pathways are related with APD: marked volume loss of the medial OFC deficits might be secondarily underdeveloped due to amygdala deficits (Blair, 2007). Indeed amygdala volume decreases are being associated with APD (Raine & Yang, 2006). APD also demonstrate cognitive pathway deficits: marked volume loss of the posterior hippocampus has been described in psychopathy (Laakso, Vaurio, et al., 2001), as well as in the STS and anterior temporal pole (Oliveira-Souza, 2008). Also here it is possible that anterior temporal and STS cortical thinning could be due to a decrease of hippocampal efferent stimuli.

The neuronal loss in the emotional pathway (amygdala + prelimbic + OFC) and cognitive pathway (hippocampus) will have functional repercussions, and accordingly in APD less affect-related activity in the amygdala/hippocampal formation, parahippocampal gyrus, ventral striatum, and in the anterior and posterior cingulate gyri has been observed (Kiehl, Smith, et al., 2001).

Thus, the structural and functional impairments found in individuals with antisocial personality disorder will most likely disrupt normal moral perception and decision-making, which in turn predisposes the individual to rule-breaking, antisocial behavior. This is supported by the fact that there is substantial overlap between areas involved in moral judgment and those related to antisocial/psychopathic behavior (Raine & Yang, 2006). Brain regions common to both include ventral and polar/medial PFC, the amygdala and angular gyrus/posterior superior temporal gyrus/sulcus.

There are however also differences: a key difference is the presence of hippocampal and anterior cingulate dysfunction in antisocial/psychopathic individuals, whereas in moral studies these structures are not consistently activated. On the other hand, the posterior cingulate is activated in moral judgment tasks, and is not implicated in antisocial behavior (Raine & Yang, 2006).

Almost all criminal and psychopathic individuals know right from wrong. It is suggested that it is predominantly the feeling of what is moral that is deficient in antisocial groups, rather than the knowing of what is moral (Raine & Yang, 2006). In other words, the most impaired aspect in APD is probably the emotional component, which is related to dysfunctional amygdala, VMPFC and OFC.

In summary, due to structural and functional impairments present in antisocial populations, which are primarily located in the emotional pathways, the integration (ACC, DLPFC) of social (anterior temporal, STS) emotional (amygdala, ACC-VMPFC, OFC) and cognitive aspects (hippocampus) in the emotional self (ACC-PFC-parietal-STS) seems to be disturbed in people demonstrating antisocial behavior.

This also explains why VMPFC lesions predispose to utilitarian moral decisions (Greene, 2007; Greene, Morelli, et al., 2007; Koenigs, Young, et al., 2007; Moll & Oliveira-Souza, 2007). With a dysfunctional emotional planning area BA32, the DLPFC which integrates emotion and cognition, only receives the cognitive component and uses this to come to a moral decision, i.e. a purely cognitive utilitarian decision.

A recent rTMS (Knoch, Pascual-Leone, et al., 2006) and tDCS (Knoch, Nitsche, et al., 2007) study of the right DLPFC have demonstrated that modulation of this area can influence moral decision making, in contrast to left sided DLPFC stimulation.

It is important to state that not all criminal acts result from a dysfunctional moral brain. Predatory murderers demonstrate marked brain differences compared with affective murderers, the latter of whom are emotionally charged. These people show prefrontal hypometabolism (Raine, Meloy, et al., 1998) which remains unaltered in time (=trait) (Anckarsater, Piechnik, et al., 2007). Predatory murderers on the other hand, who are controlled and murder in a purposeful manner demonstrate no prefrontal hypometabolism. Subcortically, however, both types of murderers are hyperactive on the right side (amygdala, hippocampus, thalamus, midbrain). These findings indicate that excessive subcortical activity predisposes to aggressive behavior, but that while predatory murderers have sufficiently good prefrontal functioning to regulate these aggressive impulses, the affective murderers lack such prefrontal control over emotion regulation (Raine, Meloy, et al., 1998).

Pedophilia

Pedophilia refers to sexual interest in prepubescent children (Fagan, Wise, et al., 2002; Blanchard & Barbaree, 2005), hebephilia, an erotic interest in pubescent children (Freund & Blanchard, 1989; Blanchard & Barbaree, 2005) and teleiophilia, in adults (Blanchard & Barbaree, 2005; Cantor, Kabani, et al., 2008). Pedophilia is a psychiatric disorder of high public concern, because 1–2 in 10 children have been sexually approached and abused by an adult (Fagan, Wise, et al., 2002; Freyd, Putnam, et al., 2005).

Neuropsychological research suggests that pedophilic and nonpedophilic men may differ in brain function and in brain structure. It has been shown, that pedophilic men have lower IQs (Cantor, Blanchard, et al., 2004), poorer visuospatial (Cantor, Blanchard, et al., 2004) and verbal memory scores (Cantor, Blanchard, et al., 2004), higher rates of non-right-handedness (Cantor, Blanchard, et al., 2004), elevated rates of having suffered childhood head injuries resulting in unconsciousness (Blanchard, Christensen, et al., 2002), and elevated rates of having failed school grades or having required placement in special education programs (Cantor, Kuban, et al., 2006). It is estimated that 47% of pedophiles demonstrate antisocial personalities (Virkkunen, 1976), and a linear relationship has been shown between victim age and psychopathology, with child offenders displaying the greatest affective and thought disturbance (Kalichman, 1991). Even though child molesters have the ability to understand moral issues they ignore these interpersonal social values (Valliant, Gauthier, et al., 2000). All these factors might contribute to high levels of recidivism, treatment drop-outs, and noncompliance which causes pedophilia to be extremely difficult to treat: in order to be effective treatment needs to be intensive, long-term, and comprehensive, possibly with lifetime follow-up (Cohen & Galynker, 2002). New treatment options are needed and therefore a pathophysiologically based neuromodulatory treatment might be welcomed.

Case reports have described changes in the right orbitofrontal region (Dressing, Obergriesser, et al., 2001; Burns & Swerdlow, 2003) and the hippocampus (Mendez,

Chow, et al., 2000). In line with these data the two main functional neuroanatomic theories of pedophilia point to (1) frontal-executive dysfunction or (2) temporo-limbic dysfunction, or a combination of both. Frontal executive dysfunction explains sexual offending by behavioral (frontal) disinhibition (Graber, Hartmann, et al., 1982), whereas temporal-limbic theories implicate either the regulation of sexual behavior (disturbed sexual urges) by deep temporal lobe structures (Hucker, Langevin, et al., 1986) or the role of such structures in behavioral disinhibition (Graber, Hartmann, et al., 1982). Dual or combined frontotemporal dysfunction theories have been offered, in which pedophilic men suffer from dysfunction both in temporal regions (disturbing sexual urges) and in frontal regions (causing behavioral disinhibition) (Cohen & Galynker, 2002; Cohen, Nikiforov, et al., 2002).

However, the idea that sexual deviance is associated with frontal and/or temporal lobe damage is only based on few investigations (Joyal, Black, et al., 2007). Recent morphological and functional imaging research may contribute to a more detailed theoretical model of pedophilia.

Pedophilia's structural brain abnormalities seem to involve both gray and white matter. Reduced gray matter is noted in pedophiles in the right amygdala, hypothalamus (bilaterally), septal regions, substantia innominata, and bed nucleus of the striae terminalis (Schiltz, Witzel, et al., 2007), as well as ventral striatum (also extending into the nucleus accumbens), the orbitofrontal cortex and the cerebellum (Schiffer, Peschel, et al., 2007). White matter abnormalities are limited to two major fiber bundles: the superior fronto-occipital fasciculus and the right arcuate fasciculus (Cantor, Kabani, et al., 2008). Because the superior fronto-occipital and arcuate fasciculi connect cortical regions that respond to sexual cues, these results suggest that pedophilia is related to a partial disconnection within a network for recognizing sexually relevant stimuli (Cantor, Kabani, et al., 2008).

Functional imaging of visual erotica in non-pedophiles has demonstrated a distributed sexual arousal network with increased neural activity in several areas, including the anterior cingulate, medial prefrontal, orbitofrontal, insular, and occipitotemporal cortices, as well as in the amygdala and the ventral striatum (Karama, Lecours, et al., 2002; Kim, Sohn, et al., 2006). Ventral striatum activation on the one hand indicates the involvement of the human reward system during the processing of visual erotica (Stark, Schienle, et al., 2005). Penile erection on the other hand correlates with activation of the right medial prefrontal cortex, the right and left orbitofrontal cortices, the insulae, the paracentral lobules, the right ventral lateral thalamic nucleus, the right anterior cingulate cortex and regions involved in motor imagery and motor preparation (supplementary motor areas, left ventral premotor area) (Moulier, Mouras, et al., 2006).

Males have significantly more activation of the amygdala (Hamann, Herman, et al., 2004), thalamus (Karama, Lecours, et al., 2002) and hypothalamus (Karama, Lecours, et al., 2002; Hamann, Herman, et al., 2004) than females. With increasing age, the hypothalamus and thalamus become less activated in males, which may be responsible for the lesser physiological arousal in response to the erotic visual stimuli associated with age (Kim, Sohn, et al., 2006). In hypoactive sexual desire disorder patients there is an abnormally maintained activity of the medial orbitofrontal cortex in contrast to controls (Stoleru, Redoute, et al., 2003). Attempted inhibition of the sexual arousal generated by visual sexual stimuli is associated with activation of the right superior frontal gyrus and right anterior cingulate gyrus (Beauregard, Levesque, et al., 2001).

In hetero- and homosexual males and females the ventral striatum, centromedian thalamus and ventral premotor cortex show a stronger neuronal response to preferred relative to non-preferred stimuli (Ponseti, Bosinski, et al., 2006), which could be similar in pedophiles.

Functional imaging in pedophiles has demonstrated hypoactivation of the hypothalamus, the periaqueductal gray, and dorsolateral prefrontal cortex during presentation of unspecific visual erotic stimuli (Walter, Witzel, et al., 2007). Presenting non-specific visual sexual stimuli furthermore activates occipitotemporal regions, the prefrontal cortex (BA 9, 11, 46 and 47) and subcortical (limbic) areas such as the striatum, globus pallidus, substantia nigra and medial temporal cortex (Schiffer, Krueger, et al., 2008), similarly to the non-specific sexual arousal network. The activation pattern in pedophiles comprises large parts of the reward system as well (Schiffer, Krueger, et al., 2008). When specifically pedophilic content was presented, activation increases in the insula, dorsolateral prefrontal cortex (BA 9, 47), parahippocampal gyrus, precuneus, some occipitotemporal regions and amygdala, suggesting these are extra processes of sexual arousal in pedophiles to a pedophile-specific stimulus (Schiffer, Krueger, et al., 2008). The parahippocampal gyrus has been associated with the recovery of perceptual relational (Prince, Daselaar, et al., 2005) information (Cabeza, Rao, et al., 2001), and the dorsolateral PFC with monitoring of the retrieved information (Cabeza, Rao, et al., 2001) which would be specific for pedophiles when looking at pedophilic content. The amygdala and insula could explain increased arousal for this specific content, and the precuneus might be related to the self-centered mental imagery (Cavanna & Trimble, 2006) associated with the pedophilic stimulus.

Based on the abovementioned available data about structural and functional abnormalities, a conceptually very simple model would suggest a reward system dysfunction in pedophiles with a wrong stimulus-reward association: the view of a child becomes associated with activation of a sexual arousal circuit. This hypothesis is supported by frontostriatal gray matter decrease in pedophiles which could reflect this dysfunctional reward reinforcement. The right sided lateralization has been suggested to predispose to aggressiveness as described for murderers (Raine, Meloy, et al., 1998).

Moreover, part of the symptoms of pedophilia resemble those of obsessive–compulsive (OC) disorder. Therefore pedophilia can be included in the OC spectrum, similar as other paraphilias, sexual compulsions, Tourette syndrome, body dysmorphic disorder, hypochondriasis, trichotillomania, eating disorders, autism, pathological gambling, kleptomania, or depersonalization disorder (Castle & Phillips, 2006). Accordingly, pedophilia might share a common etiopathological mechanism with all these obsessive-compulsive spectrum disorders (Schiffer, Peschel, et al., 2007).

Developing Neurosurgical Treatments for Moral Dysfunctions

Developing neurosurgical techniques for moral dysfunctions can be based on the principles mentioned in the introduction:

1. Morality and social behavior are products of specific brain circuit activations/deactivations.
2. These brain activations can be visualized using functional imaging techniques.
3. The cortical activations or deactivations can be transiently modulated by transcranial magnetic stimulation, the functional changes in deep brain areas by selective amytal testing.
4. Implantation of electrodes on or in these areas can subsequently modulate these areas permanently.

As explained in this chapter neurobiological, functional neuroimaging, and neuropsychological data all converge to demonstrate that psychopathy and pedophilia are nothing more than clinical expressions of specific brain circuit malfunctions.

Functional neuroimaging can be performed in multiple ways to visualize these abnormally functioning brain circuits, but specificity is most likely important. Thus a pedophile who is sexually attracted by little boys should undergo functional neuroimaging studies with these specific stimuli (e.g. pictures of naked boys) and brain activation patterns for this specific stimulus should be compared to activations to other erotic stimuli (e.g. naked women, naked men, naked girls), so that the difference in brain activation/deactivation for the specific stimulus can be traced and ideally corrected or suppressed.

Different neuromodulation techniques exist that could be used as treatment for moral dysfunction: transcranial magnetic stimulation (TMS), transcranial direct current stimulation (tDCS) and implanted electrodes. TMS or tDCS are non-invasive methods inducing temporary changes in brain functioning and thus symptoms. These non-invasive neuromodulation techniques can be used as pre-surgical tests evaluating whether a more invasive implantation of an electrode for permanent symptom suppression could be beneficial.

Neurostimulation is actually a misnomer as it could suggest that magnetic or electrical stimulation stimulates parts of the brain. Whether or not stimulation activates or inactivates the underlying brain depends on the stimulation parameters used. For 1 Hz or low frequency (LFS) TMS, stimulation results in a decreased metabolism of the underlying cortex and more widespread areas of the brain (Kimbrell, Little, et al., 1999; Speer, Kimbrell, et al., 2000) as well as in suppression of excitability of the underlying cortex (Chen, Classen, et al., 1997; Chen, 2000), whereas for higher frequency (HFS) TMS (10–20 Hz) the metabolism of the underlying cortex is increased (Kimbrell, Little, et al., 1999; Speer, Kimbrell, et al., 2000) and excitability increases as well (Chen, 2000). Therefore it is often stated that 1 Hz TMS results in inactivation of the stimulated area, whereas HFS TMS activates the underlying cortex. From a clinical point of view both LFS and HFS may have a similar result, namely creating a virtual lesion (Walsh & Rushworth, 1999) as so-called activation could represent a mere disruption of ongoing activity.

For electrical stimulation via implanted electrodes high frequency stimulation (120–130 Hz) suppresses electrical activity and induces a conduction block, arresting signal transmission (Beurrier, Bioulac, et al., 2001), whereas it is unknown what low frequency (4–40 Hz) electrical stimulation does. Note that low frequency electrical stimulation uses similar frequencies as high frequency TMS.

Pedophilia

A summary of social brain functioning suggests that the occurrence of a salient/important reinforcing stimulus may be signaled by the amygdala, which then may trigger a learning signal in the nucleus accumbens so that the reward may become better predicted in the future via formation of stimulus-reward association. The value of the reward may then be assessed (emotionally) in the OFC, to be used by the DLPFC to decide a course of action consistent with current goals after integration of simultaneous cognitive input via parietal and temporal cortices.

Five different treatments can be proposed for treating moral dysfunction in the framework of pedophilia:

1. Moral reconditioning by electrical stimulation of the nucleus accumbens with associated visual stimulus presentation.
2. Anterior cingulate stimulation for moral reconditioning and/or sexual arousal suppression.
3. Moral reconditioning by electrical stimulation of the amygdala with associated visual stimulus presentation.
4. Orbitofrontal stimulation for moral reconditioning.
5. Dorsolateral prefrontal stimulation for suppression of moral dysfunctioning.

In clinical practice this can be performed in a similar way to the previously described methodology for suppression of tinnitus and phantom pain via electrical stimulation (De Ridder et al., 2004; De Ridder, De Mulder, et al,. 2006; De Ridder et al., 2007a, 2007b, 2007c). The stimulus which leads to dysfunctional neuronal activation can be presented in a fMRI or PET scanner. This allows individual localization of the dysfunctional brain activation. The activated or deactivated areas of interest can subsequently be modulated non-invasively by transcranial magnetic stimulation. If this results either in temporary behavioral modification or in suppression of dysfunctional neuronal activity, a permanent improvement can potentially be obtained by implantation of electrodes under stereotactic guidance using the fMRI or PET as maps or by using normalized co-planar stereotactic human brain atlases such as Talairach (Talairach & Tournoux, 1988) or Schaltenbrand (Schaltenbrand & Wahren, 1977). We are aware about potential difficulties with the use of TMS as investigational tool for temporary modulation of the involved brain areas. First, it has been shown that rTMS applied over the DLPFC modulates activity in the DLPFC, the OFC, the anterior cingulate and in the nucleus accumbens, which all represent potential targets for permanent stimulation. Therefore potential behavioral changes during rTMS of the DLPFC can not exactly be attributed to functional changes in a specific structure. Another problem might be the assessment of behavioral effects of rTMS, which might be only subliminal and/or short lasting. A potential solution to the problem could be the additional use of functional imaging for the assessment of rTMS effects.

The treatments suggested are proposed for non-psychopathic pedophiles. Psychopathic pedophiles can be treated as described in the psychopathy/antisocial personality disorder treatment section.

Electrical Stimulation of the Nucleus Accumbens

In the nucleus accumbens a stimulus is paired with a reward leading to a certain response. When a child (as stimulus) becomes coupled to a reward by activation of a sexual arousal network, the continuous updating of this feedback system can be modulated in order to suppress the wrong stimulus (child)-reward (sexual arousal) association. The nucleus accumbens can be suppressed by high frequency stimulation (> 130 Hz) on the moment that stimuli with pedophilic content are presented. This should decouple the reward from the stimulus, i.e. the cost/benefit instruction is reset to non-rewarding.

In clinical practice a high resolution (3 T) fMRI can be performed presenting neutral, sexually arousing and specific pedophilic stimuli (Schiffer, Krueger, et al., 2008). The activation of the nucleus accumbens can be localized and used as a target for implanting electrodes. A non-invasive test can be performed preoperatively to evaluate potential treatment effect by transcranial magnetic stimulation of the DLPFC, which has direct connections to the nucleus accumbens (Petrides & Pandya, 1999). 20 Hz rTMS of the DLPFC in rats has been shown to increase dopamine release in the nucleus accumbens (Keck, Welt, et al., 2002; Erhardt, Sillaber, et al., 2004), 10 Hz in humans, but not in the nucleus accumbens, only in the caudate nucleus (Strafella, Paus, et al., 2001). The difference between rats and humans might be due to a variety of reasons, including species specific differences, but also frequency specificity of nucleus accumbens response could be involved.

Once implanted, the accumbens-suppressing high frequency stimulation can be triggered by the presentation of stimuli with pedophilic content, and monitored by phallometry (Freund & Blanchard, 1989; Cooper & Cernovovsky, 1992). This should result in a dissociation of the stimulus-response, in other words in moral reconditioning.

Electrical Stimulation of the Anterior Cingulate Cortex

The anterior cingulate cortex encodes whether or not an action is worth performing in view of the expected benefit and the cost of performing the action. This cost benefit analysis is dysfunctional in people with moral dysfunction: the benefit from sexually assaulting a child as perceived by the anterior cingulate is worth performing where it should be the opposite. Therefore suppressing anterior cingulate activity could normalize its function. Based on the functional imaging data that demonstrate that the right ACC is involved in sexual arousal suppression (Beauregard, Levesque, et al., 2001), the ACC modulation could benefit from this second additive working mechanism.

In clinical practice a similar approach can be taken to accumbens stimulation: high resolution fMRI can be performed presenting neutral, sexually arousing and specific pedophilic stimuli (Schiffer, Krueger, et al., 2008).

Preoperatively rTMS can be used as a non-invasive predictive test. rTMS of the DLPFC has been shown to influence the ipsilateral ACC. Slow (1 Hz) TMS increases ipsilateral metabolism on PET scan in the ACC (Ohnishi, Matsuda, et al., 2004; Knoch, Treyer, et al., 2006), as well as ipsilateral medial prefrontal cortex, contralateral ventrolateral PFC, and the contralateral ventral striatum (Ohnishi, Matsuda, et al., 2004; Knoch, Treyer, et al., 2006). Opposite effects on brain activation, including the anterior cingulate, have been shown for high frequency and low frequency rTMS (in depressed patients) (Speer, Kimbrell, et al., 2000), and 20 Hz rTMS is shown to decrease activity in the ACC (Nadeau, McCoy, et al., 2002) in responders to rTMS. Thus depending on the rTMS frequency chosen, the activity of the ACC can be increased or decreased: higher frequency (20 Hz) stimulation may increase brain glucose metabolism in a transsynaptic fashion, whereas lower frequency (1 Hz) stimulation may decrease it (Post, Kimbrell, et al., 1999). Using a double-cone coil, the ACC can also be reached by rTMS over the medial frontal cortex, as demonstrated by metabolic changes on PET scan (Hayward, Mehta, et al., 2007).

Phallometric evaluation of pedophilic stimulus presentation can subsequently be used as a criterium for definitive implantation with electrodes. Activation of the ACC on fMRI during pedophilic stimulus presentation can localize the target for implanting electrodes in a neuronavigated way, using the fMRI as guidance.

Potentially continuous stimulation without reconditioning (=presenting stimuli of pedophilic content simultaneously) can be performed, to continuously suppress sexual arousal.

Electrical Stimulation of the Amygdala

It has been demonstrated that the lateral nuclei of the amygdala predominantly subserve an input function and the central (LeDoux, 1996) nuclei an output function (116). There are however also direct pathways to the central nuclei from the orbitofrontal cortex, inhibiting the hypothalamus and autonomic nervous system activity (Ghashghaei, Hilgetag, et al., 2007). The VMPFC on the other hand excites the hypothalamus and autonomic nervous system via the basolateral nucleus (Ghashghaei, Hilgetag, et al., 2007). This suggests that based on the orbitofrontal deficiency in psychopaths the normal prefrontal input in the amygdala might be disturbed but not the output. The central nucleus of the amygdala has direct afferent connections with the brainstem, hypothalamus, basal ganglia, as well as the insula (Volz, Rehbein, et al., 1990), such that motor, autonomic and somatosensory conditioning can remain intact in psychopaths, with selective disruption of emotional conditioning.

Electrical stimulation of the basolateral amygdala increases the dopamine level in the nucleus accumbens in contrast to stimulation of the central nucleus (Howland, Taepavarapruk, et al., 2002). Thus the central nucleus exerts its tonic influence on the nucleus accumbens probably via an indirect pathway (Phillips, Ahn, et al., 2003). Incentive stimuli induce a supplementary temporary increase in dopamine in the nucleus accumbens via the direct basolateral pathway, so the VMPFC can learn. This is what is probably lacking in psychopaths: the normal tonus on the general learning process (conditioning) still functions, but emotional conditioning is diminished due to a decreased basolateral input. Therefore amygdala stimulation can be used to modulate emotional conditioning.

Clinically it has been shown that high intensity electrical stimulation at 40 Hz of the left amygdala can induce the raw feeling of extreme depression (De Ridder, unpublished data). This negative aspect of amygdala stimulation could be used as a reconditioning mechanism by activation of 40 Hz left amygdala electrical stimulation, triggered by simultaneous presentation of pedophilic stimuli. This will induce an extreme negative reward to become associated to stimuli of pedophilic content, inducing moral reconditioning. This form of treatment requires that the nucleus accumbens still functions normally to allow for the stimulus-reward to become associated.

In clinical practice an amytal test could be performed preoperatively, by selective or supraselective catheterization of the anterior choroidal artery, similarly to what is done for epilepsy (Weissenborn, Ruckert, et al., 1996; Wieser, Muller, et al., 1997) and tinnitus (De Ridder, Fransen, et al., 2006). The success of the amygdala suppression could be evaluated by phallometry (Cooper & Cernovovsky, 1992; Seto, Lalumiere, et al., 2000) during amytal infusion with simultaneous presentation of stimuli of pedophilic content. Due to its potential complications, the value of this test is limited to centers of expertise in supraselective anterior choroidal catheterization (Wieser, Muller, et al., 1997).

Potentially rTMS of the DLPFC could be used as a non-invasive preoperative test to predict treatment success, as rTMS influences ipsilateral amygdala activity: 1 Hz rTMS decreases ipsilateral amygdala activity, 20 Hz increases ipsilateral amygdala activity (Speer, Kimbrell, et al., 2000) (in depressed patients).

Using fMRI neuronavigation, an electrode can be implanted in the amygdala. The electrode can subsequently be activated at 40 Hz while presenting stimuli of pedophilic content, thereby inducing an extreme negative reward to become associated with these specific stimuli.

Electrical Stimulation of the OFC

The value of the reward which is associated to a stimulus is probably assessed by the OFC: the medial OFC is activated by positive reinforcement, approach behavior and positive feelings, whereas the lateral OFC seems to be activated by negative reinforcement, avoidance behavior and negative feelings (Elliott, Dolan, et al., 2000). The potential treatment for pedophilia lies in a modification of the activation patterns related to stimuli of pedophilic content. Activation by neurostimulation of the lateral OFC, with or without simultaneous suppression of the medial OFC on presentation of stimuli of pedophilic content should induce a negative reinforcement, avoidance behavior and negative feelings for this specific stimulus.

Potentially rTMS of the DLPFC could be used as a non-invasive preoperative test to predict treatment success, as rTMS influences orbitofrontal activity: 1 Hz rTMS decreases ipsilateral OFC activity (Knoch, Treyer, et al., 2006), 10 Hz increases OFC activity bilaterally (Knoch, Treyer, et al., 2006). The 10 Hz stimulation activates the lateral OFC (BA47) bilaterally (Knoch, Treyer, et al., 2006), and medial OFC (BA11,25) contralaterally (Knoch, Treyer, et al., 2006), thus exerting a mixed effect on the OFC.

In clinical practice, two electrodes can be implanted on the orbitofrontal cortex, guided by pedophilic activation patterns on fMRI. The medial electrode can be activated at high frequencies (>130 Hz) to have a suppressive effect and the lateral electrode at low frequencies (10–40 Hz) to have an activating effect. The electrodes can be triggered by presentation of stimuli of pedophilic content, thereby inducing a negative reward value associated with these specific stimuli, on the condition that the nucleus accumbens still functions normally.

Electrical Stimulation of the DLPFC

Based on the abovementioned theoretical model of brain functioning, the emotional component of the moral circuitry consists of the amygdala, anterior cingulate and ventromedial and orbitofrontal component, the rational component of the posterior cingulate and the integrated emotional-cognitive moral response on the dorsolateral prefrontal cortex. The internalization in the self might be mediated via the ACC-PCC and STS-angular-VLPFC circuit. The dorsolateral prefrontal cortex has extensive and direct feedback connections to all major brain areas involved in moral processing: amygdala (Aggleton, Burton, et al., 1980), ACC (Pandya & Kuypers, 1969; Pandya, Dye, et al., 1971; Goldman & Nauta, 1977; Selemon and Goldman-Rakic, 1988; Petrides & Pandya, 1999), VMPFC (Petrides & Pandya, 1999), OFC (Pandya, Dye, et al., 1971; Goldman & Nauta, 1977; Petrides & Pandya, 1999), PCC (Goldman & Nauta, 1977; Selemon & Goldman-Rakic, 1988; Petrides & Pandya, 1999) and STG/STS (Pandya & Kuypers, 1969; Pandya, Dye, et al., 1971; Selemon & Goldman-Rakic, 1988; Padberg, Seltzer, et al., 2003). Thus DLPFC stimulation can modulate almost every area involved in moral processing directly. rTMS (Knoch, Pascual-Leone et al., 2006) and tDCS (Knoch, Nitsche, et al., 2007) of the right DLPFC can influence moral decision making, possibly by modulating the right DLPFC's capacity to resist temptation (Knoch & Fehr, 2007). As the left DLPFC is more activated in pedophiles when visual pedophilic stimuli are presented, this could be suppressed via left DLPFC stimulation. Low frequency (1 Hz) DLPFC rTMS could therefore be performed as a preoperative test, evaluated by phallometry

(Cooper & Cernovovsky, 1992). As high and low frequency rTMS of the DLPFC have been suggested to have opposite effects (Kimbrell, Little, et al., 1999; Speer, Kimbrell, et al., 2000), high frequency (10–20 Hz) rightsided DLPFC rTMS could theoretically increase self-control via increased temptation resistance.

If rTMS exerts a temporary beneficial effect, the activation pattern in the DLPFC on fMRI can be elected as a target for neuronavigated electrode implantation on the DLPFC. High frequency (>130 Hz) electrical stimulation could be used for left hyperactivity seen in pedophilia, or low frequency (10–40 Hz) for right DLPFC stimulation.

Psychopathy/Antisocial Personality Disorder (APD)

Due to structural (amygdala-prelimbic-OFC and posterior hippocampus) and functional impairments present in antisocial populations, which are primarily located in the emotional pathways, the integration (ACC, DLPFC) of social input (anterior temporal, STS), emotion (amygdala, ACC-VMPFC, OFC) and cognition (hippocampus) in the emotional self (ACC-PFC-parietal-STS) seems to be disturbed in people demonstrating antisocial behavior. Thus surgical treatments should aim at restoring or compensating for the individual structural or functional deficits in these people. Due to the fact that neuromodulation techniques are more capable of suppressing overactive brain areas than augmenting inactive or hypoactive areas, this limits the treatment options for psychopaths or APD. For a neurostimulator to function it requires enough target cells to be present.

Assuming that the orbitofrontal deficiencies result in a poor reward value perception, orbitofrontal stimulation as described above coupled to simultaneous presentation of moral stimuli can only be attempted if sufficient functional orbitofrontal cortex is still present. Depending on the location of the lesion, compensatory stimulation can be performed. For example, in medial orbitofrontal lesions, where no positive reinforcement is possible for morally good stimuli, negative reinforcement for amoral stimuli can be maximized by 10–40 Hz electrical stimulation of the lateral OFC with simultaneous amoral stimulus presentation. Preoperative trials with TMS are difficult due to the lack of specificity for activating/inactivating the lateral OFC with TMS.

Amygdala stimulation has the same limitations: if structural and functional abnormalities of the amygdala are too pronounced, electrical stimulation might not be beneficial. However, high frequency (>130 Hz) electrical stimulation of the right amygdala could still be beneficial symptomatically, by functionally suppressing the right-sided subcortical hyperactivity noted in psychopathic murderers (Raine, Meloy, et al., 1998). Left sided 40 Hz stimulation might not generate the feeling of raw but extreme depression anymore with impaired OFC function, as the orbitofrontal cortex might be required for generating the (conscious) feeling of extreme depression. On the other hand, the subconscious emotional response (LeDoux, 1996) (which does not require the orbitofrontal cortex) generated by the stimulation might be strong enough to result in a reconditioning in the nucleus accumbens on a behavior level, without the associated (conscious) feeling.

Nucleus accumbens stimulation coupled to simultaneous amoral stimulus presentation might still cause moral reconditioning even with impaired OFC. The decoupling of amoral stimuli from reward in itself could be sufficient. The moral reconditioning might be subconscious, but might result in better adapted social behavior. Even though the APD patient might not consciously know why he/she doesn't do amoral things anymore, if the amoral stimuli are not rewarded anymore, the amoral behavior could cease to continue.

DLPFC might be a more promising option for the treatment of APD. Whereas normally response inhibition (Völlm, Richardson, et al., 2004) or resistance to temptations (Knoch & Fehr, 2007) is mediated via the right dorsolateral (Völlm, Richardson, et al., 2004; Knoch & Fehr, 2007), and the left orbitofrontal cortex (Völlm, Richardson, et al., 2004), the right DLPFC and left OFC are less specifically activated in patients with psychopathy, and a more bilateral DLPFC and OFC activation is present (Völlm, Richardson, et al., 2004). This is similar in pedophiles where pedophilic stimuli activate the left DLPFC (Schiffer, Krueger, et al., 2008) in stead of the right. Therefore low frequency (1 Hz) leftsided DLPFC rTMS could be performed as a preoperative test. As high and low frequency rTMS of the DLPFC have opposite effects (Cooper & Cernovovsky, 1992), high frequency (10–20 Hz) rightsided DLPFC rTMS could theoretically increase self-control via increased temptation resistance or response inhibition.

If rTMS exerts a temporary beneficial effect, the activation pattern in the DLPFC on fMRI can be elected as a target for neuronavigated electrode implantation on the DLPFC. High frequency (>130 Hz) electrical stimulation could be used for suppressing left hyperactivity or low frequency (10–40 Hz) electrical stimulation for right DLPFC stimulation.

What Surgical Treatment to Chose in Individual Cases ?

The choice of the exact surgical target will depend on the individual structural and functional brain images. In patients with important OFC or amygdala atrophy local stimulation of these structures is most likely of limited value due to the lack of substrate to modulate. Furthermore it has to be considered that a lack of grey substance comes along with a reduction of connecting nerve fibers.

Only brain areas that are visualized by activation or deactivation in fMRI or PET scan, using specific stimuli, such as naked boys for homosexual pedophiles, and not activated or deactivated in non-specific stimuli, e.g. naked women for homosexual pedophiles, are worthwhile modulating, since they represent the abnormal network involved.

Stimulation of activation areas for non-specific stimuli can be interesting if no stimulus-specific treatment is possible and if the purpose is to have a more general effect, for example impulsivity control or aggression reduction by right amygdala suppression or left DLPFC suppression.

DLPFC rTMS can be used as a preoperative test since the DLPFC influences almost all parts of the moral brain circuit directly. The combination of rTMS with functional imaging might indicate, which changes of neuronal activation correlate most with behavioral changes and thus identify the best targets for stimulation within the network.

Ethical Implications of Neurosurgical Interventions for Moral Dysfunction

A complete analysis of the ethical aspects of neuromodulation as a treatment for moral disorders is impossible within this chapter. It merits an entire book by itself. However some important aspects definitely need to be addressed.

Whether these surgeries, influencing human morality, are ethical or not clearly depends on the aim of the intervention. Suppressing morality on demand before a battle in a reversible way, in a battalion of elite soldiers, by implanted electrodes is most likely generally accepted to be unethical. However, implanting an electrode for suppressing sexual arousal or drive related to children in a pedophile who asks for help is not as clearcut, and depending on the situation it might considered ethical or unethical to do.

The ethical implications of the surgical modulation of moral dysfunction has to be addressed both by medical doctors, surgical neuroscientists, philosophers specialized in moral sciences and politicians.

Based on the abovementioned pathophysiology of moral brain circuit dysfunction in pedophilia and APD, moral dysfunction is to be considered a pathological form of brain functioning, just like Parkinson's disease, epilepsy, depression or obsessive-compulsive disorder.

Moral dysfunction has very few if any long-lasting non-invasive treatments that are efficacious, which ethically justifies the experimental study of more invasive treatments. These neuromodulation techniques should therefore aim at proving to be superior to the available treatments.

Every brain intervention has its inherent risks: inserting an electrode in or on the brain can cause bleeding, stimulation can cause epileptic seizures, foreign material in the body can become infected, surgical leads can malfunction or break, requiring reinterventions etc., all risks which are identical to other neurostimulation treatments, such as those performed for Parkinson's disease, pain, tinnitus etc.

The main advantage of these neuromodulation techniques is however that the treatment is reversible, until proven otherwise, in the sense that if the implant does not deliver the wanted outcome, or has too much side-effects, it can be turned off, or corrected by reintervention. It is however clear that in the initial experimental stage the patients should be followed up very closely in order to find out whether only the desired effect is obtained with the applied neuromodulation technique, especially so once a permanent implant is used. The advantage of TMS as a preoperative test makes the step to implanting seem smaller, as a benefit of modulating the abnormal brain circuit can be tested non-invasively before the surgery, but it should be kept in mind that this still has to be proven to be a valuable predictor of surgical success. If so, TMS can be of great help, but if not, many people could be denied a potential treatment option.

One important aspect of neuromodulation is that it could restore self-control to the patient and thereby increase her or his quality of life tremendously: "It can mean the difference between having no control and having a considerable control over one's body and life" (Glannon, 2007). This would be the same for non-psychopathic pedophiles who resent having these uncontrollable feelings, whereas for a psychopathic pedophile and psychopathic serial killer this might not apply.

Whether pedophiles, murderers and other criminals with APD should get the option to get an operation in return for less time in prison is a more difficult question, which is not a surgical problem but a political one. Due to the nature of psychopathic characteristics, it would be highly unwise to do so. Psychopathic personalities will try to answer and behave in such a way as to get the result the investigator wants, if that can benefit the psychopath. This will inherently create scientific outcome data that are impossible to trust and use. Even if so called waterproof double blind placebo controlled studies at group level demonstrate the technique is effective, once applied at an individual level it will be impossible to trust it. If the system fails for whatever reason, chances might be too high the psychopath will not report it, or he might turn off the system once freedom has been obtained. Therefore it could be considered unethical to grant prison release on the condition of being implanted.

In a recent advice on ethical aspects of modifying behavior by new technology in health care, it was concluded that as long as modifying behavior serves a widely acceptable purpose, and the results are sufficiently good, novel techniques can be a gain for health care. There are many positive aspects but some negative aspects to implementing new techniques for behavioral modification. Positive individual benefits are a potential improvement in self-control, self-support, safety and quality of life. It can benefit society by increasing safety. Risks and negative aspects are that individual freedom can also come under pressure, technology does not always deliver what it promises, and there might be side-effects and unwanted or unforeseen risks. Social implications include altered relationships, and different kinds of behavior could become medicalized. Furthermore financial gain might be involved and influence privacy, interest and wishes of the patient (Schermer, 2007).

Therefore a system of checks and balances will need to follow up on all future developments within this field of neuroethics and applied neuromorality, before it can become implemented. This kind of research needs to be performed in a very open setting, where indications, techniques and results are open to both ethical and neuroscience/neurosurgical peer review as well as scrutiny by society. Completely abolishing brain-based morality research and its clinical implications would be unwise and unethical as it might be developed then in places where control and directing the research is lacking.

Our Christianity-based society has until very recently continued to consider morality in a Cartesian dualistic way: mind and body are separate entities. Modern western neuroscience has evolved into materialitic monism, the brain equals the mind, which considers morality and therefore also moral dysfunctions as brain dysfunctions, similar to cardiac insufficiency as a heart dysfunction.

In vain of this evolution, in the future, once treatments have been proven successful, every medical doctor should be allowed to decide to treat or not to treat moral dysfunction, if requested by the patient, as moral dysfunction is no different a brain disease than Parkinson's disease, pain, epilepsy etc. Once decided by a medical doctor, *patients willing to be treated* could be referred to specialized multidisciplinary neuromodulation units making the final decisions and performing the operations. These multidisciplinary units should consist of forensic psychiatrists, criminologists, psychologists, neurosurgeons and professional neuro-ethicists.

References

Aggleton, J. P., Burton, M. J., et al. (1980). Cortical and subcortical afferents to the amygdala of the rhesus monkey (Macaca mulatta). *Brain Research, 190* (2), 347–368.

Anckarsater, H., Piechnik, S., et al. (2007). Persistent regional frontotemporal hypoactivity in violent offenders at follow-up. *Psychiatry Research, 156* (1), 87–90.

Beauregard, M., Levesque, J., et al. (2001). Neural correlates of conscious self-regulation of emotion. *Journal of Neuroscience, 21* (18), RC165.

Berridge, K. C. (2006). The debate over dopamine's role in reward: The case for incentive salience. *Psychopharmacology* (Berl). 2007; *191* (3), 391–431.

Berthoz, S., Grezes, J., et al. (2006). Affective response to one's own moral violations. *NeuroImage, 31* (2), 945–950.

Beurrier, C., Bioulac, B., et al. (2001). High-frequency stimulation produces a transient blockade of voltage-gated currents in subthalamic neurons. *Journal of Neurophysiology, 85* (4), 1351–1356.

Birbaumer, N., Veit, R., et al. (2005). Deficient fear conditioning in psychopathy: A functional magnetic resonance imaging study. *Archives of General Psychiatry, 62* (7), 799–805.

Blair, K. S., Newman, C., et al. (2006). Differentiating among prefrontal substrates in psychopathy: Neuropsychological test findings. *Neuropsychology, 20* (2), 153–165.

Blair, R. J. (2007). Dysfunctions of medial and lateral orbitofrontal cortex in psychopathy. *Annals of the New York Academy of Sciences, 1121,* 461–479.

Blair, R. J., & Cipolotti, L. (2000). Impaired social response reversal. A case of 'acquired sociopathy'. *Brain, 123* (Pt 6), 1122–1141.

Blanchard, R., & Barbaree, H. E. (2005). The strength of sexual arousal as a function of the age of the sex offender: Comparisons among pedophiles, hebephiles, and teleiophiles. *Sex Abuse, 17* (4), 441–456.

Blanchard, R., Christensen, B. K., et al. (2002). Retrospective self-reports of childhood accidents causing unconsciousness in phallometrically diagnosed pedophiles. *Archives of Sexual Behavior, 31* (6), 511–526.

Borg, J. S., Hynes, C., et al. (2006). Consequences, action, and intention as factors in moral judgments: An FMRI investigation. *Journal of Cognitive Neuroscience, 18* (5), 803–817.

Brower, M. C., & Price, B. H. (2001). Neuropsychiatry of frontal lobe dysfunction in violent and criminal behaviour: A critical review. *Journal of Neurology, Neurosurgery and Psychiatry, 71* (6), 720–726.

Burns, J. M., & Swerdlow, R. H. (2003). Right orbitofrontal tumor with pedophilia symptom and constructional apraxia sign. *Archives of Neurology, 60* (3), 437–440.

Cabeza, R., Rao, S. M., et al. (2001). Can medial temporal lobe regions distinguish true from false? An event-related functional MRI study of veridical and illusory recognition memory. *Proceedings of the National Academy of Sciences of the United States of America, 98* (8), 4805–4810.

Cantor, J. M., Blanchard, R., et al. (2004). Intelligence, memory, and handedness in pedophilia. *Neuropsychology, 18* (1), 3–14.

Cantor, J. M., Kabani, N., et al. (2008). Cerebral white matter deficiencies in pedophilic men. *Journal of Psychiatric Research, 42* (3), 167–183.

Cantor, J. M., Kuban, M. E., et al. (2006). "Grade failure and special education placement in sexual offenders' educational histories." *Archives of Sexual Behavior , 35* (6), 743–51.

Castle, D. J., & Phillips, K. A. (2006). Obsessive-compulsive spectrum of disorders: A defensible construct? *The Australian and New Zealand Journal of Psychiatry, 40* (2), 114–120.

Cavanna, A. E., & Trimble, M. R. (2006). The precuneus: A review of its functional anatomy and behavioural correlates. *Brain, 129* (Pt 3), 564–583.

Chen, R. (2000). Studies of human motor physiology with transcranial magnetic stimulation. *Muscle and Nerve*, (Suppl, 9), *23* (S9): S26–S32.

Chen, R., Classen, J., et al. (1997). Depression of motor cortex excitability by low-frequency transcranial magnetic stimulation. *Neurology, 48* (5), 1398–1403.

Cohen, L. J., & Galynker, I. I. (2002). Clinical features of pedophilia and implications for treatment. *Journal of Psychiatric Practice, 8* (5), 276–289.

Cohen, L. J., Nikiforov, K., et al. (2002). Heterosexual male perpetrators of childhood sexual abuse: A preliminary neuropsychiatric model. *The Psychiatric Quarterly, 73* (4), 313–336.

Cooper, A. J., & Cernovovsky, Z. (1992). The effects of cyproterone acetate on sleeping and waking penile erections in pedophiles: possible implications for treatment. *Canadian Journal of Psychiatry, 37* (1), 33–39.

Cote, S. M., Vaillancourt, T., et al. (2006). The development of physical aggression from toddlerhood to pre-adolescence: A nation wide longitudinal study of Canadian children. *Journal of Abnormal Child Psychology, 34* (1), 71–85.

De Ridder, D., De Mulder, G., et al. (2004). Magnetic and electrical stimulation of the auditory cortex for intractable tinnitus. Case report. *Journal of Neurosurgery, 100* (3), 560–564.

De Ridder, D., De Mulder, G., et al. (2006). Primary and secondary auditory cortex stimulation for intractable tinnitus. *ORL; Journal for Oto-Rhino-Laryngology and Its Related Specialties, 68* (1), 48–54; discussion 54–5.

De Ridder, D., De Mulder, G., et al. (2007a). Electrical stimulation of auditory and somatosensory cortices for treatment of tinnitus and pain. *Progress in Brain Research, 166*, 377–388.

De Ridder, D., De Mulder, G., et al. (2007b). Auditory cortex stimulation for tinnitus. *Acta Neurochirurgica. Supplement, 97* (Pt 2), 451–462.

De Ridder, D., De Mulder, G., et al. (2007c). Somatosensory cortex stimulation for deafferentation pain. *Acta Neurochirurgica. Supplement, 97* (Pt 2), 67–74.

De Ridder, D., Fransen, H., et al. (2006). Amygdalohippocampal involvement in tinnitus and auditory memory. *Acta oto-laryngologica. Supplementum*, (556), 50–53.

Dressing, H., Obergriesser, T., et al. (2001). Homosexual pedophilia and functional networks – an fMRI case report and literature review. *Fortschritte der Neurologie-Psychiatrie, 69* (11), 539–544.

Elliott, R., Dolan, R. J., et al. (2000). Dissociable functions in the medial and lateral orbitofrontal cortex: Evidence from human neuroimaging studies. *Cerebral Cortex, 10* (3), 308–317.

Erhardt, A., Sillaber, I., et al. (2004). Repetitive transcranial magnetic stimulation increases the release of dopamine in the nucleus accumbens shell of morphine-sensitized rats during abstinence. *Neuropsychopharmacology, 29* (11), 2074–2080.

Fagan, P. J., Wise, T. N., et al. (2002). Pedophilia. *Journal of the American Medical Association, 288* M(19), 2458–2465.

Freund, K., & Blanchard. R. (1989). Phallometric diagnosis of pedophilia. *Journal of Consulting and Clinical Psychology, 57* (1), 100–105.

Freyd, J. J., Putnam, F. W., et al. (2005). Psychology. The science of child sexual abuse. *Science, 308* (5721), 501.

Ghashghaei, H. T., Hilgetag, C. C., et al. (2007). Sequence of information processing for emotions based on the anatomic dialogue between prefrontal cortex and amygdala. *NeuroImage, 34* (3), 905–923.

Glannon, W. (2007). *Bioethics and the brain.* Oxford: Oxford University Press.

Goldman, P. S., & Nauta, W. J. (1977). Columnar distribution of cortico-cortical fibers in the frontal association, limbic, and motor cortex of the developing rhesus monkey. *Brain Research, 122* (3), 393–413.

Graber, B., Hartmann, K., et al. (1982). Brain damage among mentally disordered sex offenders. *Journal of Forensic Sciences, 27,* 125–134.

Gray, J. R., Braver, T. S., et al. (2002). Integration of emotion and cognition in the lateral prefrontal cortex. *Proceedings of the National Academy of Sciences of the United States of America, 99* (6), 4115–4120.

Greene, J. (2003). From neural 'is' to moral 'ought': what are the moral implications of neuroscientific moral psychology?" *Nature Reviews Neuroscience, 4* (10), 846–849.

Greene, J., & Haidt, J. (2002). How (and where) does moral judgment work? *Trends in Cognitive Sciences, 6* (12), 517–523.

Greene, J. D. (2007). Why are VMPFC patients more utilitarian? A dual-process theory of moral judgment explains. *Trends in Cognitive Sciences, 11* (8), 322–323; author reply 323–4.

Greene, J. D., Morelli, S. A., et al. (2007). Cognitive load selectively interferes with utilitarian moral judgment. *Cognition*. 2008; *107* (3), 1144–1154.

Greene, J. D., Nystrom, L. E., et al. (2004). The neural bases of cognitive conflict and control in moral judgment. *Neuron, 44* (2), 389–400.

Grimm, S., Schmidt, C. F., et al. (2006). Segregated neural representation of distinct emotion dimensions in the prefrontal cortex-an fMRI study. *NeuroImage, 30* (1), 325–340.

Gusnard, D. A., Raichle, M. E., et al. (2001). Searching for a baseline: Functional imaging and the resting human brain. *Nature Reviews Neuroscience, 2* (10), 685–694.

Haidt, J. (2001). The emotional dog and its rational tail: A social intuitionist approach to moral judgment. *Psychological Review, 108* (4), 814–834.

Hamann, S., Herman, R. A., et al. (2004). Men and women differ in amygdala response to visual sexual stimuli. *Nature Neuroscience, 7* (4), 411–416.

Harenski, C. L., & Hamann, S. (2006). Neural correlates of regulating negative emotions related to moral violations. *NeuroImage, 30* (1), 313–324.

Hayward, G., Mehta, M. A., et al. (2007). Exploring the physiological effects of double-cone coil TMS over the medial frontal cortex on the anterior cingulate cortex: An H2(15)O PET study. *European Journal of Neuroscience, 25* (7), 2224–2233.

Heekeren, H. R., Wartenburger, I., et al. (2003). An fMRI study of simple ethical decision-making. *Neuroreport, 14* (9), 1215–1219.

Heekeren, H. R., Wartenburger, I., et al. (2005). Influence of bodily harm on neural correlates of semantic and moral decision-making. *NeuroImage, 24*(3), 887–897.

Howland, J. G., Taepavarapruk, P., et al. (2002). Glutamate receptor-dependent modulation of dopamine efflux in the nucleus accumbens by basolateral, but not central, nucleus of the amygdala in rats. *Journal of Neuroscience, 22* (3), 1137–1145.

Hucker, S., Langevin, R., et al. (1986). Neuropsychological impairment in pedophiles. Canadian *Journal of Behavioral Science, 18,* 440–448.

Huijbregts, S. C., Seguin, J. R., et al. (2008). Maternal prenatal smoking, parental antisocial behavior, and early childhood physical aggression. *Development and Psychopathology, 20* (2), 437–453.

Joyal, C. C., Black, D. N., et al. (2007). The neuropsychology and neurology of sexual deviance: A review and pilot study. *Sex Abuse, 19* (2), 155–173.

Kalichman, S. C. (1991). Psychopathology and personality characteristics of criminal sexual offenders as a function of victim age. *Archives of Sexual Behavior , 20* (2), 187–197.

Kalsbeek, A., Buijs, R. M., et al. (1987). Effects of neonatal thermal lesioning of the mesocortical dopaminergic projection on the development of the rat prefrontal cortex. *Brain Research, 429* (1), 123–132.

Kalsbeek, A., Voorn, P., et al. (1988). Development of the dopaminergic innervation in the prefrontal cortex of the rat. *Journal of Comparative Neurology, 269* (1), 58–72.

Karama, S., Lecours, A. R., et al. (2002). Areas of brain activation in males and females during viewing of erotic film excerpts. *Human Brain Mapping, 16* (1), 1–13.

Keck, M. E., Welt, T., et al. (2002). Repetitive transcranial magnetic stimulation increases the release of dopamine in the mesolimbic and mesostriatal system. *Neuropharmacology, 43* (1), 101–109.

Kiehl, K. A., Smith, A. M., et al. (2001). Limbic abnormalities in affective processing by criminal psychopaths as revealed by functional magnetic resonance imaging. *Biological Psychiatry, 50* (9), 677–684.

Kim, S. W., Sohn, D. W., et al. (2006). Brain activation by visual erotic stimuli in healthy middle aged males. *International Journal of Impotence Research, 18* (5), 452–457.

Kimbrell, T. A., Little, J. T., et al. (1999). Frequency dependence of antidepressant response to left prefrontal repetitive transcranial magnetic stimulation (rTMS) as a function of baseline cerebral glucose metabolism. *Biological Psychiatry, 46* (12), 1603–1613.

Knoch, D., & Fehr, E. (2007). Resisting the power of temptations: The right prefrontal cortex and self-control. *Annals of the New York Academy of Sciences, 1104,* 123–134.

Knoch, D., Nitsche, M. A., et al. (2007). Studying the neurobiology of social interaction with transcranial direct current stimulation the example of punishing unfairness. *Cereb Cortex*. 2008: *18* (9), 1987–1990.

Knoch, D., Pascual-Leone, A., et al. (2006). Diminishing reciprocal fairness by disrupting the right prefrontal cortex. *Science, 314* (5800), 829–832.

Knoch, D., Treyer, V., et al. (2006). Lateralized and frequency-dependent effects of prefrontal rTMS on regional cerebral blood flow. *NeuroImage*.

Koenigs, M., Young, L., et al. (2007). Damage to the prefrontal cortex increases utilitarian moral judgements. Nature, *446* (7138), 908–911.

Koob, G. F. (2006). The neurobiology of addiction: A neuroadaptational view relevant for diagnosis. *Addiction, 101*(Suppl 1), 23–30.

Kringelbach, M. L., O'Doherty, J., et al. (2003). Activation of the human orbitofrontal cortex to a liquid food stimulus is correlated with its subjective pleasantness. *Cerebral Cortex, 13* (10), 1064–1071.

Laakso, M. P., Vaurio, O., et al. (2001). Psychopathy and the posterior hippocampus. *Behavourial Brain Research, 118* (2), 187–193.

Lacourse, E., Cote, S., et al. (2002). A longitudinal-experimental approach to testing theories of antisocial behavior development. *Development and Psychopathology, 14* (4), 909–924.

LeDoux, J. E. (1996). *The emotional brain*. New York: Simon and Schuster.

Luo, Q., Nakic, M., et al. (2006). The neural basis of implicit moral attitude–an IAT study using event-related fMRI. *NeuroImage, 30* (4), 1449–1457.

Manes, F., Sahakian, B., et al. (2002). Decision-making processes following damage to the prefrontal cortex. *Brain, 125* (Pt 3), 624–639.

McClure, S. M., York, M. K., et al. (2004). The neural substrates of reward processing in humans: The modern role of FMRI. *Neuroscientist, 10* (3), 260–268.

Mendez, M. F., Chow, T., et al. (2000). Pedophilia and temporal lobe disturbances. *The Journal of Neuropsychiatry and Clinical Neurosciences, 12* (1), 71–76.

Mitchell, D. G., Avny, S. B., et al. (2006). Divergent patterns of aggressive and neurocognitive characteristics in acquired versus developmental psychopathy. *Neurocase, 12* (3), 164–178.

Moll, J., & Oliveira-Souza, R. (2007). Moral judgments, emotions and the utilitarian brain. *Trends in Cognitive Sciences, 11* (8), 319–321.

Moll, J., Oliveira-Souza, R., et al. (2002a). Functional networks in emotional moral and nonmoral social judgments. *NeuroImage, 16* (3 Pt 1), 696–703.

Moll, J., Oliveira-Souza, R., et al. (2002b). The neural correlates of moral sensitivity: A functional magnetic resonance imaging investigation of basic and moral emotions. *Journal of Neuroscience, 22* (7), 2730–2736.
Moll, J., Oliveira-Souza, R., et al. (2005). The moral affiliations of disgust: A functional MRI study. *Cognitive and Behavioral Neurology, 18* (1), 68–78.
Moll, J., Eslinger, P. J., et al. (2001). Frontopolar and anterior temporal cortex activation in a moral judgment task: Preliminary functional MRI results in normal subjects. *Arquivos de neuro-psiquiatria, 59* (3-B), 657–664.
Moll, J., Krueger, F., et al. (2006). Human fronto-mesolimbic networks guide decisions about charitable donation. *Proceedings of the National Academy of Sciences of the United States of America, 103* (42), 15623–15628.
Moll, J., Zahn, R., et al. (2005). Opinion: The neural basis of human moral cognition. *Nature Reviews Neuroscience, 6* (10), 799–809.
Montague, P. R., Dayan, P., et al. (1996). A framework for mesencephalic dopamine systems based on predictive Hebbian learning. *Journal of Neuroscience, 16* (5), 1936–1947.
Moulier, V., Mouras, H., et al. (2006). Neuroanatomical correlates of penile erection evoked by photographic stimuli in human males. *NeuroImage, 33* (2), 689–699.
Nadeau, S. E., McCoy, K. J., et al. (2002). Cerebral blood flow changes in depressed patients after treatment with repetitive transcranial magnetic stimulation: Evidence of individual variability. Neuropsychiatry, Neuropsychology, and Behavioral Neurology, *15* (3), 159–75.
Narayan, V. M., Narr, K. L., et al. (2007). Regional cortical thinning in subjects with violent antisocial personality disorder or schizophrenia. *American Journal of Psychiatry, 164* (9), 1418–1427.
Newman, J. P., & Kosson, D. S. (1986). Passive avoidance learning in psychopathic and nonpsychopathic offenders. *Journal of Abnormal Psychology, 95* (3), 252–256.
Ohnishi, T., Matsuda, H., et al. (2004). rCBF changes elicited by rTMS over DLPFC in humans. *Supplements to Clinical Neurophysiology, 57,*715–720.
Oliveira-Souza, R., Hare, R. D., Bramati, I. E., Garrido, G. J., Azevedo Ignácio, F., Tovar-Moll, F., et al. (2008). Psychopathy as a disorder of the moral brain: fronto-temporo-limbic grey matter reductions demonstrated by voxel-based morphometry. *NeuroImage, 40* (3), 1202–1213.
Padberg, J., Seltzer, B., et al. (2003). Architectonics and cortical connections of the upper bank of the superior temporal sulcus in the rhesus monkey: An analysis in the tangential plane. *Journal of Comparative Neurology, 467* (3), 418–434.
Pandya, D. N., Dye, P., et al. (1971). Efferent cortico-cortical projections of the prefrontal cortex in the rhesus monkey. *Brain Research, 31* (1), 35–46.
Pandya, D. N., & Kuypers, H. G. (1969). Cortico-cortical connections in the rhesus monkey. *Brain Research, 13* (1), 13–36.
Petrides, M., & Pandya, D. N. (1999). Dorsolateral prefrontal cortex: comparative cytoarchitectonic analysis in the human and the macaque brain and corticocortical connection patterns. *European Journal of Neuroscience, 11* (3), 1011–1036.
Phillips, A. G., Ahn, S., et al. (2003). Amygdalar control of the mesocorticolimbic dopamine system: parallel pathways to motivated behavior. *Neuroscience & Biobehavioral Reviews, 27* (6), 543–554.
Ponseti, J., Bosinski, H. A., et al. (2006). A functional endophenotype for sexual orientation in humans. *NeuroImage, 33* (3), 825–833.
Post, R. M., Kimbrell, T. A., et al. (1999). Repetitive transcranial magnetic stimulation as a neuropsychiatric tool: Present status and future potential. *Journal of ECT, 15* (1), 39–59.
Prince, S. E., Daselaar, S. M., et al. (2005). Neural correlates of relational memory: Successful encoding and retrieval of semantic and perceptual associations. *Journal of Neuroscience, 25* (5), 1203–1210.
Quartz, S., & Sejnowski, T. J. (2002). *Liars, lovers and heroes. What the new brain science reveals about how we become who we are*. New York: HarperCollins.

Quay, H. C. (1965). Psychopathic personality as pathological stimulation-seeking. *American Journal of Psychiatry, 122,*180–183.

Raine, A., Lencz, T., et al. (2000). Reduced prefrontal gray matter volume and reduced autonomic activity in antisocial personality disorder. *Archives of General Psychiatry, 57* (2), 119–127; discussion 128–9.

Raine, A., Meloy, J. R., et al. (1998). Reduced prefrontal and increased subcortical brain functioning assessed using positron emission tomography in predatory and affective murderers. *Behavioral Sciences & The Law, 16* (3), 319–332.

Raine, A., & Yang, M. (2006). Neural foundations to moral reasoning and antisocial behavior. *Social Cognitive and Affective Neuroscience, 1* (3), 203–213.

Rilling, J. K., Glenn, A. L., et al. (2007). Neural correlates of social cooperation and non-cooperation as a function of psychopathy. *Biological Psychiatry, 61* (11), 1260–1271.

Rilling, J. K., Sanfey, A. G., et al. (2004). Opposing BOLD responses to reciprocated and unreciprocated altruism in putative reward pathways. *Neuroreport, 15* (16), 2539–2543.

Robertson, D., Snarey, J., et al. (2007). The neural processing of moral sensitivity to issues of justice and care. *Neuropsychologia, 45* (4), 755–766.

Rolls, E. T., Hornak, J., et al. (1994). Emotion-related learning in patients with social and emotional changes associated with frontal lobe damage. *Journal of Neurology, Neurosurgery and Psychiatry, 57* (12), 1518–1524.

Schaltenbrand, G., & Wahren, W. (1977). *Atlas for stereotaxy of the human brain*. Stuttgart: Thieme Medical Publishers.

Schermer, M. (2007). Gedraag je! Ethische aspecten van gedragsbeinvloeding door nieuwe technologie in de gezondheidszorg. Utrecht.

Schiffer, B., Krueger, T., et al. (2008). Brain response to visual sexual stimuli in homosexual pedophiles. J Psychiatry Neurosci, *33* (1), 23–33.

Schiffer, B., Peschel, T., et al. (2007). Structural brain abnormalities in the frontostriatal system and cerebellum in pedophilia. *Journal of Psychiatric Research, 41* (9), 753–762.

Schiltz, K., Witzel, J., et al. (2007). Brain pathology in pedophilic offenders: Evidence of volume reduction in the right amygdala and related diencephalic structures. *Archives of General Psychiatry, 64* (6), 737–746.

Schultz, W. (2004). Neural coding of basic reward terms of animal learning theory, game theory, microeconomics and behavioural ecology. *Current Opinion in Neurobiology 14* (2), 139–147.

Selemon, L. D., & Goldman-Rakic, P. S. (1988). Common cortical and subcortical targets of the dorsolateral prefrontal and posterior parietal cortices in the rhesus monkey: Evidence for a distributed neural network subserving spatially guided behavior. *Journal of Neuroscience, 8* (11), 4049–4068.

Seto, M. C., Lalumiere, M. L., et al. (2000). The discriminative validity of a phallometric test for pedophilic interests among adolescent sex offenders against children. Psychological Assessment, *12* (3), 319–327.

Speer, A. M., Kimbrell, T. A., et al. (2000). Opposite effects of high and low frequency rTMS on regional brain activity in depressed patients. *Biological Psychiatry, 48* (12), 1133–1141.

Stark, R., Schienle, A., et al. (2005). Erotic and disgust-inducing pictures–differences in the hemodynamic responses of the brain. *Biological Psychology, 70* (1), 19–29.

Stoleru, S., Redoute, J., et al. (2003). Brain processing of visual sexual stimuli in men with hypoactive sexual desire disorder. *Psychiatry Research, 124* (2), 67–86.

Strafella, A. P., Paus, T., et al. (2001). Repetitive transcranial magnetic stimulation of the human prefrontal cortex induces dopamine release in the caudate nucleus. *Journal of Neuroscience, 21* (15), RC157.

Talairach, J., & Tournoux, P. (1988). *Co-planar stereotaxic atlas of the human brain*. New York: Thieme.

Toro, R., Leonard, G., et al. (2008). Prenatal exposure to maternal cigarette smoking and the adolescent cerebral cortex. *Neuropsychopharmacology, 33* (5), 1019–1027.

Tremblay, R. E. (2000). The development of aggressive behavior during childhood: What have we learned from the past century? *International Journal of Behavioral Development, 24,* 129–141.
Valliant, P. M., Gauthier, T., et al. (2000). Moral reasoning, interpersonal skills, and cognition of rapists, child molesters, and incest offenders. *Psychological Reports, 86* (1), 67–75.
Verney, C., Berger, B., et al. (1982). Development of the dopaminergic innervation of the rat cerebral cortex. A light microscopic immunocytochemical study using anti-tyrosine hydroxylase antibodies. *Brain Research, 281* (1), 41–52.
Virkkunen, M. (1976). The pedophilic offender with antisocial character. *Acta Psychiatrica Scandinavica, 53* (5), 401–405.
Völlm, B., Richardson, P., et al. (2004). Neurobiological substrates of antisocial and borderline personality disorder: Preliminary results of a functional fMRI study. *Criminal Behaviour and Mental Health, 14* (1), 39–54.
Völlm, B., Richardson, P., et al. (2007). Neuronal correlates of reward and loss in Cluster B personality disorders: A functional magnetic resonance imaging study. *Psychiatry Research, 156* (2), 151–167.
Volz, H. P., Rehbein, G., et al. (1990). Afferent connections of the nucleus centralis amygdalae. A horseradish peroxidase study and literature survey. *Anatomy and Embryology* (Berl), *181* (2), 177–194.
Walsh, V., & Rushworth, M. (1999). A primer of magnetic stimulation as a tool for neuropsychology. *Neuropsychologia, 37* (2), 125–135.
Walter, H., Abler, B., et al. (2005). Motivating forces of human actions. Neuroimaging reward and social interaction. *Brain Research Bulletin, 67* (5), 368–381.
Walter, M., Witzel, J., et al. (2007). Pedophilia is linked to reduced activation in hypothalamus and lateral prefrontal cortex during visual erotic stimulation. *Biological Psychiatry, 62* (6), 698–701.
Weissenborn, K., Ruckert, N., et al. (1996). A proposed modification of the Wada test for presurgical assessment in temporal lobe epilepsy. *Neuroradiology, 38* (5), 422–429.
Wieser, H. G., Muller, S., et al. (1997). The anterior and posterior selective temporal lobe amobarbital tests: Angiographic, clinical, electroencephalographic, PET, SPECT findings, and memory performance. *Brain and Cognition, 33* (1), 71–97.
Yang, Y., Raine, A., et al. (2005). Volume reduction in prefrontal gray matter in unsuccessful criminal psychopaths. *Biological Psychiatry, 57* (10), 1103–1118.

Does It Pay to be Good? Competing Evolutionary Explanations of Pro-Social Behaviour

Matthijs van Veelen

Brain, Behaviour, Evolution

Our brain makes us behave. If we want to understand why we have the brains that we have, we need to figure out how we benefit from the behaviour it induces. Insofar as our brain is moral, the natural evolutionary question is then how behaving morally promotes our survival and procreation. Does it pay to be good? Doesn't it pay more to be bad? This question seems to preoccupy many of us already in everyday life. It has inspired many writers to produce novels and film scripts that revolve around the question whether or not the bad guy gets a taste of his own medicine. The imminent tragic downfall of a good man – while the villain seems to get away with whatever evil he is up to – makes us hope for a few twists and turns that make everyone get what they deserve. These exciting ingredients are not always processed with subtlety and realism, but even in books and films that appeal to our sense of tragedy through a bad ending, there is the same tension. Recently scientists also tried to answer the question whether or not it pays to be good – which is exciting too, but in a different way – and they have produced a considerable amount of studies in doing so. In this chapter I will describe some of those evolutionary models and try to indicate how close we are to an answer.

In this short overview I will focus on two things in particular. The first is whether or not the behaviour as it is modelled captures the essentials of actual human moral behaviour. One of the focal problems in the literature is for instance the evolution of altruism. While the supply of models has been growing, we might want to ask ourselves if the altruism that features in these models accurately describes our behaviour. "What Moral Behaviour?" discusses this, and a few other elements that are present in different models, such as reciprocity and altruistic punishment. It also indicates their benefits and shortcomings as descriptions of actual human (moral) behaviour.

As a second point I would like to stress the importance of deriving testable implications. There has been a wave of models of the evolution of altruism (or

---

M. van Veelen (✉)
Department of Economics (AE), Amsterdam University, Amsterdam, Netherlands
e-mail: C.M.vanVeelen@uva.nl

J. Verplaetse et al. (eds.), *The Moral Brain*, DOI 10.1007/978-1-4020-6287-2_8,

cooperation, or sociality), some of which come with rather strong claims concerning their explanatory power. With quite different models being proposed as explanations of human altruism, it is natural to ask which one is correct, since one would think that they cannot all be right. Deriving testable predictions that set the different explanations apart is essential for answering this question. Ideally such testable implications would also concern characteristics of the brain.

What Moral Behaviour?

Altruism

Most papers in this field have aimed at explaining altruism, which is regularly taken to be an act that increases the fitness of some other individual, at the expense of one's own fitness. This is a defendable choice, because such behaviour exists and obviously poses an evolutionary puzzle. Modelling always requires simplifying in order to capture the essentials, and if there is one aspect of our social behaviour that seems wise to distill, then it is giving up ones own fitness for the sake of the survival or procreation of someone else. Consequently the peculiarities of our altruistic behaviour have not attracted too much attention, that is, not in the models that are supposed to explain its evolution. While there are quite a few models that aim at explaining altruism in this simple definition, it is clear that the complexity of our moral behaviour goes beyond just being altruistic. It is to be expected that further progress needs to involve more accurate and more complex descriptions of our pro-social behaviour.

A first aspect that is absent in most evolutionary models – but present in some models of pro-social behaviour – is that whether or not an individual performs an altruistic act may very well depend on the status quo. If person A has a lot to eat, and person B has almost nothing, then we expect B not to be too inclined to share food with person A. If endowments would be reversed, however, and B lives in abundance while A has trouble making ends meet, then B might be much more willing to give. Such dependence on the status quo was introduced by Fehr and Schmidt (1999) and Bolton and Ockenfels (2000) and, in a different setting, tested in the lab by Andreoni and Miller (2002). In order to capture the dependence on the status quo, *altruistic preferences* were introduced, which is a tool from mathematical economics (see Box 1 Mas-Colell, Whinston, & Green (1995)). One element of both the Fehr–Schmidt and the Bolton–Ockenfels model is that, for reasonable parameters, individuals will behave altruistically when they are ahead, but spiteful when behind. (Behaviour is called spiteful if a person is willing to give up money, food or fitness of its own to reduce the amount of money, food or fitness of another individual). While there are evolutionary models that explain altruism, and there are models that explain spite, there are no models that explain both of them at the same time. Present models explaining altruism predict that if there is altruism, it should be there regardless of whether one is ahead or behind. The same holds for models that explain spite: if they predict spite, they predict nothing but spite. (See van Veelen (2006a) for a more formal version of this statement and a mathematical proof). Better and improved models should therefore explain behaviour that is conditional on the status quo, which can sometimes lead to altruistic, but sometimes also to spiteful behaviour. This is a challenge, and if we would find such models, then that would bring altruistic behaviour in evolutionary models a step closer to actual human other-regarding behaviour.

Box 1 Other-Regarding Utility Functions

How exactly other-regarding (or pro-social) behaviour may depend on the status quo, can be described with a utility function, assuming that there is some consistency in this dependence. A utility function returns a number, when fed with an alternative from which the individual can choose. Those alternatives are, in this setting, combinations of fitness for the individual itself and for the other. An example of an (altruistic) utility function is $u_{0.5}(x_{\text{self}}, x_{\text{other}})= x_{\text{self}}+ 0.5x_{\text{other}}$. If the individual itself would get 1 and the other would get 2, the utility function returns the value 2. If the individual itself would get 2 and the other would get 1, the utility function would return the value 2.5. If this utility function describes the behaviour of individual A, then that implies that individual A prefers the second alternative to the first, as 2.5 is larger than 2.

If the choice between two such alternatives does not differ between different status quo's, then the utility function that describes such behaviour is linear. This gives us a first family of utility functions that has members that are determined by a parameter α and are defined by:

$$u_{\alpha}\left(x_{\text{self}}, x_{\text{other}}\right) = x_{\text{self}} + \alpha x_{\text{other}}$$

An intuitive interpretation of α is that it is the weight that one individual attaches to the interest of the other person. (Edgeworth (1881) called α a coefficient of efficient sympathy). To visualize a utility function, one can draw pictures with iso-utility curves. An iso-utility curve is a set of points that represent alternatives that yield the same value of the utility function, implying that the person who's behaviour it describes is indifferent between them. The pictures below represent a very altruistic person, that weights the interests of the other as much as its own – α = 1 – and a less, but still considerably altruistic person, that cares for the interests of the other half as much as for its own – α = 0.5. Note that the sharper ascend of the iso-utility curves in the second picture implies that it takes a larger gain of the other individual to make an individual with this preference relation go for the more altruistic alternative.

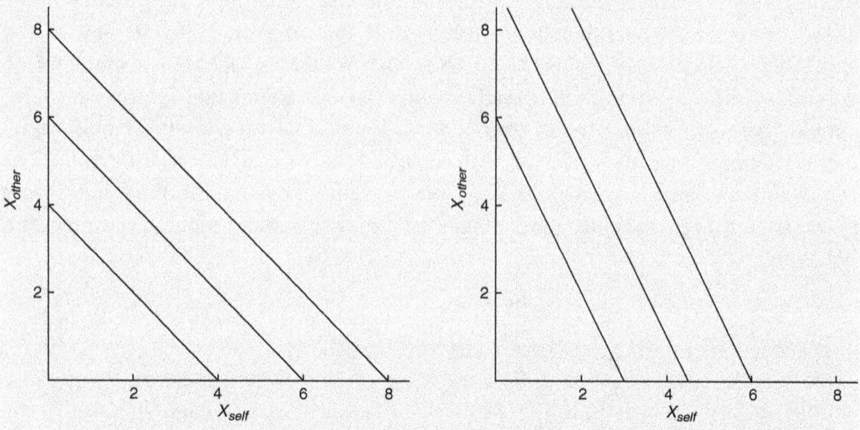

$u\,(x_{\text{self}}, x_{\text{other}}) = x_{\text{self}} + x_{\text{other}}$ $u\,(x_{\text{self}}, x_{\text{other}}) = x_{\text{self}} + \frac{1}{2} x_{\text{other}}$

A more general family of preference relations, and one that allows for true dependence on the status quo, has parameters α and ρ and is represented by utility functions

$$u_{a,\varrho}\,(x_{\text{self}}, x_{\text{other}}) = \left(x_{\text{self}}^{\varrho} + a x_{\text{other}}^{\varrho}\right)^{1/\varrho}$$

For α = 1, the next pictures show iso-utility curves for different values of ρ. This parameter represents the dislike of inequality; the lower ρ, the more the iso-utility curves are actually curved, which means that it matters more how equal or unequal the alternatives are.

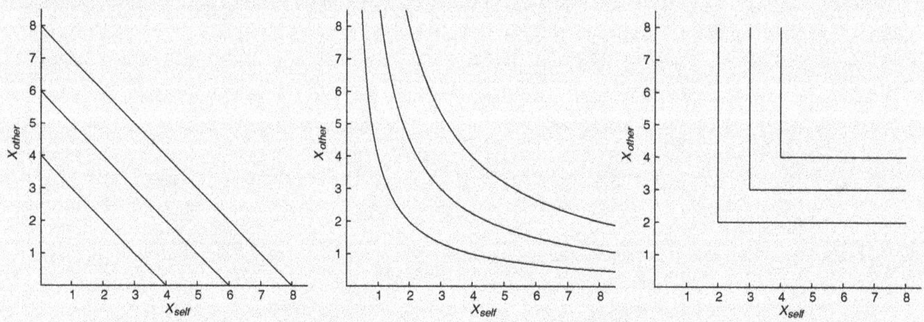

$\alpha = 1, \rho = 1$ $\alpha = 1, \rho \to 0$ $\alpha = 1, \rho \to -\infty$

There are two prominent example of utility functions that encompass both altruism and spite. The formulas can be found in Fehr and Schmidt (1999) and Bolton and Ockenfels (2000). We will only draw two pictures with iso-utility curves.[1]

[1]For Fehr and Schmidt (1999) α = 2/3 and β= 1/3 is chosen and for Bolton and Ockenfels (2000) $a = 1$ and $b = 15$.

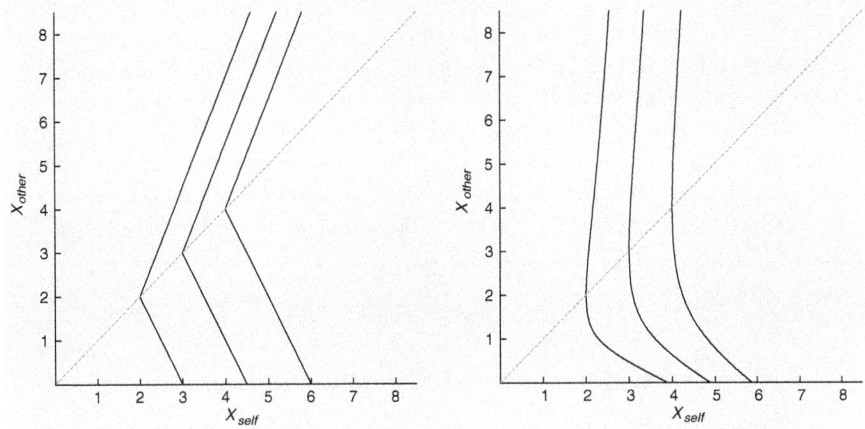

Fehr-Schmidt (1999) Bolton-Ockenfels (2000)

A slightly more formal introduction can also be found in van Veelen (2006a).

A second aspect that is worth discussing concerns the use of the word "altruism". The standard usage of the word prescribes that it refers to behaviour that increases the fitness of some other individual, at the expense of one's own fitness. This is a perfectly valid and clear-cut definition. Nonetheless, even though there are good reasons for sticking to this way in which the word is used, we will see that strictness also comes at a cost here. In "Different Types of Models", we see that if one individual gives away food, thereby increasing someone else's survival probability and decreasing its own, there are different models that could explain such behaviour. One is kin selection, in which case relatedness between donor and recipient should be high enough for selfish genes to build altruistic individuals. In this model, the loss of survival probability of the donor should be offset by the gain of survival probability of the other, weighted by a measure of relatedness. Here the act is altruistic according to the strict definition. Another explanation however could be sexual selection, where altruistic behaviour signals good health or good genes and attracts mates. In such a model the loss in survival probability should be offset by the gain in opportunities to reproduce. Here the act of giving is not altruistic in the strict sense of the definition, because reproductive opportunities also determine fitness. The behaviour under study, however, may be the same; in both cases the donor might genuinely care for the recipient, and, more importantly, in both cases we observe one individual reducing its own survival probability and exchange it for an increase in the probability of survival of another. This at first sight surprises us, and any explanation will have to indicate how this is offset or outweighed by something else. By what it is outweighed, in this case also determines whether or not the behaviour deserves the label "altruistic" in the strict interpretation. It can be seen as a bit inconvenient to let the label for the behaviour that we want to explain, depend on the explanation. Although it certainly is defendable to restrict the term altruism to the behaviour in the first explanation, and relabel the behaviour if we find that the second explanation applies, another approach is also possible. One could let the word altruism refer to any behaviour where the donor sacrifices something (for instance survival probability) to serve the interests of another individual. This would imply that, when describing behaviour, we consciously narrow our focus to, for instance, the effect of the behaviour on the survival probabilities of what usually are called donor and recipient(s), while broadening our view again when we are looking for the explanation, for which, as we will see, there are quite a few candidates. But even if we do choose to stick to the strict definition, it is worth realizing that behaviour is usually observed is this narrow sense, and that labeling it as altruistic then implies a choice for a subset of the possible explanations.

Reciprocity, Indirect Reciprocity, Altruistic Punishment and Third Party Punishment

A second ingredient that at least plays a role in morality is reciprocity. In general, reciprocity means that an individual conditions its behaviour on the behaviour of the individual it interacts with. Reciprocity is also clearly present in human social interactions. A natural setting for studying the evolution of reciprocity is one with repeated interaction, where it might pay to behave reciprocally. (The reason why it pays is that it can seize opportunities for mutual cooperation with others that are also cooperative and reciprocal, and avoids exploitation by others that are not).

While reciprocity first of all applies to dyadic interactions, humans may also behave differently depending on how the individuals they interact with treat others. This is called indirect reciprocity, or third party punishment. Furthermore punishment is called altruistic if the punisher incurs more costs than benefits from punishing, and other population members benefit, for instance from a change in behaviour of the punished. These definitions are the natural next step towards moral behaviour, because they go beyond dyadic interactions. (See Fehr & Gächter, 2002, and Nowak & Sigmund, 2005, for a review. The latter paper also contains further references to relevant papers).

Our moral behaviour is built from a few ingredients. There is empathy in there, because most moral arguments cease to work if one does not at least care a bit about others. The ability to put oneself in someone else's position therefore is a vital part of moral judgements. Morality also contains some logic concerning symmetry arguments, as most moral statements derive their strength from a claim to being more or less impartial judgements. Rawls (1971) and Harsanyi (1977) have tried to formalise the structure of moral judgements by introducing the imaginary veil of ignorance, behind which no-one knows which role it will be assigned. Their approach is normative (or axiomatic) rather than descriptive, but it is undeniably true that we are not insensitive for symmetry arguments, which indicates that impartiality is an important ingredient of moral judgements too.

The review on moral psychology by Haidt (2007) on the other hand is completely descriptive. This review contains four "principles" that concern moral behaviour. The first one is that moral reasoning regularly follows moral intuition, rather than precede it, although moral reasoning can sometimes override moral intuition[1]. The second is that the capacity for moral reasoning seems to serve less as a guide to individual decision making than as a way to be able to influence the moral judgements by others of your own behaviour. Moral judgements in practice therefore show a large degree of self-serving bias. The third is the human capacity to form communities within which selfishness is punished and virtue rewarded. The fourth is that morality is about more than harm and fairness. The extra ingredients are intuitions about ingroup–outgroup dynamics and loyalty to the group, intuitions about authority, and intuitions about bodily and spiritual purity.

As we will see below, most formal (mathematical) models restrict themselves to altruism, reciprocity or indirect reciprocity. The question why we are sensitive to symmetry or impartiality arguments is not really addressed there, nor do the four principles feature in there as characteristics of human behaviour that are under selection pressure. This does not mean that we do not have an idea where to look for explanations. Some aspects of our moral judgements even have more or less obvious evolutionary reasons that do not need a mathematical model to be explained. For some of these elements, however, there are considerable discrepancies between the behaviour we can explain with our models, and the behaviour we observe and should be explaining. While many models focus on altruism - not even on other-regarding utility functions, or preferences – our moral behaviour is much more complex than that. This obviously does not at all mean that the modelling so far was not interesting and useful – on the contrary – but further progress will have to involve increased realism concerning the behaviour to be explained.

Different Types of Models

The multitude of models can roughly be divided into four categories. First one can consider natural selection at three different levels: gene, individual and group. That makes three types of selection: kin selection, natural selection at the individual level, and group selection. The fourth category is sexual selection.

Kin Selection

The insight that genes are the basic unit of reproduction is essential for understanding kin selection. If a gene can increase the number of copies of itself by helping a related individual survive or procreate – where a related individual is likely to carry copies of the same gene – then it will do so, even if that means that such help reduces the chances to survive or procreate for the individual that the gene itself is in. Doing what is best for the reproduction of the gene can therefore include

[1]Moral intuition is sometimes referred to as being in the domain of affect, and moral reasoning as cognitive, but Haidt (2007) follows Bargh & Chartrand (1999) for terminology, in order to avoid creating the impression that affective reactions do not involve information processing.

performing behaviour that decreases the fitness of one particular carrier, under the condition that it will have to increase the fitness of other, related individuals enough to offset this decrease. How much of an increase in the fitness of related individuals will do to offset the decrease of ones own fitness is summarized by Hamilton's rule, that states that benefits have to weighted with relatedness. When reduced to interactions between two individuals, Hamilton's rule predicts that behaviour whereby an individual raises the fitness of another individual by b and lowers its own fitness by c will be selected for if $rb - c > 0$ and will be selected against if $rb - c < 0$, where r is a measure of relatedness (Hamilton, 1964). After Hamilton's groundbreaking article, a whole literature on kin selection has developed; see for instance Lehmann and Keller (2006) or van Veelen (2007) and references therein. The idea of kin selection is explained very clearly in Dawkins' (1976) *The Selfish Gene*, which also more or less holds the view that any altruism or pro-social behaviour that is there must be the result of kin selection.

Natural Selection at the Individual Level

Although kin selection is an adequate explanation of a considerable share of human other regarding behaviour, such as parental care, there still is a wide range of altruistic, fair and moral acts that can never be understood if we only think in terms of genetic relatedness. After all, we are not just nice to our family members, but we are also kind towards the neighbours and we do have friends that we help when they need a hand. There are differences, but humans exhibit behaviour towards others they are not genetically related to and that, at least at first sight, shows some similarities with the behaviour towards their kin. One way or another, it seems that evolution has extended our warm feelings across the family boundaries. The same goes for some other animals, for instance for chimpanzees. Every once in a while, chimps get into fights, and when they do, they can surely use some help. Assistance in a fight can be seen as an act of altruism, as fighting is a risky business, and it is one that is performed too, as chimps sometimes do come to each others aid. These supportive interventions in aggressive encounters are also not restricted to kin, although there are differences (see for instance Harcourt & de Waal, 1992).

The explanation of such helping behaviour might be that what goes around, comes around. If it does, then there are two issues that we can discuss. The first is: if the cost of helping is offset by the benefit of the favour when returned, can we still call this helping behaviour altruistic? If we follow the strict definition of altruism, then the answer is obviously no. Helping then is nothing but a good investment, that does not decrease, but increases the fitness of the individual that performs it. Again, this is a more than defendable way of using the word. I would like to mention though that an alternative is to use it in a less stringent way. Sticking to the chimp example, there is by itself no difference between assistance to a brother in a fight and assistance to a "friend" in a fight. In both cases the helping individual increases the survival probabilities of the helped at the expense of its own. The explanation might however be different, for which we have to zoom out, and understand that in the second case it pays for the individual to help, while in the first case it only pays for the gene.

The second issue is: why would things come around? It is clear that being helpful can pay, but why would others reciprocate? There are two cases to be discerned here. The first is the dyadic situation, in which case we would speak of direct reciprocity. The way to model this situation is in a repeated game setting, which was done in a classic computerized tournament by Axelrod (1984). It turns out that reciprocal strategies perform well; tit-for-tat did best in his tournament, and also in the following literature, reciprocal strategies come out on top regularly; see for instance Bendor and Swistak (1995), Binmore and Samuelson (1992, 1997), Imhof, Fudenberg, and Nowak (2005) or van Veelen (2006b) and references therein.

How indirect reciprocity evolved is an even harder question. The best known models are the ones by Nowak and Sigmund (1998a, 1998b) and Leimar and Hammerstein (2001), which feature "image scoring strategies", that are indirectly reciprocal. These models also fall in the category of natural selection at the individual level, because here it still is in the best interest of the individual itself to reciprocate indirectly.

Group Selection

The debate on group selection and morality has followed a peculiar pattern including remarkable shifts in the consensus. The following quote suggests that Darwin did think of groups as a possible unit of selection, especially when it comes to morality.

> It must not be forgotten that although a high standard of morality gives but a slight or no advantage to each individual man and his children over the other man of the same tribe, yet that an increase in the number of well-endowed men and advancement in the standard of morality will certainly give an immense advantage to one tribe over another. There can be no doubt that a tribe including many members who, from possessing in a high degree the spirit of patriotism, fidelity, obedience, courage, and sympathy, were always ready to aid one another, and to sacrifice themselves for the common good, would be victorious over most other tribes; and this would be natural selection. At all times throughout the world tribes have supplanted other tribes and as morality is one important element in their success, the standard of morality and the number of well-endowed men will thus everywhere tend to rise and increase.
>
> (Charles Darwin, *The Descent of Man*, page 166).

If he would have said that a high standard of morality would be a *disadvantage* to the individual, rather than "a slight or no advantage", then this would have been an even better reason for group selectionists to call upon Darwin himself for their case, but even as it is now it seems pretty obvious that Darwin uses a group selection argument as an explanation for the existence of morality.

Later, others have been a lot less sparing than Darwin in invoking this argument. Sober and Wilson (1998) name a few examples of unjustified group selection claims.

> According to Allee (1951), dominance hierarchies exist to minimize within group conflict, so that the entire group could be more productive. According to Wynne-Edwards (1962),

> individual organisms restrain themselves from consuming food and from reproducing, so that the population can avoid crashing to extinction. And according to Dobzhansky (1937), whole species maintain genetic diversity to cope with new environmental challenges; like savvy investors, they diversify their portfolios because the future is uncertain. (Sober & Wilson, *Unto Others*, page 4)

In the 1960s however, such arguments became a heresy. In his book *Adaptation and Natural Selection* (1966) G. C. Williams definitively did away with the group selectionists, and everything that even smelled of group selection was sentenced to the stake. In order to guard the congregation from this dangerous heterodoxy, all biologists were taught a new First Commandment: Thou shalt not invoke any group selection argument. In *The Extended Phenotype*, Richard Dawkins (1982) gives the following definition of group selection: "A hypothetical process of natural selection among groups of organisms. Often invoked to explain the evolution of altruism. Sometimes confused with kin selection."

If we leave the religious analogy for a moment, we might try to understand what made this discussion so fierce and at times even unpleasant. One of the most appealing aspects of the theory of evolution is that it explains why so many things are so very good at carrying out particular tasks. An arm is a perfect instrument to reach out and grab things and an eye is a very effective device to scan the surroundings. This attractive side of the theory of evolution can however lead us into temptation to cook up a story about adaptive advantages for everything that seems to have a function. "For the good of the group" or even "for the good of the species" are arguments that seem natural, but which lots of time are just wrong, because in most realistic models it is not groups that die or survive, but individuals. Good for the group is all very well, but if it is the individual who pays, then selection will catch up with all pro-group behaviour. Yet the temptation to abuse the evolutionary vocabulary remains strong. I imagine that it is the flood of improper use of evolutionary arguments that made the refutation of group selection so vehement and categorical. Group selection, however, is not always bad.

The comeback of group selection is for a large part due to D.S. Wilson, who claims that not all group selection models are necessarily unrealistic. In *Unto Others*, Sober and Wilson (1998) therefore argue that group selection has been an important force in the evolution of altruism. While the book, as well as most of the literature indiscriminately lumps together a whole bunch of models under the label "group selection", it is worthwhile noticing that this set really contains models of two rather different types (see Van Veelen & Hopfensitz, 2007; and (Van Veelen, 2009)).

The first type of group selection models could be called the standard group selection model (see Price, 1972; Hamilton, 1975; Frank, 1998, and Sober & Wilson, 1998). What is essential for this type of model, is that groups are not formed randomly, but in an assortative way, which is realistic in many group-structured populations. This assortment can come in degrees, depending on assumptions of the model. If groups form slightly assortative, then altruists on average end up in groups where they are teamed up with a slightly larger number of altruists than egoists do.

Groups can also form completely assortative, in which case altruists and egoists are completely segregated and find themselves in homogeneous groups. The more assortative the group formation process is, the larger the differences in composition between groups.

There are two equivalent ways in which we can see why such a model can result in altruism being selected. From the individual point of view the argument is as follows: altruists incur more costs than egoists do, because they are altruists, which is bad for them. They do however also enjoy more benefits than egoists do, because they interact more with altruists than egoists do, as a result of the assortative group formation. Whether a given form of altruism can evolve, depends on how these costs and benefits are balanced. From the "gene for altruism" point of view the result is the same; an altruistic act is of course costly, but the benefits are enjoyed by a disproportional share of altruists, which then promotes the share of altruistic genes in the total population. Again, whether a given form of altruism can evolve, depends on how these costs and benefits are balanced.

This first type of group selection model can also be seen as a kin selection model. Because of the assortative group formation, group members are (slightly) related to each other; two random group members share more genes than two random population members. This particular type of group selection model therefore could also be classified under kin selection. In the literature, an effort is made by some to restate group selection models as kin selection models (see for instance Grafen, 2007, and Lehmann, Keller, West, & Roze, 2007).

The essential feature of the second type of model is that the fates of group members are, to some extent, aligned. This really makes it a different kind of model. A good example is a simulation model by Bowles, Choi, and Hopfensitz (2003). This model of the co-evolution of altruism and norms includes violent conflicts, and these wars are survived together or not at all. In war, group members therefore share their fate completely. In such a situation helping behaviour (or moral behaviour) can evolve, because the cost that an individual faces for helping a fellow group member can be offset by the increase in chances to survive a conflict with another group together. More general, if the fates of group members are, to some extend, tied together, then behaviour that serves the interests of the group, also serve the interests of the individual, thereby allowing for a cooperation and labour division to evolve. The second type of group selection model therefore is in line with seeing the emergence of social groups as another "major transition" in biology, such as the transition from single-celled to multicellular life. (See also Weibull and Salomonsson (2006), whose group selection models falls under this category).

Again, for the second type of model, the behaviour would not be called altruistic in the strict definition, but one could choose to call helping behaviour within the group altruistic, whether it is explained by the first type of group selection model, or the second. Summarizing, the difference between the two types of group selection model is that the working of the first one depends on the answer to the question "*is the probability that you also have the gene large enough*", while the other needs an affirmative answer to "*are our interests enough in line*".

Sexual Selection

Yet another evolutionary explanation of helping behaviour is that it signals fitness, and therefore attracts partners. Those partners can increase the evolutionary payoff of their own parental investment by doing this shared investing together with someone fit and making sure that their genetic contribution to the offspring is tied to a good set of genes provided by the other. Here helping behaviour may be costly on the survival side, but it is beneficial on the reproduction side, and therefore it does not necessarily reduce fitness. Again helping behaviour then is not altruistic in the strict sense of the word, but one could also be less strict and call the possibly genuinely felt need to care for others altruistic, even though the individual that does the helping does get – possibly unanticipated – gains in another domain.

There are two distinct concepts that are important within sexual selection. The first is Fisher's runaway process, where a trait (such as altruism) and preference for a trait (a preference to mate with altruistic individuals) play leapfrog and each causes the other to increase. The second is Zahavi's handicap principle (Zahavi, 1997), where only costly signals can convey the true quality of the sender. This is formalised in Grafen (1990a, 1990b). *The Mating Mind* by Geoffrey Miller (2001) strongly advocates the strength and importance of sexual selection, and includes altruism and morality as important examples.

Which One Is Right?

Given this multitude of suggested explanations, it makes sense to think how we should deal with it. Surely they can be thought of as competing explanations, and they can first of all be treated as such. The task we then face is to derive empirical implications of the different models and focus on where they differ. This way we can, in classic Popperian style, reject explanations that are wrong. One such example is van Veelen (2006a), where it is shown that altruism, when explained by kin selection or by the first type of group selection model, should be independent of the status quo, as described above in the section "What Moral Behaviour?". This is not in line with observations of human other-regarding behaviour, which clearly is dependent on status quo (see also the paper for a formal version of this claim and a precise version of the contrast between prediction and empirical evidence).

The different models are in the first place competing explanations, but it would be too simple only to see them as mutually exclusive possibilities fighting for our approval. To illustrate that, I would like to draw a parallel between the evolution of altruism and the evolution of the mouth. The mouth of our earliest ancestors that had one, supposedly served a single purpose: the efficient intake of food. Since then however, millions of years have passed and in the meantime mutation and selection have equipped us with a mouth that does much more than only eating. We also use our present pair of jaws for verbal communication, we sing with it, we express our happiness by forming a smile and our tenderness through a kiss. Since evolution tends to take the line of the least resistance, one can imagine that adapting an eating instrument for speech is much easier than developing a separate device to make us talk. The same might just go for altruism. Primarily it may have been there towards relatives only, optimizing the gene's fitness (kin selection). Suitably adapted however, it might very well serve as a fitness indicator (sexual selection) or an optimal strategy to play all kinds of "games" with, that arise with the growing complexity of human interactions (natural selection). Therefore it is certainly possible that there are a few different models that all make sense and that they are explanations for different sides of the same complex of altruism, fairness and morality. The great challenge is therefore to see to what extent the different possible selection processes have shaped (different aspects of) our moral behaviour.

Our task is therefore is to increase the precision and detail with which we describe actual human moral behaviour, and contrast that with predictions of combinations of models. The fact that so much of the current theoretical evolutionary research is only focussed on explaining altruism indicates that this is only the beginning. One example of a challenge is that we think that humans are capable of good as well as bad, or, in particular, that they can be host to altruistic as well as spiteful sentiments, as for instance Fehr and Schmidt (1999) and Bolton and Ockenfels (2000) suggest. If so, then we need more complex and detailed models of population structure to explain that. Evolutionary models now predict either altruism or spite, regardless of the status quo, but not both, depending on the initial situation. Such complex models might also include interactions between different existing models.

Verbal Versus Formal

There are many formal, mathematical models that aim at explaining altruism, reciprocity and indirect reciprocity. There are some that aim at explaining norms. I do not know of formal models that go far beyond that. This does of course not at all mean that there are no explanations, or ideas where to look for explanations of the more complex aspects of morality. An inspiring idea is that morality "binds and builds" (see Haidt, 2007, and references therein). It "constrains individuals and ties them to each other to create groups that are emergent entities with new properties", which suggests a group selection model of the second type (see "Group Selection"). This might also include religion, that also typically unites people in a moral community (Durkheim, 1915). So far, however, such a suggested explanation is incomplete, as it does not specify why it would not be better for individual reproduction to loosen up a bit from this group creation process, or free-ride on the enforcement of the norms within the group by others. It therefore makes sense to explore those suggestions in formal models, which forces us to be precise and allows the explanation to undergo the test whether behaviour is evolutionary stable, which means that mutations cannot invade. The better descriptions of moral behaviour, as well as more verbal explanations are however a necessary input for formal modelling, that otherwise risks cooking up models that explain behaviour that we do not observe in humans, or overlook ingredients in the explanation. This also applies to the human capacity for groupishness and linking our emotions to those of other people, including larger groups of them (Newberg, D'Aquili, & Rause, 2001, Gallese, Keysers, & Rizzolatti, 2004). We more or less see what it does, and we can also imagine how it serves the group, but in order to be a proper explanation, we should also understand how it is cheater-proof.

Brain Scanning: Proximate and Ultimate Causes

A blunt, outsider's way of describing a large share of current brain research is that it locates mental activities and describes their relative intensities. This is a fascinating thing to do; Sanfey, Rilling, Aronson, Nystrom, and Cohen (2003) for instance let subjects play an ultimatum game, and finds that responding to unfair offers involves emotions as well as cognition or reasoning. Similar studies have looked at what parts of the brain are active when players reciprocate (Knoch, Pascual-Leone, Meyer, Treyer, & Fehr, 2006), punish altruistically (Fehr & Gächter, 2002, de Quervain et al., 2004) and donate to charities (Moll et al., 2006), to name a few fMRI scans therefore are a relatively new tool that helps us get at the proximate causes of behaviour, which is a welcome complement for more traditional lab experiments (Glimcher & Rustichini, 2004).

According to the classic distinction, evolutionary research aims at finding ultimate, rather than proximate explanations; it figures out, not how the mind works, but why it works the way it does. Finding out how our brains make us behave, along with an accurate description of our behaviour itself, therefore can be seen as a prerequisite for the search for evolutionary explanations. There are however reasons why the interaction between the work on proximate and ultimate causes has its limitations.

What a theorist would dream of, is the following. Suppose there are two competing explanations for altruistic behaviour, one from the domain of group selection and one from the domain of sexual selection. The group selection explanation implies that altruistic behaviour should be brought about by activity in the left half of the brain, the sexual selection model would imply activity in the right half of the brain. To settle the issue, we put subjects in an fMRI scan, look at the data or the pictures, and discard one of the two models, depending on the results.

Unfortunately, models of the evolution of morality are, as far as I know, not really near such a scenario. The behaviour in current models is not even complex enough to allow for differences in actual behaviour itself, let alone in how it is brought about by the brain. This of course does not mean that it is by nature not possible to find evidence for one theory or against another in brain scans, but it does indicate that there still remains a lot to be done. It does also make one expect that the search for ultimate causes will lag behind the search for proximate ones and that evolutionary theorists will benefit from ever better images of how the actual moral brain works.

References

Allee, W. (1951). *Cooperation among animals*. New York: Henry Shuman.

Andreoni, J., & Miller, J. (2002). Giving according to GARP: An experimental test of the consistency of preferences for altruism. *Econometrica, 70*, 737–753.

Axelrod, R. (1984). *The evolution of cooperation*. New York: Basic Books.

Bargh, J. A., & Chartrand, T. L. (1999). The unbearable automaticity of being. *American Psychologist, 54*, 462–479

Bendor, J., & Swistak, P. (1995). Types of evolutionary stability and the problem of cooperation. *Proceedings of the National Academy of Sciences, 92*, 3596–360.

Binmore, K. G., & Samuelson, L. (1992). Evolutionary stability in repeated games played by finite automata. *Journal of Economic Theory, 57*, 278–305.

Binmore, K. G., & Samuelson, L. (1997). Muddling through: Noisy equilibrium selection. *Journal of Economic Theory, 74*, 235–265.

Bolton, G. E., & Ockenfels, A. (2000). ERC: A theory of equity, reciprocity, and competition. *American Economic Review, 90*, 166–193.

Bowles, S., Choi, J.-K., & Hopfensitz, A. (2003). The co-evolution of individual behaviours and social institutions. *Journal of Theoretical Biology, 223*, 135–147.

Dawkins, R. (1976). *The selfish gene*. Oxford: Oxford University Press.

Dawkins, R. (1982). *The extended phenotype: The long reach of the gene*. Oxford: Oxford University Press.

Dobzhansky, T. (1937). *Genetics and the origin of species*. New York: Columbia University Press.

Durkheim, E. (1915). *The elementary forms of religious life*. New York: The Free Press(Reprint: 1965).

Edgeworth, F. D. (1881). *Mathematical Physics*, Londan, P. Kegan.

Fehr, E., & Gächter, S. (2002). Altruistic punishment in humans. *Nature, 415*, 137–140.

Fehr, E., & Schmidt, K. (1999). Fairness, competition, and inequality. *Quarterly Journal of Economics, 114*, 17–868.

Frank, S. A. (1998). *Foundations of social evolution*. Princeton, NJ: Princeton University Press,.

Gallese, V., Keysers, C., & Rizzolatti, G. (2004). A unifying view of the basis of social cognition. *Trends in Cognitive Sciences, 8*, 396–403.

Glimcher, P. W., & Rustichini, A. (2004). Neuroeconomics: The consilience of brain and decision. *Science, 306*, 447–452.

Grafen, A. (1990a). Sexual selection unhandicapped by the Fisher process. *Journal of Theoretical Biology, 144*, 473–516.

Grafen, A. (1990b). Biological signals as handicaps. *Journal of Theoretical Biology, 144*, 517–546.

Grafen, A. (2007). Detecting kin selection at work using inclusive fitness. *Proceedings of the Royal Society B, 274*, 713–719.

Haidt, J. (2007). The new synthesis in moral psychology. *Science, 316*, 998–1002.

Hamilton, W. D. (1964). The genetical theory of social behaviour (I and II). *Journal of Theoretical Biology, 7*, 1–16, 17–32.

Hamilton, W. D. (1975). Innate social aptitudes of man: Approach from evolutionary genetics. In R. Fox (Ed.), *Biosocial anthropology* (pp. 133–155). New York, Wiley.

Harcourt, A. H., & de Waal, F. B. M. (Eds.). (1992). *Coalitions and alliances in humans and other animals*. Oxford: Oxford University Press.

Harsanyi, J. C. (1977). *Rational behavior and bargaining equilibrium in games and social situations*. Cambridge UK: Cambridge University Press.

Imhof, L. A., Fudenberg, D., & Nowak, M. A. (2005). Evolutionary cycles of cooperation and defection. *Proceedings of the National Academy of Sciences, 102*, 10797–10800.

Knoch, D., Pascual-Leone, A., Meyer, K., Treyer, V., & Fehr, E. (2006). Diminishing reciprocal fairness by disrupting the right prefrontal cortex. *Science, 314*, 829–832.

Lehmann, L., & Keller, L. (2006). The evolution of cooperation and altruism – a general framework and a classification of models. *Journal of Evolutionary Biology, 19*, 1365–1376.

Lehmann, L, Keller, L., West, S., & Roze, D. (2007). Group selection and kin selection: Two concepts but one process. *Proceedings of the National Academy of Sciences, 104*, 6736–6739.

Leimar, O., & Hammerstein, P. (2001). Evolution of cooperation through indirect reciprocity. *Proceedings of the Royal Society of London, 268*, 748–753.

Mas-Colell, A., Whinston, M. D., & Green, J. R. (1995). *Microeconomic theory*. New York/Oxford: Oxford University Press.

Miller, G., 2001. *The mating mind: How sexual choice shaped the evolution of human nature*. New York: Anchor Books.

Moll, J., Krueger, F., Zahn, R., Pardini, M., Oliveira-Souza, R., & Grafman, J. (2006). Human fronto-mesolimbic networks guide decisions about charitable donation. *PNAS, 103*, 15623–15628.

Newberg, A., D'Aquili, E., & Rause, V. (2001). *Why god won't go away: Brain science and the biology of belief*. New York: Ballantine Books,

Nowak, M. A., & Sigmund, K. (1998a). The dynamics of indirect reciprocity. *Journal of Theoretical Biology, 194*, 561–574.

Nowak, M. A., &. Sigmund, K. (1998b). Evolution of indirect reciprocity by image Scoring. *Nature, 393*, 573–577.

Nowak, M. A., & Sigmund, K. (2005). Evolution of indirect reciprocity. *Nature, 437*, 1291–1298.

Price, G. R. (1972). Extension of covariance selection mathematics. *Annals of human Genetics, 35*, 485–489

de Quervain, D. J.-F., Fischbacher, U., Treyer, V., Schellhammer, M., Schnyder, U., Buck, A., et al. (2004). The neural basis of altruistic punishment. *Science, 305*, 1254–1258.

Rawls, J. (1971). *A theory of justice*. Cambridge, MA: Harvard University Press.

Sanfey, A. G., Rilling, J. K., Aronson, J. A., Nystrom, L. E, & Cohen, J. D. (2003). *The neural basis of economic decision-making in the ultimatum game. Science, 300*, 1755–1758.

Sober, E., & Wilson, D. S. (1998). *Unto others: The evolution and psychology of unselfish behavior*. Cambridge, MA: Harvard University Press.

van Veelen, M. (2006a). Why kin and group selection models may not be enough to explain human other-regarding behaviour. *Journal of Theoretical Biology, 242*, 790–797.

van Veelen, M. (2006b). Evolution of strategies in repeated games with discounting Tinbergen Institute Discussion Paper TI 2006-115/1

van Veelen, M. (2007a). Hamilton's missing link. Journal of Theoretical Biology, 246, 551–554.

van Veelen, M. (2009). Group selection, kin selection, altruism and cooperation: When inclusive fitness is right and when it can be wrong. *Forthcoming in Journal of Theoretical Biology.*

van Veelen, M., & Hopfensitz, A. (2007b). In love and war: Altruism, norm formation, and two different types of group selection. *Journal of Theoretical Biology, 249*, 667–680.

Weibull, J. W., & Salomonsson, M. (2006). Natural selection and social preferences. *Journal of Theoretical Biology, 239*, 79–92.

Wynne-Edwards, V. C. (1962). *Animal dispersion in relation to social behavior*. Edinburgh: Oliver and Boyd.

Zahavi, A. (1997). *The handicap principle: A missing piece of Darwin's puzzle*. Oxford: Oxford University Press.

How Can Evolution and Neuroscience Help Us Understand Moral Capacities?

Randolph M. Nesse

Trying to understand morality has been a central human preoccupation for as far back as human history extends, and for very good reasons. The core phenomenon is readily observable: we humans judge each other's behaviour as right or wrong, and each other's selves as moral or immoral. If others view you as moral, you will thrive in the bosom of a human group. If, however, others view you as immoral, you are in deep trouble; you may even die young, either at the hands of others, or alone in the bush. These are very good reasons indeed for close attention to morality.

There are, however, two problems. The first is how to distinguish right from wrong, the second is inhibiting temptations to do what is wrong (that temptations to do right are not a problem is most interesting). The first problem poses few concerns for most people – they are confident that they know what is right, based on their intuitive emotional responses. However, different people have different emotional responses, and intuition makes a poor argument. Finding a general principle that explains the individual instances would be incredibly valuable. So, for several thousands of years, philosophers have tried to find general moral principles. They have also argued about where they come from, why they have normative force, and how they are best applied to individual instances (Darwall, Gibbard, & Railton, 1997). Thousands of books chronicle the human quest for moral knowledge.

Now, in a mere eye blink of history, the scene has changed. Completely new kinds of knowledge are being brought to bear. Neuroscience is investigating the brain mechanisms involved in moral decisions, moral actions, and responding to moral and immoral actions by self and others. Evolutionary biology is investigating why those brain mechanisms exist, how they give a selective advantage, and why there is genetic variation that influences moral tendencies. This is an exciting time for those of us curious about morality.

R.M. Nesse (✉)
The University of Michigan, Ann Arbor, MI, USA
e-mail: nesse@umich.edu

This chapter introduces Chapter 10, which is reprinted from: Nesse, R. M. (2007). Runaway social selection for displays of partner value and altruism, *Biological Theory*, 2(2), 143–155.

J. Verplaetse et al. (eds.), *The Moral Brain*, DOI 10.1007/978-1-4020-6287-2_9,

Before jumping in with new theory and data, however, it is worth stepping back to see how existing explanations fit what we know already. Otherwise, our efforts to help may be received very much like those of a child who holds up his portable video game in a symphony concert and yells excitedly, "Everyone has to try this!" It is impossible to summarize the accomplishments of moral philosophy over even the Western tradition, but three generalizations may be useful. First, general agreement has been reached on some issues. Second, a remarkable amount of disagreement persists. Third, much of the disagreement comes from confusion about what kind of question is at issue, and much of the agreement is about the need to frame questions carefully.

The most fundamental distinction is between descriptive statements and normative statements. One set of questions is descriptive – what are moral capacities and moral behaviors like, and where do they come from? They are about what is. They are, for the most part, science. The other set of questions is normative. They are about what is right and what is wrong. They are about what we ought to do.

The importance of keeping these questions separate is a major recent (as these things go) advance in moral philosophy, attributed usually to David Hume in the 18th century (Hume, 1985 [1740]). Knowing what the world is like cannot directly tell us what is right. Framed in a more familiar form, you cannot get an Ought from an Is. Normative principles about what we should do cannot be derived from knowledge about what the world is like, not even from complete scientific knowledge about the brain and how its moral capacities were shaped by natural selection. (Subtle arguments apply, of course, but those are for another time).

The attempt to derive normative principles from facts is often called the naturalistic fallacy, but that is not exactly accurate. GE Moore, in his 1903 book *Principia Ethica* (Moore, 1903), defined the naturalistic fallacy as attempts to prove a moral claim based on a definition of "good" based on natural properties such as what is desirable or pleasant. Moore's point is that "good" cannot be defined in such terms, because it is not an object in the natural world. This is obviously closely related to the Is-Ought barrier, so, like others, I will use the phrase "naturalistic fallacy" in the more general sense.

While nearly every article on evolution and morals notes the importance of avoiding the naturalistic fallacy, many then go ahead and nonetheless draw moral guidance from observations about the world. Or, if they don't, readers do, often with blissful naïveté. To observe the phenomenon for yourself, explain the naturalistic fallacy to undergraduate students, then explain how natural selection has shaped male mammals to compete to get as many matings as possible, and open the discussion. The fact of increased competition among males leads some students to conclude that such behavior is natural and therefore right. Others will disagree, and the resulting animated exchanges demonstrate how people draw normative implications from the most abstract principles in biology, all warnings notwithstanding.

The naturalistic fallacy is, for humans, remarkably natural. For instance, when an evolutionary biologist describes forced mating as a potential adaptive strategy, he can warn against the naturalistic fallacy page after page, but few readers will even notice; most will be outraged because they think he approves of rape. His protests that he said no such thing will be ignored.

While such readers may be illogical, their reactions reflect important human tendencies. They intuitively recognize that describing a human behavior as a "normal" adaptation will lead many to conclude that the behavior is right, or at least not wrong. For instance, some young men, upon learning that natural selection has shaped organisms to maximize reproduction, change their personal sexual behaviour dramatically. I have also observed several people who changed their behavior after learning that relationships are mutually beneficial exchanges in which the maximum payoff goes those who can best deceive others. They changed not only their view of relationships, but their actual relationships; previously secure close partnerships became much more difficult. And then, of course, there is the dire history of people leaping from the fact of natural selection to justifications for eugenics and even genocide. Human views about the origins and functions of the moral capacities have tangible effects on behavior. Caution is warranted.

With this background, we can ask what neuroscience and evolution offer to understanding morality and immorality. The simultaneous focus in this book on psychopathy and normal moral capacities is particularly useful. Medicine has consistently found it difficult to explain abnormal conditions until both the normal mechanisms and their functions are clear. Conversely, studies of pathology often offer the best evidence about the functional significance of a trait. If you want to know what the thyroid gland is for, clues come from observing what happens when the thyroid gland is missing or malfunctioning. If you want to understand the benefits of moral capacities, study individuals who lack them.

Understanding the moral capacities requires two kinds of knowledge, evolutionary and proximate. They address fundamentally different questions (Dewsbury, 1999; Tinbergen, 1963). A proximate explanation is about how a mechanism works. Neuroscience offers proximate explanations at a low level. Psychology offers proximate explanations about morality at a higher level. Evolutionary explanations are different. They address why the mechanisms exist at all, in terms of selection and other evolutionary forces that account for the mechanism being the way it is. This is usually described as "the function" of a trait (although that often turns out to be too simple). Proximate and evolutionary investigations can inform each other, but they are about different questions.

Evolution

An evolutionary explanation of how moral capacities have increased fitness is the essential foundation for understanding morality. For the purposes of this book, the most important conclusion is that this foundation is still under construction. I have written much about this (Nesse, 2006), others have devoted their lives to it (Hammerstein, 2003; Katz, 2000) and this volume contains a comprehensive review (van Veelen in this volume). While much remains to be done, a rough framework is in place.

As most readers will know, naïve group selection seemed sufficient to explain altruism and morality until Williams pointed out its deficiencies (Williams, 1966), and Hamilton (Hamilton, 1964) and Trivers (Trivers, 1971) offered alternative explanations of kin selection and reciprocity, respectively. A vast amount of research since then has framed a general solution (Hammerstein, 2003). In very broad brush strokes, the vast majority of cooperative behavior in animals can be explained by kin selection or mutualisms. Well-documented examples of reciprocity in animals turn out to be rare (Stevens, Cushman, & Hauser, 2005), with the exception of our species. For humans, trading favors is at the center of life. We have emotions shaped to cope with the situations that routinely arise in reciprocity relationships (Nesse, 1990). Extraordinary social institutions enforce agreements, thus allowing vast social complexity.

Despite dozens of issues still on the table, a general explanation for moral capacities that facilitate exchange relationships is within reach. Skill in managing such relationships brings a net gain, and so selection should shape tendencies to do what works. That means paying close attention to who you are dealing with, and it usually means following rules. Explaining altruism beyond reciprocity and kinship (for instance, helping a dying friend), is more difficult. One approach is to explain such altruism away as self-deception or mistakes, another is to attribute it to group selection or cultural influences. I have previously argued that a capacity for commitment (in the economic game-theory sense of the word) can shape capacities for communal relationships that explain some aspects of altruism and our moral capacities. I still think that is important, but it leaves much unexplained.

Finally, I found articles by Mary Jane West-Eberhard (West-Eberhard, 1979; West-Eberhard, 1983) that offered a new perspective. In the late nineteen seventies she discovered that extraordinary social traits can result from the same kind of runaway selection that shape extraordinary sexually dimorphic traits, like peacock tails. The only difference is that the fitness benefits come not from being chosen as a mate, but from being chosen as a social partner. In the course of evolutionary history, once personal relationships began yielding a selective advantage, individuals who chose better partners began to gain an advantage. However, getting the best partners is not merely a matter of choosing, it depends more on being preferred as a partner. Individuals therefore display resources they can offer to their partners, and personal and moral characteristics that make them desirable partners, such as generosity and honesty. My paper, reprinted in this volume, unites this fundamental idea with modern mathematical models and findings from human social science, to argue that runaway selection based on partner choice can explain human moral and other social capacities that are otherwise inexplicable (Nesse, 2007).

One particularly interesting aspect, not developed in the article, is how social selection can explain why we value certain diverse personal characteristics as "virtue." Recent sophisticated social science methods have revisited and confirmed long-recognized virtues such as bravery, creativity, wisdom, persistence, integrity, vitality, love, kindness, social intelligence, fairness, forgiveness, humility, gratitude, hope, and humor (Peterson & Seligman, 2004). How do all of these characteristics come together to be recognized as virtues? I suspect it is because they are the characteristics we want in our partners. And, because we want to be valued as partners, they are also the traits we seek to display, and to live up to. They are products of social selection. Their unity may arise, not from a grand unifying philosophical principle, but from their origins in social selection for tendencies to choose, and to be, excellent relationship partners.

Evolution and Pathology

The framework of Darwinian medicine can be useful to analyze the evolutionary origins of presumably abnormal states, such as antisocial personality disorder (Nesse & Williams, 1994). Is it a disorder created only in modern environments, a product of infection, a tradeoff, a constraint, or is it an adaptation? The first question is whether it has had some kind of utility, but it is essential to keep all the alternative hypotheses on the table and not to jump to one conclusion.

Linda Mealey has argued that sociopathy might be a frequency dependent alternative strategy that gives a selective advantage when it is rare (Mealey, 1995). She notes a variety of supportive evidence, including the additional matings garnered by some psychopaths, however it is important to note several reasons why the hypothesis is not widely accepted.

Most alternative strategies are mating strategies, such as those used by dominant and subordinate male orangutans. Other morphs, such as benthic and limnic morphologies in fish, are alternatives for living in different ecological niches. These alternative strategies need not be associated with genetic differences. For instance, many fish change sex depending on the social environment. Such facultative adaptations, shaped by natural selection, monitor the environment and express one body type or another, depending on the situation. Locusts change from solitary to swarming morphs depending on the circumstances. The difference need not be categorical; early exposure to heat in infancy increases the number of sweat glands.

These principles may be useful for understanding antisocial personality disorder. It is hard to see the benefits of a genetically determined psychopath behavioral morph, when more flexible regulation would be more efficient. Why be stuck playing one strategy when flexibility is superior? Furthermore, substantial evidence now shows that genetic tendencies to sociopathy are not deterministic, they interact with early events to cause the disorder in some individuals. For instance, Caspi et al. found that rates of conduct disorder and criminal conviction increased dramatically with exposure to child abuse in all genotypes, but the increase was greater in those with low MAOA activity (Caspi et al., 2002). This suggests considering antisocial personality as an alternative strategy that emerges depending on early experience.

A more basic question is whether antisocial personality disorder is one condition, multiple related conditions, or positions on a continuum. Experience talking with psychopaths reveals their enormous diversity. Some use violence to get what they want. Others never use violence, but manipulate others, and get their greatest pleasure from deceiving others, not from what they get out of the deception. Others have extraordinary seduction skills that allow them to get what they want, and then abscond. Others are simply socially incompetent; they cannot manage relationships, and they flounder in all kinds of ways. And, of course, others are simply shrewd political manipulators who become powerful leaders. Despite this observed variation, a review of five studies of antisocial personality disorder, finds four that support a view of antisocial personality as a distinct category instead of a dimension (Haslam, 2003).

This brings up the question of why such diverse characteristics as low empathy, impulsivity, using violence, and inability to have close relationships should occur together so reliably. The most obvious possibility is that all aspects arise from a common proximate cause, perhaps impulsiveness, or a low ability to learn from punishment. Another possibility is that they occur together because they are all useful aspects of a strategy for social influence. This could reflect a frequency dependent genetic strategy, as Mealey proposed, but there are several alternatives. They could also occur together as parts of a facultative adaptation that emerges in response to certain early experiences. However, they could also merely reflect what people fall back on when some defect in the cognitive and emotional apparatus makes normal social relationships impossible. Etiological heterogeneity is likely (Silberg, Rutter, Tracy, Maes, & Eaves, 2007) and there is no need to posit just one kind of deficit. Some might lack empathy, others may be unable to learn from punishment, others may simply be too impulsive to be reliable relationship partners, others may simply believe that others are untrustworthy. If a person is, for whatever reason, unable to create and benefit from ordinary social relationships, he or she will fall back to simpler strategies. Early experience with using violence and deception to manipulate others soon results, by simple learning, in more and more effective psychopaths who are locked into one strategy of social influence. The coherence of the syndrome may arise not from within, but from doing what works when you are incapable of maintaining and benefitting from enduring social relationships.

Neuroscience

Several chapters in this book demonstrate that moral decision making and moral emotions arise from brain mechanisms. Of course, we have long known that this had to be true. Nonetheless, because we are all innate dualists (Bloom, 2004), the simple fact can still seem shocking.

The next task is to find out what parts of the brain carry out moral tasks. Is one locus specialized for the task? Is there a circumscribed module to take care of it? As has been the case for other capacities, from language, to pain, to emotion, moral tasks are not processed in any one location. They may even be handled in different loci by different people. Several chapters in this book take on the challenge of trying to discover where the brain processes moral information. They amply demonstrate that these tasks are not carried out everywhere, but also that they are not carried out by a specific locus; they are carried out in diverse regions that are hard to specify.

It is disappointing that we cannot point to one brain locus, and say "morality happens here." However, what we observe is exactly what an evolutionary perspective leads us to expect. A few very specific responses, such as vomiting and panic, have functions so specific and universally essential that they have been conserved for tens of millions of years; they have specific loci devoted largely to managing their expression. Equally old tasks that are not so tightly constrained, such as balance, are more distributed. New tasks, such as language, have been grafted onto existing structures in whatever way works, resulting in a hodge-podge of loci and connections.

This evolutionary view is very different from the massive modularity that is often associated with evolutionary psychology (De Schrijver in this volume). From my perspective, it is important to recognize that specific kinds of situations have posed the adaptive challenges that shaped brains with capacities for moral reasoning. The resulting mechanisms deal effectively with those situations. However, this by no means implies that selection would shape separate mechanisms to deal with each situation. On the contrary, we should expect modules with massive overlap in their arousal, processing, and output, and in the brain loci that mediate their functioning.

Moral capacities are very recent and very nonspecific. They require input of many kinds from many sources and outputs to many effectors. Selection shaped them by acting on relevant variation wherever it was available to co-opt old structures to new uses. For instance, disgust is ancient and has obvious adaptive utility – it motivates avoidance of pathogens (Curtis & Biran, 2001). Individuals with even a slight tendency to experience disgust after betrayal would avoid the betrayer. In just a few tens of thousand more years, it should not be surprising that moral violations arouse the same brain areas aroused by disgust (Greene & Haidt, 2002). Similarly, see Chapter 2 about how communal relationships and the moral emotions that sustain them seem to be have arisen from the mother infant attachment system (Moll & Oliveira-Souza in this volume). Both examples seem likely to be correct. They illustrate how evolutionary thinking can help inhibit tendencies to think of the brain as a machine with newly minted modules for each challenge, and how it can help us to accept the messy reality of functions carried out by multiple interconnected multifunctional loci.

Conclusion

Evolutionary analyses to understand the origins of moral capacities are coming along, but no one thesis is dominant at this point except for the general conclusion that natural selection has shaped capacities for coping with the situations that arise in reciprocity relationships, and additional moral capacities that make communal relationships possible.

Even at this stage, however, an evolutionary perspective can help to guide neuroscience research about antisocial personality disorder by encouraging attention to how new functions have been grafted onto structures with multiple other functions, and attention to likely constraints that make such systems vulnerable to failure. Moral capacities are evolutionarily new and almost completely restricted to humans. This means that the substantial genetic variation in moral traits may best be explained because the phenotype is in transition. A related possibility is that optimum may vary markedly across groups and times. Even in a stable setting, fitness may be about the same across a wide range of the distribution.

What trait? Social selection shapes extraordinary concern about what others think about us, and motivations to please others. The benefits of such tendencies may explain why social anxiety disorders are vastly more common than psychopathy. The costs of excess social sensitivity are, however, large. It seems entirely possible that reproductive success will be roughly the same for individuals across a wide range of the distribution. Our expectation that there is some sharp peak that defines "normal" may be incorrect. This does not sit well with our human wish to define categories and declare some normal and some abnormal. However, it may reflect a more realistic view that can help us better understand morality and immorality.

Acknowledgments Preparation of this manuscript was made possible by a Fellowship from the Berlin Institute for Advanced Study.

References

Bloom, P. (2004). *Descartes' baby: How the science of child development explains what makes us human*. New York: Basic Books.

Caspi, A., McClay, J., Moffitt, T. E., Mill, J., Martin, J., Craig, I. W., et al. (2002). Role of genotype in the cycle of violence in maltreated children. *Science, 297*(5582), 851–854.

Curtis, V., & Biran, A. (2001). Dirt, disgust, and disease. Is hygiene in our genes? *Perspectives in Biology and Medicine, 44*(1), 17–31.

Darwall, S. L., Gibbard, A., & Railton, P. A. (1997). *Moral discourse and practice: Some philosophical approaches*. New York: Oxford University Press.

Dewsbury, D. A. (1999). The proximate and the ultimate: Past, present and future. *Behavioural Process, 46*, 189–199.

Greene, J., & Haidt, J. (2002). How (and where) does moral judgment work? *Trends in Cognitive Sciences, 6*(12), 517–523.

Hamilton, W. D. (1964). The genetical evolution of social behavior i, and ii. *Journal of Theoretical Biology, 7*, 1–52.

Hammerstein, P. (2003). *Genetic and cultural evolution of cooperation*. Cambridge, MA: MIT Press in Cooperation with Dahlem University Press.

Haslam, N. (2003). The dimensional view of personality disorders: A review of the taxometric evidence. *Clinical Psychology Review, 23*(1), 75–93.

Hume, D. (1985 [1740]). *An treatise of human nature*. London: Penguin Classics.

Katz, L. (2000). *Evolutionary origins of morality : Cross disciplinary perspectives*. Devon: Imprint Academic.

Mealey, L. (1995). Sociopathy. *Behavioral and Brain Sciences, 18*(3), 523–599.

Moore, G. E. (1903). *Principia ethica*. Cambridge, UK: Cambridge University Press.

Nesse, R. M. (1990). Evolutionary explanations of emotions. *Human Nature, 1*(3), 261–289.

Nesse, R. M. (2006). Why so many people with selfish genes are pretty nice-except for their hatred of *the selfish gene*. In A. Grafen & M. Ridley (Eds.), *Richard dawkins* (pp. 203–212). London: Oxford University Press.

Nesse, R. M. (2007). Runaway social selection for displays of partner value and altruism. *Biological Theory, 2*(2), 143–155.
Nesse, R. M., & Williams, G. C. (1994). *Why we get sick: The new science of Darwinian medicine*. New York: Vintage Books.
Peterson, C., & Seligman, M. E. P. (2004). *Character strengths and virtues: A handbook and classification*. New York: Oxford University Press.
Silberg, J. L., Rutter, M., Tracy, K., Maes, H. H., & Eaves, L. (2007). Etiological heterogeneity in the development of antisocial behavior: The Virginia twin study of adolescent behavioral development and the young adult follow-up. *Psychological Medicine, 37*(8), 1193.
Stevens, J. R., Cushman, F. A., & Hauser, M. D. (2005). Evolving the psychological mechanisms for cooperation. *Annual Review of Ecology, Evolution, and Systematics, 36*(1), 499–518.
Tinbergen, N. (1963). On the aims and methods of ethology. *Zeitschrift für Tierpsychologie, 20*, 410–463.
Trivers, R. L. (1971). The evolution of reciprocal altruism. *Quarterly Review of Biology, 46*, 35–57.
West-Eberhard, M. J. (1979). Sexual selection, social competition, and evolution *Proceedings of the American Philosophical Society, 123*(4), 222–234.
West-Eberhard, M. J. (1983). Sexual selection, social competition, and speciation. *Quarterly Review of Biology, 58*(2), 155–183.
Williams, G. C. (1966). *Adaptation and natural selection: A critique of some current evolutionary thought*. Princeton, N.J.: Princeton University Press.

Runaway Social Selection for Displays of Partner Value and Altruism

Randolph M. Nesse

The discovery of evolutionary explanations for cooperation is one of the great achievements of late 20th century biology. As most readers know, benefits to the group rarely explain tendencies to help others (Williams, 1966; Dawkins, 1976), benefits to kin explain altruism in proportion to the coefficient of relatedness (Hamilton, 1964), and mutual benefits and reciprocal exchanges explain much cooperation between nonrelatives (Trivers, 1971). Subsequent theoretical and empirical studies have blossomed into a body of knowledge that can explain much social behavior (Wilson, 1975; Trivers, 1985; Dugatkin, 1997; Alcock, 2001; Hammerstein, 2003).

Controversies continue, however. Some arise from a profusion of models for cooperation that use inconsistent terminology and that tend to emphasize one explanation when several may apply (Frank, 1998; Hirshleifer, 1999; Hammerstein, 2003; Lehmann & Keller, 2006; Nowak, 2006; West, Griffin, & Gardner, 2007). Other controversies reflect impassioned debates about human nature (Midgley, 1994; Wright, 1994; Ridley, 1997; Segerstråle, 2000; de Waal, Macedo, Ober, & Wright, 2006; Dugatkin, 2006). However, some controversies persist because no explanation seems entirely satisfactory for some phenomena, especially human capacities for altruism and complex sociality.

While kin selection and variations on reciprocity explain most human capacities for cooperation, some observations don't fit the usual models. In behavioral economics laboratory experiments, and in everyday life, people tend to be more altruistic than predicted (Gintis, 2000; Fehr & Rockenbach, 2004; Brown & Brown, 2006; de Waal et al., 2006). They also tend to punish defectors even when that is costly (Henrich & Boyd, 2001; Boyd, Gintis, Bowles, & Richerson, 2003). People follow rules and they are preoccupied with morals and mores, monitoring and gossiping about even minor deviations (Axelrod, 1986; Katz, 2000; Krebs, 2000;

R.M. Nesse (✉)
The University of Michigan, Ann Arbor, MI, USA
e-mail: nesse@umich.edu

Reprinted from: Nesse, R. M. (2007). Runaway social selection for displays of partner value and altruism. *Biological Theory, 2*, 143–155.

J. Verplaetse et al. (eds.), *The Moral Brain*, DOI 10.1007/978-1-4020-6287-2_10,

de Waal et al., 2006). Perhaps most interesting of all, close friends take pains to avoid making exchanges explicit because calling attention to them harms relationships (Batson, 1991; Mills & Clark, 1994; Dunbar, 1996; Tooby & Cosmides, 1996; Nesse, 2001, 2006; Brown & Brown, 2006). Friendships exist, but they remain in want of a satisfactory evolutionary explanation (Smuts, 1985; Silk, 2003).

This article argues that well-established models of social selection may explain how partner choice could shape extreme prosocial traits in humans. It begins by reviewing early descriptions of social selection (West-Eberhard, 1979, 1983) and more recent formal models that illustrate the value of calculating the fitness components from social selection separately from those that arise from the rest of natural selection (Tanaka, 1996; Wolf et al., 1999; Frank, 1998, 2006). Next, it reviews the recent recognition of the power of partner choice (Noë & Hammerstein, 1994) and connects these insights with recent models of how covariance of partner phenotypes can lead to runaway social selection (Tanaka, 1996; Breden & Wade, 1991). These lines of work come together with recent work on human altruism to suggest that the fitness benefits of being chosen as a partner may shape extreme displays of partner value, including capacities for genuine altruism, that are otherwise difficult to explain.

Social Selection

Social selection is the subtype of natural selection in which fitness is influenced by the behavior of other individuals (West-Eberhard, 1979, 1983; Wolf et al., 1999; Frank, 2006). Although well-established in biology, the term social selection is slightly problematic because epidemiologists use the same phrase to describe the entirely different phenomenon of some social groups having a higher proportion of individuals with some condition. For instance, the proportion of people with schizophrenia is higher in inner cities simply because many cannot afford to live elsewhere. Also potentially confusing is the idiosyncratic use of social selection as an alternative to sexual selection (Roughgarden, Oishi, & Akcay, 2006), when in fact it is a subtype. These potential confusions aside, social selection is the standard term for fitness changes resulting from the social behaviors of other individuals.

Sexual selection by female choice is the best known subtype of social selection. Female biases for mating with ornamented males select for more elaborate male displays, and the advantages of having sons with extreme displays (and perhaps advantages from getting good genes) select for stronger preferences (Grafen, 1990; Kokko, Brooks, Jennions, & Morley, 2003). The resulting positive feedback makes displays and preferences more and more extreme until genetic variation is exhausted, or until the fitness increase from more matings equals the fitness decrease from lowered competitive ability and earlier mortality (Andersson, 1994; Kokko, Jennions, & Brooks, 2006). Sexual selection is social selection because individual fitness is influenced by the choices and behaviors of other individuals. West-Eberhard made the point succinctly it in one of the first papers on the topic:

> Sexual selection refers to the subset of social competition in which the resource at stake is mates. And social selection is differential reproductive success (ultimately, differential gene replication) due to differential success in social competition, whatever the resource at stake." (1979, p. 158).

Social selection arising from conspecific choices and behaviors has been described in detail (Crook, 1972; West-Eberhard, 1975, 1983; Tanaka, 1996; Wolf et al., 1999; Frank, 2006). Suprisingly, however, its full power is just now being recognized. The perspective of social selection shifts attention away from individual strategies in iterated exchanges, and towards the prior and larger fitness challenges of identifying the best available partners and doing whatever will get them to choose one as a partner. In formal models, this means partitioning the force of social selection resulting from the covariance of partner's phenotypes separately from other forces of natural selection. Following Queller (1992) and Frank (1997), Wolf, et al. (1999) describe social selection by saying "factors other than one's own phenotype may affect an individual's fitness. . .individual variation in fitness can be attributed to variation in the value of traits expressed by an individual's social partners." (1999, pp. 255–256).

Building from Lande and Arnold's model of sexual selection (1983), Wolf, et al. (1999, p. 256) partition relative fitness ω into one component from social selection and a separate component from the rest of natural selection: $\omega = \alpha + \beta_N z_i + \beta_S z'_j + \varepsilon$ where β_N is the natural selection gradient, β_S is the social selection gradient, z_i is the trait in the individual and z'_j is a covarying trait in the partner (the prime sign indicates that the trait is in the partner, α is fitness uncorrelated with the traits, and ϵ is error). They then derive a generalized phenotypic version of Hamilton's rule to show that selection favors an altruistic trait z_{ij} whenever $\frac{C^{ij}}{P_{ii}}\beta_S + \beta_N > 0$, where C^{ij} is the phenotypic covariance between the trait in the individual and the partner, and P_{ii} is the character's variance. Here, β_N is the selection cost for an altruistic trait (and will therefore be <0), and β_S is the benefit to the partner (which will be >0), so the altruistic trait will be selected only if its covariance with the associated trait is large compared to the trait's variance. The model, very similar to Frank's (1997, 1998), and also drawing on Fisher and Price, is based on phenotypes and does not require covariance of genes within individuals. Partner choice creates phenotypic covariance that can shape extreme traits such as displays of one's value as a partner. How far will social selection push such traits at the expense of other components of natural selection? An answer to this important question requires detailed analysis of social selection by partner choice.

While all social behavioral tendencies can be interpreted as products of social selection because they involve choice by other individuals (Wolf et al., 1999; Frank, 2006), the emphasis here is on forces of selection that arise from choices about relationship partners and group membership. If potential partners or group members vary in resources and tendencies to reliably bestow them on close partners, then a preference for resource-rich selectively-altruistic partners will give a selective advantage. Being preferred as a partner gives fitness advantages because it gives more possible partners to choose from (Noë & Hammerstein, 1994). This will select for displays of resources and selective altruism that reflect an individual's potential value as a partner.

The nonrandom association of individuals with extreme displays and those with strong preferences can result in runaway social selection that increases both traits to extremes that decrease other fitness parameters (Breden & Wade, 1991; Tanaka, 1996; Wolf et al., 1999). This model differs from sexual selection because in most cases preferences and displays will both be present in the same individuals. Also, benefits to others pay off not only directly, but also because benefits to partners eventually result in benefits to the self via interdependence (Rothstein, 1980; Humphrey, 1997; Brown & Brown, 2006). At equilibrium, many individuals will be presenting and assessing expensive displays in a competition that results in partnerships between individuals of similar partner value.

In sexual selection, runaway occurs only when the covariance of the trait and the display is greater than the viability decrease from the display. At equilibrium, further increases in female preference would lower fitness because of decreased viability of sons (Kokko et al., 2006). However, "even small changes in female behavior (which cost little) can generate strong selection when a male's fitness depends primarily on his mating success." (Kokko et al., 2006, p. 59). In selection for social partners, the cost of choosing partners with extremely high value has little or no disadvantage comparable to the disadvantage experienced by females who chooses mates with the most extreme displays. Displays of partner value will, therefore, continue under directional selection until their marginal benefits impose equal costs to other fitness components, such as ability to accumulate material resources. Thus, social selection for partners can, like sexual selection, explain extremely costly traits.

In a model of social selection that emphasizes signaling submission and real fighting ability, Tanaka (1996) addresses the possibility of runaway social selection more directly. As in the Wolf, et al. model, fitness is partitioned into components from social selection that are distinct from the rest of natural selection in order to assess where the equilibrium for a signal lies. That point often is reached, he concludes, by runaway selection that quickly arrives at the equilibrium where the marginal benefits of further increasing the signal are balanced by its direct costs. Crespi (2004) has argued that such positive feedback cycles are much more common in nature than is usually recognized. Deception and cheating have been major themes in reciprocity research, and they apply in social selection models, but their effects are limited by inexpensive gossip about reputations and by the difficulty of faking expensive resource displays (Tanaka, 1996).

Social Selection in Nature

If the above models are correct, then examples of non-sexual social selection should be observed in the natural world. Some examples of traits shaped by preferences in one species for displays in another species illustrate runaway selection without genetic covariation in the same genome. As Darwin noted (1871), flowers have elaborate and diverse forms because they compete to satisfy pollinator preferences. Flowers preferred by pollinators contribute more genes to future generations, so floral displays become increasingly extravagant until the marginal benefits from attracting more pollinators are matched by costs to other aspects of fitness, such as investment in leaves and roots (Armbruster, Antonsen, & Pélabonb, 2005). Benefits can also come from not being chosen. Staying near the center of a selfish herd is shaped by predator preferences. Stotting protects gazelles because it is an honest signal of vigor that discourages predators from useless chases.

Signals between members of the same species are shaped by the same mechanisms (Grafen, 1984; Bradbury & Vehrencamp, 1998). Social coordination signals are ubiquitous. For instance, a bird on a nest makes distinctive movements to signal to its partner that it is ready to trade roles. The signal benefits both parties, so there is no selection for an extreme signal. In competitive situations, amplified signals are common (Tanaka, 1996). When a wolf bares its throat to signal yielding in a fight, both parties benefit by avoiding the danger of an escalated fight; a prominent submission display that creates real vulnerability pays off by avoiding useless fighting. Status displays in lieu a fight are likely to be extreme because only expensive honest signals will influence the competitor. Note that such signaling behaviors give benefits only because they interact with the phenotypes of other individuals who have been primed by selection to be influenced.

Some examples, such as males competing for a territory, blur the boundary between sexual and other social selection. Others arise more clearly from nonsexual social selection, such as the huge brightly colored beaks of both male and female toucans. They do not result from sexual selection; non-social toucan species have less exaggerated and more sexually dimorphic beaks. They are more likely honest signals of ability to defend a nesting territory (West-Eberhard, 1983). Bright coloration in both sexes is also prominent in territorial lizards and some mammals, especially lemurs. Social selection has also been proposed as the explanation for bright coloration of reef fish. West-Eberhard offers a wealth of examples, and reasons why species recognition hypotheses are insufficient (1983). She also notes that Wynne-Edwards (1962) provides additional examples, even if he was wrong about how selection shaped them. This is especially important because it highlights the power of social selection to account for phenomena that might otherwise appear to be products of group selection. While the sources of female ornamentation remain an active research focus, a recent review endorses the importance of social selection:

> Almost 20 years ago, West-Eberhard argued that monomorphic showy plumage was associated with aggressive social displays (over territories or other resources) by both sexes. Her argument was supported by examples from several taxa including toucans, parrots and humming birds. West-Eberhard's suggestions resulted in surprisingly little empirical research in the following years. However, among published studies, most seem to support her view (Amundsen, 2000, p. 151).

Domestication

Domestication illustrates how social preferences can shape profoundly prosocial traits. It requires no conscious breeding, only preferences that influence fitness among individuals from the other species who vary on traits that matter to humans

(Price, 1984; Diamond, 2002). For instance, wolves with less fear of humans and lower levels of aggression were able to, and allowed to, stay closer to ancestral human camps where the fitness value of food scraps was a domesticating selection force. In turn, those humans who had tendencies to be altruistic towards dog-progenitors received fitness benefits – initially warnings of danger, but later, help in the hunt and protection. This process selected for genes that increase human altruism towards dogs, and it shaped dogs who behave in ways that please humans enormously.

Humans also show many characteristics of being domesticated – low rates of aggression, increased cooperation, eagerness to please others, and even changes in bone structure similar to those characteristic of domesticated animals (Leach, 2003). It seems plausible that humans have been domesticated by the preferences and choices of other humans. Individuals who please others get resources and help that increase fitness. Aggressive or selfish individuals get no such benefits and are at risk of exclusion from the group, with dire effects on fitness. The result is thoroughly domesticated humans, some of whom can be enormously pleasing.

This process does not depend on the success of the group. Instead, individuals constantly make small self-interested social choices that shape the behaviors of others who learn to do whatever works. The resulting effects on fitness shape the species by social selection. This process offers a dramatic example of a Baldwin effect, in which learning shapes adaptive behavior patterns that create new selection forces that rapidly facilitate better ability to exploit the new niche (Dennett, 1995; Laland, Odling-Smee, & Feldman, 2000; Weber & Depew, 2003; West-Eberhard, 2003; Ananth, 2005). Once the benefits of relationships increased above a crucial threshold, they created a newly complex social environment where individuals with special social skills got increasing fitness advantages shaped more extreme cognitive and prosocial traits (Humphrey, 1976; Byrne & Whiten, 1988; Alexander, 2005).

Herbert Simon, in a 1990 article on "social selection and successful altruism," described how selection for "docility" could give rise to behaviors that benefit others more than the self. Simon defined docility as, "persons who are adept at social learning who accept well the instruction society provides them" (p. 1666). His model is based on the fitness benefits of general social learning, and the assumption that "limits on rationality in the face of environmental complexity" result in individuals behaving altruistically for the good of society without recognizing the "tax" they are paying. In contrast, the model developed in this article views altruism as a result of the fitness benefits of social selection, not as a result of cognitive constraints.

Social Selection for Cooperation

Indirect benefits to kin are one powerful force that shapes conspecific cooperation. Ability to recognize kin, and preferences for helping them, give benefits to genes in kin that are identical by descent to those in the helper (Hamilton, 1964; Dugatkin, 1997; Frank, 1998; Queller & Strassmann, 1998; West, Pen, & Griffin, 2002). This process has been described and studied so extensively that there is no need to repeat the details here. One subtype, "green-beard effects" has been controversial, but it now appears that selection does sometimes shape kinship cues that facilitate kin altruism (Queller, Ponte, Bozzaro, & Strassmann, 2003). Phenotype variability can also be shaped by social interactions involved in reproductive competitions, at least in wasps (Tibbetts, 2004).

Preferences for helping nonrelatives who will help in return are also obviously valuable (Trivers, 1971). The challenge is how to get the benefits of trading favors without being exploited (Krebs & Dawkins, 1984; Alexander, 1987; Cosmides, 1989; Fehr & Fischbacher, 2003). Following Price (Frank, 1997) and Queller (1992), Frank (1997, 2006) points out that such cooperation can be modeled as correlated behaviors, an information problem equivalent to that of kin selection. In kin selection, a behavior increases inclusive fitness if its cost to the self is less than the benefit to the other times the coefficient of relatedness, r. In correlated behaviors, the cost is the direct effects of the behavior on the individual's fitness, the benefit is the indirect benefit from others (holding constant individual behavior), and r reflects the similarity of other's behavior, that is, the information an individual has about benefits others will likely offer. Both kin selection and correlated behavior can thus be analyzed by partitioning fitness into direct costs, indirect benefits, and a scaling factor that reflects relatedness in the former case, and information about other's anticipated behavior in the latter (Frank, 1997; Wolf et al., 1999).

The iterated prisoner's dilemma has long been the dominant model for cooperation based on reciprocity (Axelrod, 1984; Sigmund, 1993; Axelrod, 1997). In this model, the maximum joint benefit for two players comes from repeated cooperation, but an individual can get a greater payoff from defecting on any move when the other cooperates. Tit-For-Tat (starting with a cooperative act and then doing what the other person did on the previous move) is a remarkably robust strategy that nicely models some human interactions. The tractability of models based on the prisoner's dilemma has fostered scores of valuable studies (Axelrod, 1997).

It is less clear, however, that prisoner's dilemma models accurately reflect the kinds of trait variation on which selection acted to create capacities for social cognition. In most studies, anonymous agents are randomly paired, information is only about prior behavior with one agent or the sum of all agent's behavior, the same algorithm is used for interactions with all other players, and only two outcomes are possible, cooperate or defect. Reputation and punishment have increasingly been added to such models (Fehr & Fischbacher, 2003; Axelrod, Hammond, & Grafen, 2004; Henrich et al., 2006). However, few reciprocity models have all of the ingredients that are important to human cooperation in close relationships: reputation, communication, agreements, promises, threats, third party enforcement, and especially, opportunities to use extensive information to choose partners from a selection of possibilities (Kitcher, 1993; Hammerstein, 2001; Nesse, 2001; Noë, 2001). While variations in tendencies to cooperate or defect in discrete interactions with rapidly shifting partners certainly create selection forces, they explain only some aspects of some human relationships (Fehr & Henrich, 2003; Barclay & Willer, 2007. Nonetheless, such models have been a boon for the study of cooperation.

Another difficulty is that the kinds of reciprocal exchange modeled by the prisoner's dilemma seem to be rare in nature. Most apparent examples of reciprocity identified by field research now appear to be better explained by kinship or mutual benefits (Connor, 1995; Stevens, Cushman, & Hauser, 2005). Cooperative hunting is a prime example. Participants all gain, so defection does not pay. Impala grooming is a reciprocal exchange, but of the most minimal kind. Grooming bouts are traded back and forth in parcels so small that the example blurs the border between reciprocity and mutualism (Connor, 1995), although grooming may be tradable for other resources (Manson, Navarrete, Silk, & Perry, 2004). Another example, previously thought to exemplify reciprocal exchange between nonrelatives, is vampire bats sharing blood with others who did not succeed in that night's hunt (Wilkinson, 1984). However, it turns out the sharing almost always is between kin. Coalitions of male baboons were also thought to demonstrate reciprocity, but on reexamination, the males do not share mating opportunities to any great extent. A review by Stevens, et al. (2005) assesses the evidence for reciprocity in nature and concludes that there are few examples, perhaps, they say, because most animals have severe capacity constraints for memory and cognition.

Where reciprocal helping does exist, it is usually maintained by systems for assessing potential partners or withdrawing resources from defectors (Sachs, Mueller, Wilcox, & Bull, 2004). Parceling, as in reciprocal grooming, distributes resources in small packets so defection is not an issue (Connor, 1995). Another strategy is to distribute resources selectively depending on the behavior of others. For instance, yucca plants abandon flowers with too many moth larvae. This can be viewed as a punishment that selects for moths who limit egg deposition. However, abandoning the flowers with too many larvae is in the direct self-interest of the yucca plant, and this makes it advantageous for moths to limit the number of eggs laid in any one flower.

Image scoring (Nowak & Sigmund, 1998; Wedekind & Milinski, 2000) and other reputation-based strategies such as indirect reciprocity (Alexander, 1987), offer information about an individual's reliability as a partner and can lead to mutually profitable exchanges even in the absence of repeated interactions (Riolo, Cohen, & Axelrod, 2001). This is not the place to analyze the diversity of cooperation models, but it is important to recognize that delayed reciprocal exchange of resources is as rare in other animals as it is ubiquitous in humans. Furthermore, human cultures vary substantially in their levels of individual cooperation, with much of the variance attributable to variations in the patterns of economic exchange (Henrich et al., 2005), further demonstrating that human cooperation strategies are marshaled to suit the circumstances.

The role of partner choice in facilitating cooperation has long been recognized (Bull & Rice, 1991), but has been emphasized only recently (Noë & Hammerstein, 1994; Noë, 2001; Sachs et al., 2004). When there is choice, potential partners must compete in markets that change the dynamics of cooperation. Between-species partner choice is illustrated by symbioses in which the slower-evolving organism selects among individuals in a faster evolving species to get the most valuable partners, for instance, the plant symbioses with bacteria and fungi (Simms & Taylor, 2002;

Kummel & Salant, 2006). Choice of conspecific partners may be far more powerful (Roberts, 1998; Noë & Hammerstein, 1994).

Social Selection for Prosocial Traits in Humans

The possibility that social selection shaped human capacities for altruism and complex sociality was suggested in West-Eberhard's seminal publication on the topic (1979, p. 228):

> It is tempting to speculate that the explosive evolutionary increase in the proto hominid brain size, which had the appearance of a "runaway" process, was associated with the advantage of intelligence in the maneuvering and plasticity associated with social competition in primates.

The complexity of the social environment is widely recognized as a selection force likely to be important for explaining human social abilities (Humphrey, 1976; Alexander & Borgia, 1978; Alexander, 1979; Byrne & Whiten, 1988). The full implications for human prosocial traits have yet to be developed, although one wide-ranging treatment suggests that social selection may have enormous scope for explaining human capacities for art and literature, as well as capacities for intelligence and cooperation (Alexander, 2005).

A closely related model for the evolution of human altruism is based on sexual selection. Geoffrey Miller (2000, 2007) has suggested that sexual selection may account for many extreme human cognitive and behavioral traits that are otherwise difficult to explain, especially altruism. He cites evidence that both women and men prefer to marry kind reliable partners, giving a fitness advantage via sexual selection to individuals of both sexes with these heritable personality traits. Sexual selection could thus shape extreme altruism. This potentially important hypothesis has not been emphasized in recent literature, perhaps because it is difficult to study. Miller (2000) acknowledges that other forms of social selection may be important, but mostly, he says, "because they change the social scenery behind sexual selection."

Mate choices create potent selection forces, but so do choices of relationship partners. The fitness benefits from choosing social partners are more distant from direct reproduction, but they can influence fitness nearly every day and at all ages. If partnerships yield a net gain for both parties, then fitness increases with the increase in the number of others who want you as a partner, at least for the first few partners. If partners vary in value, then fitness will be increased by behaving in ways that increase the number of others who want you as a partner (Noë & Hammerstein, 1994). A good way to increase the number of available number of partners is to advertise, and to usually provide, more benefits than others can or will provide (Roberts, 1998; Barclay & Willer, 2007; Hardy & Van Vugt, 2006).

Such "competitive altruism" has been the topic of several descriptions and studies (Roberts, 1998; Barclay & Willer, 2007; Hardy & Van Vugt, 2006). The latter two studies are especially germane because they model and provide data that demonstrate competitive altruism in humans. Competitive altruism gives an advantage when extreme generosity results in disproportionate payoffs from pairing with the best partners. In Barclay and Willer's study using a prisoner's-dilemma-like task, generosity levels increased dramatically when participants knew their behaviors were observable and could be used by others choosing partners. The effect was robust even though the experiment was anonymous. Hardy and Van Vugt also demonstrated increased altruism when behavior is observed, and they found that the most altruistic individuals gained the highest status and were preferred as partners, thus gaining benefits.

The resulting positive feedback process can shape costly displays, and preferences for partners who present such displays. Displays of resources, talent and other indicators of partner value are prominent aspects of human cultures (Barkow, 1989; Dunbar, Knight, & Power, 1999; Miller, 2000; Schaller & Crandall, 2003; Alexander, 2005). Conspicuous consumption, from potlatches to Rolexes, has been interpreted as wasteful status displays (Veblen, 1899), but such displays not only entice mates, they also advertise an individual's desirability as a relationship partner or a group member. Competitions in such displays reward only the most extreme and remarkable performances and creations (Veblen, 1899; Frank, 1999; Alexander, 2005).

People advertise their reputations as much as their resources, and displays of moral character are an equally impressive aspect of human cultures (Katz, 2000). Reputation display competitions may be important for explaining human moral capacities and altruistic behaviors that are not reliably reciprocated. Recent models suggest that altruism itself may be an honest advertisement based on the handicap principle (Gintis, Smith, & Bowles, 2001; Pilot, 2005; Hardy & Van Vugt, 2006; Barclay & Willer, 2007).

Strong reciprocity is closely related (Gintis et al., 2001). As Fehr and Henrich put it (2003, p. 57), "The essential feature of strong reciprocity is a willingness to sacrifice resources in both rewarding fair behavior and punishing unfair behavior, *even if this is costly and provides neither present nor future economic rewards for the reciprocator.*" They argue that this apparently "excess" altruism is not a mistake, but an adaptation that arises because even small amounts of conformist transmission give advantages to cooperate-punish strategies that result in their spread in cultural groups. The previous argument about the Baldwin effect and emergent forces of selection in groups is similar, but focuses more attention on behaviors at the level of the individual. Prior work on the evolution of capacities for commitment (Nesse, 2001) is also related, although commitment strategies rely more on intensive communication of threats and promises, and ways to make them believable even when fulfilling the commitment would not be in the actor's interests. As noted already, research on cooperation is vulnerable to confusion because it probably is shaped by multiple selection pressures that are hard to disentangle.

The assortment that brings cooperators together need not be based on recognition or identity tags; simple environmental or partner preferences are sufficient (Pepper & Smuts, 2002; Pepper, 2007). Any mechanism that associates cooperators gives advantages to those with prosocial traits (Wolf et al., 1999; Frank, 2006). The results of such selective association of cooperators can be framed as trait-group selection (Wilson & Sober, 1994), but such models are very different from old group selection, so to prevent confusion "an alternative is to state as simply as possible what they are – models of nonrandom assortment of altruistic genes" (West et al., 2007, p. 11).

The opportunity to choose from a variety of partners, and the possibility of negotiating contracts and prices, suggests applying market models to the problem of cooperation (Noë & Hammerstein, 1995; Hammerstein, 2001; Noë, Hooff, & Hammerstein, 2001). Consumers and producers, whether humans, other animals, plants or fungi, select among available partners based upon their utility, availability, and price. Replacement of cheaters with more profitable partners exerts a powerful selection force for transaction quality and the ability to conceal and detect defection (Frank, 1988, Trivers, 2000). This shapes market efficiency and integrity, even to the apparently maladaptive extreme of guarantees that "the customer is always right." Such guarantees are exploitable and costly, but competition for customers keeps them prevalent.

The argument that social selection shapes extreme traits for winning competitions for relationship partners can be readily expanded to encompass parallel processes at the group level. Individuals in groups assess the qualities of potential future members and admit those who offer the most while demanding the least. Conversely, prospective new members assess which group offers them the most at the least cost. The result is a sorting of individuals by their abilities to contribute resources, creating groups readily ranked in quality. However, because being a big fish in a small pond can payoff better than being a small fish in a big pond, the partner value of members will overlap between groups (Frank, 1985).

Skew theory (Reeve & Shen, 2006) may clarify the dynamics of individuals competing for resources other than access to reproduction in social groups. Individuals in groups should value new members proportional to their effects on group members' ability to get resources. Potential members display both their resources and their willingness to share them. After an individual joins a group, the dynamics shift to those based on the costs and benefits of allowing a member to stay, and competition for allies and position within a group. Social selection from competitions to join the best groups may be more powerful than competition to be chosen as an individual partner, but the complexities make it wise to focus here on simpler partnerships.

It is important to note that the behaviors of individuals groups can create emergent forces of natural selection in groups that shape otherwise inexplicable traits such as genuine altruism, group loyalty and boundaries that define the in-group and devalue out-groups (Alexander & Borgia, 1978; Boyd & Richerson, 1985). Such forces may emerge reliably from individuals and partnerships pursing their own interests. While such emergent selection forces would not exist without the group, they are very different from group selection in that they do not depend on the success of the group.

Models

Most models partition fitness effects into social selection and natural selection components, and describe how covariance between traits in associated partners can account for the strength of social selection (West-Eberhard, 1975; West-Eberhard,

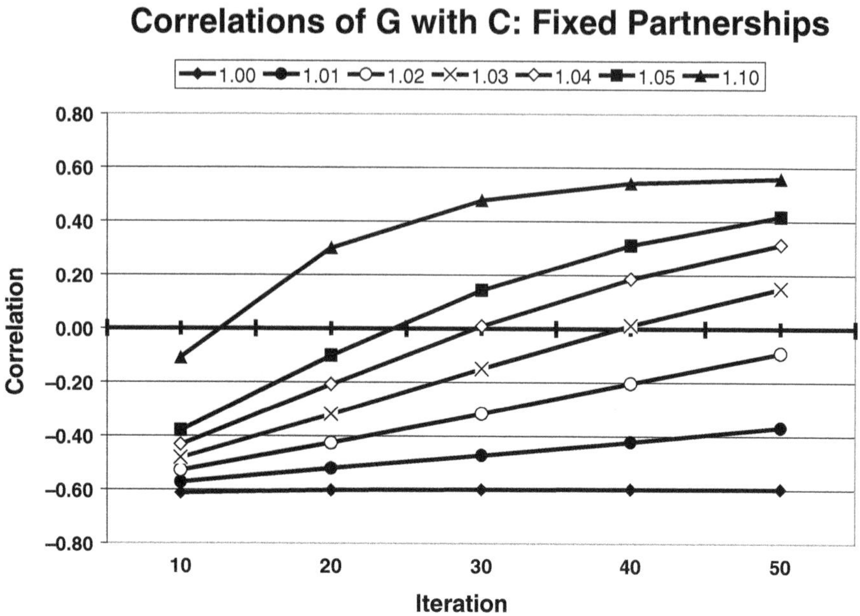

Fig. 1 Correlation of G with C for fixed partnerships over 50 iterations for six levels of R

1983; Tanaka, 1996; Wolf et al., 1999; Frank, 2006). It is difficult, however, for such models to describe the dynamic process of choosing repeatedly among many possible partners as a function of behaviors that change over time.

An agent-based shared-investment model may help to illustrate some of these processes. A simple initial model assigns each agent a randomly distributed generosity parameter, G, that ranges from 0.0 to 1.0. Each of 100 agents is endowed with capital, C=100. In each iteration, pairs of agents invest a percentage of their total resources, (G*C) and (G*C') respectively (the prime mark indicates the partner's parameters). Both partners receive a payoff equal to half of their total joint investment times R, the rate of return: Payoff = R*((G*C)+(G'*C'))/2.

If this model is run without sorting, agents remain in fixed pairs. The agent with a higher G does worse because it invests more than the partner on each move, but they share the payoffs equally. Despite the higher payoffs for the less generous agent in each pair, when all 100 agents are considered, more generous agents on average have superior payoffs as reflected in increasing correlations of G with C with each iteration. The correlation of G with C increases with each iteration. How fast it becomes positive depends on R. As shown in Fig. 1, when R=1.03, correlations become positive by iteration 40. For R=1.05 the correlation becomes positive at iteration 25, but reaches only 0.40 at iteration 50. When R=1.10, the correlation becomes positive at iteration 12 and approaches an asymptote of 0.60.

Model 2 is the same except that at each iteration the agents are sorted according to GxC, the total investment made on the previous move. This increasingly pairs

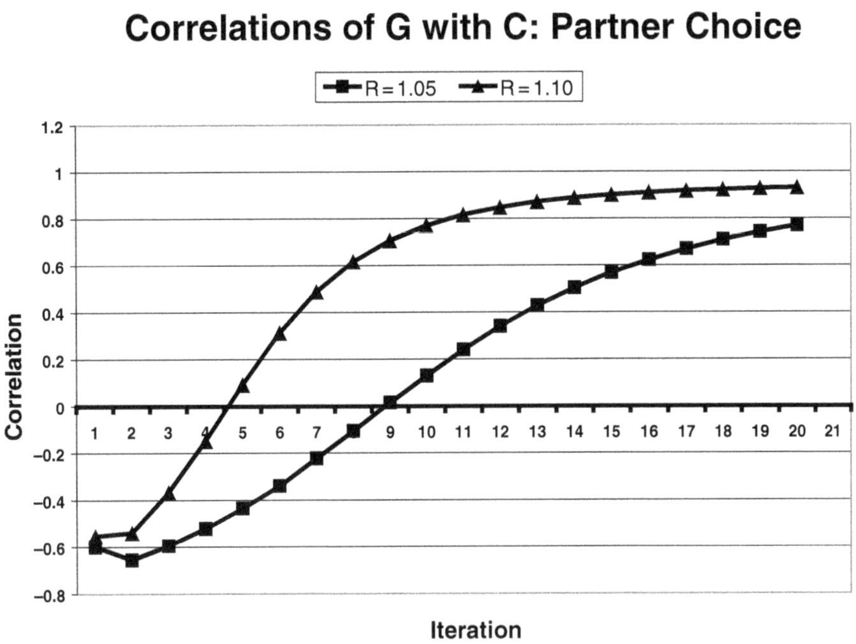

Fig. 2 Correlation of G with C given partner choice over 50 iterations for two levels of R

more generous agents as if each one were watching all others and pairing with the available partner who offers the best combination of resources and generosity. More generous agents still accumulate capital more slowly than their less generous partners, but the sorting process greatly increases the maximum correlation and how quickly it becomes positive. As illustrated in Fig. 2, G×C becomes positive at the 9th iteration if R=1.05, and at the 5th iteration if R=1.10. Both continue on to correlations much higher than in the model without partner choice.

These simple models illustrate how partner choice can shape increased generosity. The model could easily be elaborated by allowing reproduction as a function of capital accumulation, or by using a genetic algorithm to see what parameters are optimal and whether different subtypes of agents find evolutionarily stable alternative strategies. Such models could also use random normal distributions of R in order to study the influence of stochastic payoffs. It will be interesting to discover the optimal levels of generosity across different levels of other parameters and whether populations of agents go to a stable equilibrium or if they cycle. Future models also need to incorporate the possibility of deception, although continuing choice among known potential partners makes deception less important than in most reciprocity models. Social selection models lend themselves to investigations of how hierarchy influences cooperation.

The Invisible Hand

Adam Smith was preoccupied with finding explanations for sympathy (Smith, 1976 [1759]), and his followers argue that he would be dissatisfied with current evolutionary theories of altruism (Kahil, 2004). In his book on the moral passions, Smith mentioned the invisible hand only once, and this was with respect to the division of resources. The idea of the invisible hand seems equally germane, however, to the origins of moral emotions. Individuals pursue their interests by trying to attract the best possible partners. To succeed, they must offer to fulfill the wishes and expectations of potential partners at the lowest possible price. This usually requires carrying out many expensive actions that help and please others. Self-interested partner choice may be the invisible hand that shaped human capacities for sympathy.

Social exchange with partner choice gives rise to emergent forces of natural selection that can shape social traits far more sophisticated than generic sympathy. These forces should give fitness advantages to those who pay close attention to what others want, something very much like theory of mind. They could also shape empathic concern for the welfare of partners and strong motives to make reparations, not only for actual defections, but for even hints of possible lack of attention to the other's needs (Wu & Axelrod, 1995). And, they can shape love, spite, contempt and the whole range of social emotions (Nesse, 1990). Most globally, these social market forces shape desires to please others in general, and desires to avoid any cause of displeasure. Indeed, powerful internal mechanisms reward us for helping others (Brown, Nesse, Vinokur, & Smith, 2003), and cause guilt and shame when we cause others pain or disappointment (Gibbard, 1990).

Caveats and Conclusions

Several caveats and limitations should be kept in mind. First, as already noted, multiple mechanisms of selection shape capacities for cooperation. While this article emphasizes the effects of runaway social selection resulting from social partner choice, several other forces are involved, including sexual selection, the benefits of mutualisms, and plain reciprocity.

Second, and closely related, the fitness benefits of social selection are intimately involved with reciprocity and kin selection. In one sense this is not an issue. Other different perspectives, such as reciprocity and kin selection, can be modeled in a common framework. The social selection perspective is distinctive, however, because it shifts the focus of attention away from decisions to cooperate or defect and abilities to detect cheating, and towards the quite different tasks of selecting carefully among a variety of potential partners, trying to discern what they want, and trying to provide it, so one is more likely to be chosen and kept as a partner.

Reciprocity and social selection models of cooperation differ not only because they partition fitness effects differently, but also because social selection gives rise to runaway processes that can account for traits that decrease survival or competitiveness, such as extreme altruism. For the same reason, the benefits of socially selected traits may come at the cost of increased vulnerability to serious mental disorders. For instance, rapid selection for complex social capacities may have pushed some traits close to a fitness "cliff-edge" beyond which lies catastrophic cognitive failure of the sort seen in schizophrenia (Nesse, 2004).

Social selection calls attention to the locus of selection's action: heritable variations in social traits that influence abilities to get and maintain relationships with preferred social partners. Empathy, self-esteem, guilt, anger and tendencies to display moral traits and to judge others may be shaped directly by social selection. Instead of describing a stable equilibrium, social selection focuses attention on the dynamic process that shapes social traits.

Third, human nature is not unitary. Some people are profoundly prosocial, others lack all sympathy. Do individuals who lack sympathy have a genetic defect? Or, did they miss some early experience necessary to development of the capacity? Or, is sympathy a facultative trait expressed only in certain social circumstances? Or is selection for such capacities so recent that gene frequencies are changing rapidly? Or are they maintained in some frequency dependent equilibrium? (Mealey, 1995). These are important questions, as yet unanswered. While finding the mean values and distributions for any trait in any species is valuable, attempts to essentialize human nature are at odds with both observation of human variation and an evolutionary view of how human nature came to be.

Forces of social selection may also vary significantly between different groups (Henrich et al., 2005). Even within one society, different subgroups show different social patterns. It also seems possible that the benefits of partner choice may be much larger in some settings compared to others. For instance, if most economic activity requires little cooperation and no trading, then attending closely to other's needs will be of little value as compared to a situation in which competitive presentations of self influence fitness strongly. High rates of narcissism may be a reliable product of certain social and economic structures (Lasch, 1979).

A related concern is whether the opportunities for partner choice have influenced fitness long enough to create forces of social selection sufficient to shape complex social traits. To find out will require anthropological data interpreted in this framework. The possibility that capacities for profound sociality arose from culture without influences from natural selection seems unlikely. Humans clearly have social capacities that are qualitatively different from other animals (Kitcher, 1993; Dunbar, 1998; Tomasello, 1999).

Finally, words hide all manner of imprecision that is revealed only by transforming them into mathematical statements. The mathematical models in this paper are rudimentary. Among other factors that need exploration are deception, different parameters for payoffs and noise, and the possibility that viscosity or other grouping mechanisms may maintain different equilibria.

No definitive experiment is likely to prove the role of social selection in shaping human capacities for cooperation, and, for the reasons just noted, cross-species comparisons will not be very useful. Nonetheless, just as a reciprocity models suggested looking for specialized cheater detection capacities, social selection models suggest looking for specialized capacities for determining what others want, for monitoring whether one is pleasing them, and for presenting a social self that will make one desirable as a social partner. Of course, we already know quite a lot about theory of mind and the evolution of self-esteem (Leary & Baumeister, 2000), so to demonstrate that they were shaped by social selection will require predicting unnoticed aspects and looking to see if they are there.

In sum, partner choice can create runaway forces of social selection that may have shaped human prosocial tendencies and capacities for advanced social cognition that are otherwise difficult to explain. Whether this turns out to be correct awaits additional modeling, experiments, field studies, and further syntheses with the principles of microeconomics.

Acknowledgments Thanks for very helpful comments from two anonymous reviewers, and to members of my laboratory group, and to colleagues who offered valuable advice along the way including Robert Axelrod, Lee Dugatkin, Steve Frank, Kern Reeve, Bobbi Low, Richard Nisbett, Peter Railton, Mary Rigdon, Stephen Salant, Stephen Stearns, Barbara Smuts, Mary Jane West-Eberhard, and to Lucy for inspiration.

References

Alcock, J. (2001). *The triumph of sociobiology*. New York: Oxford University Press.

Alexander, R. D. (1979). *Darwinism and human affairs*. Seattle, WA: University of Washington Press.

Alexander, R. D. (1987). *The biology of moral systems*. New York: Aldine de Gruyter.

Alexander, R. D. (2005) Evolutionary selection and the nature of humanity. In V. Hösle & C. Illies (Eds.), *Darwinism and philosophy* (pp. 301–348). Notre Dame: University of Notre Dame Press.

Alexander, R. D., & Borgia, G. (1978). Group selection, altruism, and the levels of organization of life. *Annual Review of Ecology and Systematics, 9*, 449–474.

Amundsen, T. (2000). Why are female birds ornamented? *Trends in Ecology and Evolution, 15*(4), 149–155.

Ananth, M. (2005). Psychological altruism vs. biological altruism: Narrowing the gap with the Baldwin effect. *Acta Biotheoretica, 53*(3), 217–239.

Andersson, M. (1994). *Sexual selection*. Princeton, NJ: Princeton University Press.

Armbruster, W. S., Antonsen, L, & Pélabonb, C. (2005). Phenotypic selection on *Dalechampia* blossoms: Honest signaling affects pollination success. *Ecology, 86*(12), 3323–3333.

Axelrod, R. (1984). *The evolution of cooperation*. New York: Basic Books, Inc.

Axelrod, R. (1986). An evolutionary approach to norms. *American Political Science Review, 80*, 1095–1111.

Axelrod, R., Hammond, R. A., & Grafen, A. (2004). Altruism via kin-selection strategies that rely on arbitrary tags with which they coevolve. *Evolution, 58*(8), 1833–1838.

Axelrod, R. M. (1997). *The complexity of cooperation: Agent-based models of competition and collaboration*. Princeton, NJ: Princeton University Press.

Barclay, P., & Willer, R. (2007). Partner choice creates competitive altruism in humans. *Proceedings of the Royal Society B: Biological, 274*(1610), 749–753.

Barkow, J. H. (1989). *Darwin, sex, and status: Biological approaches to mind and culture*. Toronto: University of Toronto Press.

Batson, C. D. (1991). *The altruism question: Toward a social psychological answer*. Hillsdale, NJ: L. Erlbaum, Associates.

Boyd, R., Gintis, H., Bowles, S., & Richerson, P. J. (2003). The evolution of altruistic punishment. *Proceedings of the National Academy of Sciences United States of America, 100*(6), 3531–3535.

Boyd, R., & Richerson, P. J. (1985). *Culture and the evolutionary process*. Chicago: University of Chicago Press.
Bradbury, J. W., & Vehrencamp, S. L. (1998). *Principles of animal communication*. Sunderland, MA: Sinauer Associates.
Breden, F., & Wade, M. J. (1991). "Runaway" social evolution: Reinforcing selection for inbreeding and altruism. *Journal of Theoretical Biology, 153*(3), 323–337.
Brown, S. L., & Brown, R. M. (2006). Selective investment theory: Recasting the functional significance of close relationships. *Psychological Inquiry, 17*(1), 1–29.
Brown, S. L., Nesse, R. M., Vinokur, A. D., & Smith, D. M. (2003). Providing social support may be more beneficial than receiving it: Results from a prospective study of mortality. *Psychological Science, 14*(4), 320–327.
Bull, J. J., & Rice, W. R. (1991). Distinguishing mechanisms for the evolution of co-operation. *Journal of Theoretical Biology, 149*(1), 63–74.
Byrne, R. W., & Whiten, A. (1988). *Machiavellian intelligence: Social expertise and the evolution of intellect in monkeys, apes, and humans*. Oxford: Oxford University Press.
Connor, R. C. (1995). Altruism among non-relatives: Alternatives to the 'Prisoner's Dilemma.' *Trends in Ecology and Evolution, 10*(2), 84–86.
Cosmides, L. (1989). The logic of social exchange: Has natural selection shaped how humans reason? Studies with the Wason selection task. *Cognition, 31*(3), 187–276.
Crespi, B. J. (2004). Vicious circles: Positive feedback in major evolutionary and ecological transitions. *Trends in Ecology and Evolution, 19*(12), 627–633.
Crook, J. H. (1972). Sexual selection, dimorphism, and social organization in the primates. In B. Campbell (Ed.), *Sexual selection and the descent of man* (pp. 231–281). Chicago: Aldine.
Darwin, C. (1871). *The descent of man and selection in relation to sex*. London: John Murray.
Dawkins, R. (1976). *The selfish gene*. Oxford: Oxford University Press.
de Waal, F. B. M., Macedo, S., Ober, J., & Wright, R. (2006). *Primates and philosophers: how morality evolved*. Princeton, NJ: Princeton University Press.
Dennett, D. (1995). *Darwin's dangerous idea*. New York: Simon and Schuster.
Diamond, J. (2002). Evolution, consequences and future of plant and animal domestication. *Nature, 418*(6898), 700–707.
Dugatkin, L. A. (1997). *Cooperation among animals: An evolutionary perspective*. New York: Oxford University Press.
Dugatkin, L. A. (2006). *The altruism equation: Seven scientists search for the origins of goodness*. Princeton: Princeton University Press.
Dunbar, R. I. (1998). The social brain hypothesis. *Evolutionary Anthropology, 6*, 178–190.
Dunbar, R. I. M. (1996). *Grooming, gossip, and the evolution of language*. Cambridge, MA: Harvard University Press.
Dunbar, R. I. M., Knight, C., & Power, C. (1999). *The evolution of culture: An interdisciplinary view*. New Brunswick, NJ: Rutgers University Press.
Fehr, E., & Fischbacher, U. (2003). The nature of human altruism. *Nature, 425*(6960), 785–791.
Fehr, E., & Henrich, J. (2003). Is strong reciprocity a maladaptation? On the evolutionary foundations of human altruism. In P. Hammerstein (Ed.), *Genetic and cultural evolution of cooperation* (pp. 55–82). Cambridge, MA: MIT Press in Cooperation with Dahlem University Press.
Fehr, E., & Rockenbach, B. (2004). Human altruism: Economic, neural, and evolutionary perspectives. *Current Opinions in Neurobiology, 14*(6), 784–790.
Frank, R. H. (1985). *Choosing the right pond: Human behavior and the quest for status*. New York: Oxford University Press.
Frank, R. H. (1988). *Passions within reason: The strategic role of the emotions*. New York: W.W. Norton.
Frank, R. H. (1999). *Luxury fever: Why money fails to satisfy in an era of excess*. New York: Free Press.
Frank, S. A. (1997). The Price equation, Fischer's fundamental theorem, kin selection and causal analysis. *Evolution, 51*(6), 1712–1729.

Frank, S. A. (1998). *Foundations of social evolution*. Princeton, NJ: Princeton University Press.
Frank, S. A. (2006). Social selection. In C. W. Fox & J. B. Wolf (Eds.), *Evolutionary genetics: Concepts and case studies* (pp. 350–363). New York: Oxford University Press.
Gibbard, A. (1990). *Wise choices, apt feelings: A theory of normative judgment*. Oxford: Oxford University Press.
Gintis, H. (2000). Strong reciprocity and human sociality. *Journal of Theoretical Biology, 206*, 169–179.
Gintis, H., Smith, E. A., & Bowles, S. (2001). Cooperation and costly signaling. *Journal of Theoretical Biology,. 119*(1), 103–119.
Grafen, A. (1984). Natural selection, kin selection, and group selection. In J. R. Krebs & N. B. Davies (Eds.), *Behavioural ecology: An evolutionary approach* (pp. 62–84). Sunderland, MA: Sinauer.
Grafen, A. (1990). Biological signals as handicaps. *Journal of Theoretical Biology, 144*, 517–546.
Hamilton, W. D. (1964). The genetical evolution of social behavior I, and II. *Journal of Theoretical Biology, 7*, 1–52.
Hammerstein, P. (2001). Games and markets: Economic behavior in humans and other animals. In R. Noë, J. A. R. A. M. v., Hooff, & P. Hammerstein (Eds.), *Economics in nature: Social dilemmas, mate choice and biological markets* (pp. 1–22). Cambridge: Cambridge University Press.
Hammerstein, P. (2003). *Genetic and cultural evolution of cooperation*. Cambridge, MA: MIT Press in Cooperation with Dahlem University Press.
Hardy, C. L., & Van Vugt, M. (2006). Nice guys finish first: The competitive altruism hypothesis. *Personality and Social Psychology Bulletin, 32*(10), 1402–1413.
Henrich, J., & Boyd, R. (2001). Why people punish defectors. Weak conformist transmission can stabilize costly enforcement of norms in cooperative dilemmas. *Journal of Theoretical Biology, 208*(1), 79–89.
Henrich, J., Boyd, R., Bowles, S., Camerer, C., Fehr, E., Gintis, H., et al. (2005). "Economic man" in cross-cultural perspective: Behavioral experiments in 15 small-scale societies. *Behavioral and Brain Sciences, 28*(6), 795–815.
Henrich, J., McElreath, R., Barr, A., Ensminger, J., Barrett, C., Bolyanatz, A., et al. (2006). Costly punishment across human societies. *Science, 312* (5781), 1767–1770.
Hirshleifer, J. (1999). There are many evolutionary pathways to cooperation. *Journal of Bioeconomics, 1*, 73–93.
Humphrey, N. (1997). Varieties of altruism – and the common ground between them. *Social Research, 64*, 199–209.
Humphrey, N. K. (1976). The social function of intellect. In P. G. Bateson & R. A. Hinde (Eds.), *Growing points in ethology* (pp. 303–318). London: Cambridge University Press.
Kahil, E. L. (2004). What is altruism? *Journal of Economic Psychology, 25*, 97–123.
Katz, L. (2000). *Evolutionary origins of morality: Cross disciplinary perspectives*. Devon: Imprint Academic.
Kitcher, P. (1993). The evolution of human altruism. *The Journal of Philosophy, 90*(10), 497–516.
Kokko, H., Brooks, R., Jennions, M. D., & Morley, J. (2003). The evolution of mate choice and mating biases. *Proceedings of the Royal Society B: Biological, 270* (1515), 653–664.
Kokko, H., Jennions, M. D., & Brooks, R. (2006). Unifying and testing models of sexual selection. *Annual review of Ecology and Evolutionary Systematics, 37*, 43–66.
Krebs, D. L. (2000). The evolution of moral dispositions in the human species. *Annals of the New York Academy of Sciences, 907*, 132–148.
Krebs, J., & Dawkins R (1984). Animal signals: Mind-reading and manipulation. In J. R. Krebs & N. B. Davies, (Eds.), *Behavioral ecology: An evolutionary approach* (pp. 380–402). Sunderland, MA: Sinauer.
Kummel, M., & Salant, S. W. (2006). The economics of mutualisms: Optimal utilization of mycorrhizal mutualistic partners by plants. *Ecology, 87* (4), 892–902.

Laland, K. N., Odling-Smee, J., & Feldman, M. W. (2000). Niche construction, biological evolution, and cultural change. *Behavioral and Brain Sciences, 23* (1), 131–146; discussion 146–175.

Lande, R., & Arnold, S. J. (1983). The measurement of selection on correlated characters. *Evolution, 37*, 1210–1226.

Lasch, C. (1979). *The culture of narcissism: American life in an age of diminishing expectations*. New York: Warner Books.

Leach, H. M. (2003). Human domestication reconsidered. *Current Anthropology, 44* (3), 349–368.

Leary, M. R., & Baumeister, R. F. (2000). The nature and function of self-esteem: Sociometer theory. In M. P. Zanna, (Ed.), *Advances in experimental social psychology* (Vol. 32, pp. 2–51). San Diego, CA: Academic Press.

Lehmann, L., & Keller, L. (2006). The evolution of cooperation and altruism – A general framework and a classification of models. *Journal of Evolutionary Biology, 19* (5), 1365–1376.

Manson, J. H., Navarrete, C. D., Silk, J. B., & Perry, S. (2004). Time-matched grooming in female primates: New analyses from two species. *Animal Behavior, 67*, 493–500.

Mealey, L. (1995). Sociopathy. *Behavioral and Brain Sciences, 18* (3), 523–599.

Midgley, M. (1994). *The ethical primate: Humans, freedom, and morality*. New York: Routledge.

Miller, G. (2007). Sexual selection for moral virtues. *Quarterly Review of Biology, 82* (2), 97–126.

Miller, G. F. (2000). *The mating mind: How sexual choice shaped the evolution of human nature*. New York: Doubleday.

Mills, J., & Clark, M. S. (1994). Communal and exchange relationships: Controversies and research. In R. Erber & R. Gilmour (Eds.), *Theoretical frameworks for personal relationships* (pp. 29–42). Hillsdale, NJ: Lawrence Erlbaum.

Nesse, R. (2004). Cliff-edged fitness functions and the persistence of schizophrenia (commentary). *Behavioral and Brain Sciences, 27* (6), 862–863.

Nesse, R. M. (1990). Evolutionary explanations of emotions. *Human Nature, 1* (3), 261–289.

Nesse, R. M. (2001). *Evolution and the capacity for commitment*. New York: Russell Sage Foundation.

Nesse, R. M. (2006). Why so many people with selfish genes are pretty nice – except for their hatred of *The Selfish Gene*. In A. Grafen & M. Ridley (Eds.), *Richard Dawkins* (pp. 203–212). London: Oxford University Press.

Noë, R. (2001). Biological markets: Partner choice as the driving force behind the evolution of mutualisms. In R. Noë, J. A. R. A. M. v. Hooff, & P. Hammerstein (Eds.), *Economics in nature: Social dilemmas, mate choice and biological markets* (pp. 93–118). New York: Cambridge University Press.

Noë, R., & Hammerstein, P. (1994). Biological markets: Supply and demand determine the effect of partner choice in cooperation, mutualism and mating. *Behavioral Ecology and Sociobiology, 35*, 1–11.

Noë, R., & Hammerstein, P. (1995). Biological markets. *Trends in Ecology and Evolution, 10* (8), 336–339.

Noë, R., Hooff, J. A. R. A. M. v., & Hammerstein, P. (2001). *Economics in nature: Social dilemmas, mate choice and biological markets*. New York: Cambridge University Press.

Nowak, M. A. (2006). Five rules for the evolution of cooperation. *Science, 314* (5805), 1560–1563.

Nowak, M. A., & Sigmund, K. (1998). Evolution of indirect reciprocity by image scoring. *Nature, 393*, 573–577.

Pepper, J. W. (2007). Simple models of assortment through environmental feedback. *Artificial Life, 13* (1), 1–9.

Pepper, J. W., & Smuts, B. B. (2002). A mechanism for the evolution of altruism among nonkin: Positive assortment through environmental feedback. *The American Naturalist, 160*, 205–213.

Pilot, M. (2005). Altruism as advertisement – a model of the evolution of cooperation based on Zahavi's handicap principle. *Ethology, Ecology and Evolution, 17* (3), 217–231.

Price, E. O. (1984). Behavioral aspects of animal domestication. *Quarterly Review of Biology, 59*(1), 1–32.

Queller, D. C. (1992). A general model for kin selection. *Evolution, 46*(2), 376–380.

Queller, D. C., Ponte, E., Bozzaro, S., & Strassmann, J. E. (2003). Single-gene greenbeard effects in the social amoeba Dictyostelium discoideum. *Science, 299*(5603), 105–106.
Queller, D. C., & Strassmann, J. E. (1998). Kin selection and social insects. *BioScience, 48*(3), 165–175.
Reeve, H. K., & Shen, S. F. (2006). A missing model in reproductive skew theory: the bordered tug-of-war. *Proceedings of the National Academy of Sciences USA, 103*(22), 8430–8434.
Ridley, M. (1997). *The origins of virtue: Human instincts and the evolution of cooperation.* New York: Viking.
Riolo, R. L., Cohen, M. D., & Axelrod, R. (2001). Evolution of cooperation without reciprocity. *Nature, 414*(6862), 441–443.
Roberts, G. (1998). Competitive altruism: From reciprocity to the handicap principle. *Proceedings of the Royal Society B: Biological, 265*(1394), 427–431.
Rothstein, S. I. (1980). Reciprocal altruism and kin selection are not clearly separable phenomena. *Journal of Theoretical Biology, 87*, 255–261.
Roughgarden, J., Oishi, M., & Akcay, E. (2006). Reproductive social behavior: Cooperative games to replace sexual selection. *Science, 311*(5763), 965–969.
Sachs, J. L., Mueller, U. G., Wilcox, T. P., & Bull, J. J. (2004). The evolution of cooperation. *Quarterly Review of Biology, 79*(2), 135–160.
Schaller, M., & Crandall, C. S. (2003). *The psychological foundations of culture*. Mahwah, NJ: Lawrence Erlbaum.
Segerstråle, U. C. O. (2000). *Defenders of the truth: The battle for science in the sociobiology debate and beyond*. New York: Oxford University Press.
Sigmund, K. (1993). *Games of life: Explorations in ecology, evolution, and behaviour*. New York: Oxford University Press.
Silk, J. B. (2003). Cooperation without counting: The puzzle of friendship. In Hammerstein, P. (Ed.), *Genetic and cultural evolution of cooperation: Dahlem workshop report* (Vol. 29, pp. 39–54). Cambridge, MA: MIT Press.
Simms, E. L., & Taylor, D. L. (2002). Partner choice in nitrogen-fixation mutualisms of legumes and rhizobia1. *Integrative and Comparative Biology, 42*(2), 369–380.
Simon, H. A. (1990). A mechanism for social selection and successful altruism. *Science, 250*, 1665–1668.
Smith, A. (1976 [1759]). *The theory of moral sentiments*. Oxford: Clarendon Press.
Smuts, B. B. (1985). *Sex and friendship in baboons*. New York: Aldine Publishing Company.
Stevens, J. R., Cushman, F. A., & Hauser, M. D. (2005). Evolving the psychological mechanisms for cooperation. *Annual Review of Ecology, Evolution, and Systematics, 36*(1), 499–518.
Tanaka, Y. (1996). Social selection and the evolution of animal signals. *Evolution, 50*, 512–523.
Tibbetts, E. A. (2004). Complex social behaviour can select for variability in visual features: A case study in Polistes wasps. *Proceedings of the Royal Society B: Biological, 271*(1551), 1955–1960.
Tomasello, M. (1999). *The cultural origins of human cognition*. Cambridge, MA: Harvard University Press.
Tooby, J., & Cosmides, L. (1996). Friendship and the Banker's paradox: Other pathways to the evolution of adaptations for altruism. In J. M. Smith, W. G. Runciman, & R. I. M. Dunbar (Eds.), Evolution of social behaviour patterns in primates and man (pp. 119–143). London: The British Academy: Oxford University Press.
Trivers, R. (2000). The elements of a scientific theory of self-deception. *Annals of the New York Academy of Sciences, 907*(1), 114–131.
Trivers, R. L. (1971). The evolution of reciprocal altruism. *Quarterly Review of Biology, 46*, 35–57.
Trivers, R. L. (1985). Social evolution. Menlo Park, CA: Benjamin/Cummings.
Veblen, T. (1899). *The theory of the leisure class: An economic study in the evolution of institutions*. New York: Macmillan.
Weber, B. H., & Depew, D. J. (2003). *Evolution and learning: The Baldwin effect reconsidered.* Cambridge, MA: MIT Press.
Wedekind, C., & Milinski, M. (2000). Cooperation through image scoring in humans. *Science, 288*(5467), 850–852.

West-Eberhard, M. J. (1975). The evolution of social behavior by kin selection. *Quarterly Review of Biology, 50*(1), 1–33.
West-Eberhard, M. J. (1979). Sexual selection, social competition, and evolution *Proceedings of the American Philosophical Society, 123*(4), 222–234.
West-Eberhard, M. J. (1983). Sexual selection, social competition, and speciation. *Quarterly Review of Biology, 58*(2), 155–183.
West-Eberhard, M. J. (2003). *Developmental plasticity and evolution*. New York: Oxford University Press.
West, S. A., Griffin, A. S., & Gardner, A. (2007). Social semantics: Altruism, cooperation, mutualism, strong reciprocity and group selection. *Journal of Evolutionary Biology, 20*(2), 415–432.
West, S. A., Pen, I., & Griffin, A. S. (2002). Cooperation and competition between relatives. *Science, 296*(5565), 72–75.
Wilkinson, G. S. (1984). Reciprocal food sharing in the vampire bat. *Nature, 308*, 181–184.
Williams, G. C. (1966). *Adaptation and natural selection: A Critique of some current evolutionary thought*. Princeton, NJ: Princeton University Press.
Wilson, D. S., & Sober, E. (1994). Reintroducing group selection to the human behavioral sciences. *Behavioral and Brain Sciences, 17*(4), 585–607.
Wilson, E. O. (1975). *Sociobiology*. Cambridge, MA: Harvard University Press.
Wolf, J. B., Brodie, E. D. III, & Moore, A. J. (1999). Interacting phenotypes and the evolutionary process. II. Selection resulting from social interactions. *American Naturalist, 153*, 254–266.
Wright, R. (1994). *The moral animal: The new science of evolutionary psychology*. New York: Pantheon Books.
Wynne-Edwards, V. C. (1962). *Animal dispersion in relation to social behavior*. Edinburgh: Oliver and Boyd.
Wu, J., & Axelrod, R. (1995). How to cope with noise in the iterated prisoner's dilemma. *Journal of conflict resolution, 39*(1), 183–189.

The Evolved Brain: Understanding Religious Ethics and Religious Violence

John Teehan

Introduction

We are living in the midst of one of the greatest periods of intellectual discovery in the history of religious studies. Scholars from anthropology, psychology, evolutionary biology, neuroscience and philosophy are developing a cognitive science of religion which promises to revolutionize and profoundly deepen our understanding of religion. Scholars have tried for centuries to lay bare the empirical bases for religious beliefs but, without disparaging those efforts, it is only with the development of the cognitive sciences that we can move beyond arm-chair speculation or merely phenomenal descriptions of religion and develop an empirically grounded explanation for religious belief and behavior.

The work of Pascal Boyer, Scott Atran, Stewart Guthrie, Justin Barrett, William Irons, and a host of others sets a compelling account of religious belief as a natural outgrowth of basic cognitive functioning.The main lesson to be taken from this work is that evolutionary processes have shaped our emotional and cognitive predispositions in such a way that the brain channels "human experience toward religious paths" (Atran, 2002, p. 11). Or as Pascal Boyer puts it "the explanation for religious beliefs and behaviors is to be found in the way all human minds work" and the workings of the human mind bear the stamp of its evolutionary history (Boyer, 2001, p. 2). No divine revelation is required for belief in gods.

Of concern here is how neuroscience and evolutionary psychology – both part of the larger paradigm of cognitive science – may together, or separately, contribute to our understanding of human morality, and more specifically religious morality. With the development of ever more sophisticated brain scanning techniques our understanding of how the brain works is increasing dramatically. With

J. Teehan (✉)
Department of Religion, Hofstra University, Hempstead, NY, USA
e-mail: John.P.Teehan@hofstra.edu

This is an elaboration of views set out in print as "The Evolutionary Basis of Religious Ethics" in *Zygon: Journal of Religion and Science*, September 2006, Vol. 41/3, pp. 747–774, and which itself developed the ideas first published as "The Evolution of Religious Ethics" in Free Inquiry, June/July 2005, Vol. 25, No. 4.

J. Verplaetse et al. (eds.), *The Moral Brain*, DOI 10.1007/978-1-4020-6287-2_11,

technology such as PET scans, SPECT scans and fMRI, we can observe the brain functioning during a variety of tasks. Recently researchers have begun using these technologies to explore the neural bases of religious experiences and have come up with varying though intriguing results. Several researchers have zeroed in on the involvement of the temporal lobes. Michael Persinger of Laurentian University has found that electro-magnetic stimulation of the temporal lobes induces subjects to have the experience of a "presence" which they interpret according to culturally available religious idioms –i.e. Christians tend to experience Christian images, Buddhists' experiences match Buddhist expectations, etc. (an atheist subject experienced images of his old girlfriend – make of that what you will, Persinger, 1987, 1999).

Neurologist V. Ramachandran also identifies the temporal lobes as significant (Ramachandran, 1997) and the NY Times recently reported on research indicating that out of body experiences may be caused by stimulation of the temporal-parietal junction (NY Times, 10/3/06). Then there is the work of Andrew Newberg and his colleagues on the neurology of peak religious experiences. This research has garnered the most public attention, due largely to a series of popular publications by Newberg, et al, but also due to the more extensive research project they have established at the University of Pennsylvania. Newberg's focus has been on the neurological correlates of "mystical" experiences generated during deep meditation, as that is practiced by experienced individuals from a variety of religious backgrounds. The results of these studies point to the involvement of the parietal lobe and the thalamus areas. Newberg reports that he has found an anatomical asymmetry in the thalamus in long time, experienced meditators that is rare in the general population. Whether this asymmetry is caused by or the cause of the "mystical" states is, according to Newberg, an open question. Relevant to this is a report by The Society of Neuroscience on studies showing that meditation practices can physically alter the brain, leading to an increase in the cortical areas associated with attention and sensory processing (www.sfn.org/news). Now I do not intend this list to be anything more than suggestive of the work being done to uncover the neurological bases of experiences connected with religious beliefs. What I want to consider is just what conclusions can be drawn from such research.

One possible response is to adopt a reductionistic view of religion. If stimulation of the brain can generate religious experiences then it may follow that such experiences are nothing more than neuro-chemical reactions which are then explained in terms of available cultural idioms. This approach creates a great deal of anxiety for religious believers and generates both suspicion and hostility toward scientific approaches to religion. By reducing god and faith to by-products of brain chemistry, neuroscience threatens to undermine these belief systems. This reading of the research just confirms what religious critics and skeptics have suspected about the natural origins of religion, and furnishes them with empirical arguments bolstered by cutting edge neuroscience.

However, neuroscience need not lead to religious skepticism and in fact has been used to argue the case for the existence of a spiritual dimension to reality. This is the position, for example, of Andrew Newberg who has asserted that his research into what he terms "the mystical mind" supports a belief in Absolute Unitary Being, without connecting this to any particular notion of god (Newberg, D'Aquili, & Rause, 2001). This neuroscientific work is being readily integrated into a religious world view that posits that god has constructed the brain so that it is capable of experiencing him and that neuroscience is simply identifying the structures god created to accomplish this purpose. In fact there is a growing field that identifies itself as Neurotheology developing theories along such lines.

Newberg's move, however, is not original to him. In fact it is quite similar to a view set out in one of the earliest psychological accounts of religion, *On the Varieties of Religious Experience* (1902) by American philosopher/psychologist, William James. In that work James traced the source of religious experiences to the workings of the unconscious mind. Once we have grounded religious experience in the unconscious, he claimed science could go no farther, given the state of late 19th century psychology. The question that remained for James was whether this origin in the unconscious was a naturalistic one or was the unconscious, perhaps, a doorway to a transcendent level of existence. In customary honesty, James admitted that he believed the unconscious to be a doorway to the beyond, but to his credit he also asserted that neither possibility could be supported by the evidence presently available. His reading of the evidence as pointing toward a "something more" was termed by James an "overbelief," i.e. an intellectual pre-disposition toward a particular worldview. James recognized that overbeliefs do not constitute evidence and so do not constitute an answer for anyone not sharing that particular worldview. As a scientist a more conclusive answer must await future advances in the science of the mind (James, 507–519).

Over one hundred years have passed from James' work to the work of Newberg. One might have hoped that with the dramatic growth of our knowledge of the brain, and all the work done by neuroscientists such as Newberg, that the more definitive evidence James sought might be at hand. However, from reading Newberg's analysis one might think that nothing new was gained that could inform our assessment of the nature of religious experiences; that we are still in a position where one is free to bring whatever religious or skeptical interpretation one is inclined toward.

Now this is not meant as a criticism of Newberg but rather as an indication of a limitation of the neuroscience of religion. Neuroscience provides invaluable insight into the neuro-chemical bases of religious experience and is making vital discoveries about the anatomical correlates of such experience, but by itself this work leaves us to speculate on the meaning of these results for an assessment of religion. In other words, it shows us how religion is expressed in the brain but it does not provide an interpretation of those experiences that speaks to religion's ultimate origins, or allow us to evaluate religious beliefs. Of course, neuroscientists may rightly complain that this is not their job; that the goal of neuroscience is to do just what I have claimed it does – uncover the workings of the brain correlated with human experiences and activities. But my concern is to see what neuroscience can contribute to an understanding of human morality, and specific to my work, religious morality. And in this endeavor neuroscience plays a role but is not sufficient, in itself. What is missing from the psychological work of William James and from the neuroscientific work of Newberg is an evolutionary perspective.

Neuroscience sets out the neurological substrate of religious/moral experience but leaves open an explanation of the origin of those physical systems. Evolution, on the other hand, provides an explanation for the development of those systems that can answer, in theory, not only how they came to be but why they came to be – that is, what environmental challenges lead to the evolution of the cognitive responses that constitute religious experience. Evolutionary psychology can shed light on the origin of religion and religious moral systems. What I would like to do now is to demonstrate just how an evolutionary perspective might shed light on our understanding of religious moral traditions.

The thesis I wish to defend is that we may best understand religious ethical traditions as cultural expressions of underlying evolved cognitive and emotional predispositions. If we begin with (1) an understanding of how evolution has shaped our moral psychology and (2) an understanding of the moral function of religion we can then see that religious moralities, despite their variety and their ostensible transcendent origins, are grounded in evolutionary strategies that have their origin in promoting reproductive fitness. And understood this way the vexing question of religious violence may be understood in a different light.

Evolutionary Models for Morality

From an evolutionary perspective, morality is a means to resolve social conflict and thereby make social living, and cooperative action, possible. Humans are social beings, descended from a long line of social beings. As our ancestors faced the rigors and dangers of daily existence group living allowed them to better confront these challenges. The benefits of cooperative behavior in hunting and gathering and in defense against predators are obvious, but there were, of course, costs. Cooperation requires an individual to share the dangers and the costs necessary to promote the good of the group. This raises a problem from an evolutionary perspective. Typically, those individuals who most successfully promote their own interests will have an advantage in the struggle for survival. Sacrificing my interest for the good of the group does not seem to make sense from an evolutionary perspective. This is the problem of altruism. From an evolutionary approach the challenge is to understand how behavior that lowers an agent's fitness in order to raise the fitness of another can arise from a process driven by, so called, "selfish" genes (Dawkins, 1976).

While work continues on this issue, it is no longer a problem for evolutionary theory. A vigorous and growing sub-discipline has been developed, providing ever more sophisticated models to explain the evolutionary bases of morality. For the purposes of this essay I will focus on two of the classic models which provide an explanation of altruistic behavior consistent with Darwinian mechanisms. The first is the theory of kin selection, which was rigorously established by William Hamilton, in 1964. Evolutionary theory holds that behavior that increases reproductive success (i.e. fitness) will be selected for, and passed on, with reproductive success measured strictly in the number of genes passed from one generation to the next. Given this, sacrificing for your children makes sense then because you are protecting your genetic investment. People who do not care for children will not have many descendants. This type of self-sacrifice is really long-term self-interest and poses no problem for evolution. Hamilton realized, however, that child-bearing is not the only way to get copies of an individual's genes into the next generation.

My child contains copies of 50% of my genes, but my full siblings also contain copies of 50% of my genes, and so their children, my nieces and nephews, contain copies of 25% of my genes, and so on through the various degrees of familial relationships. Sacrificing my immediate interests for my kin can also be consistent with long-term self interest. If I sacrifice some of my resources to increase the fitness of my nieces, then when they reproduce they will be passing on copies of my genes to the next generation, and so I am in fact enhancing my own fitness. This broader conception of genetic self-interest is termed, by Hamilton, "inclusive fitness." While this may not coincide with philosophical notions of altruism it does create the possibility for at least limited self-sacrificial behavior in a way consistent with evolutionary processes.

Hamilton's theory of kin selection is one model for explaining the evolution of morality, but it is not sufficient. However powerful kin selection may be it is restricted to relatively small groups where likelihood of relatedness is high. This model explains why it makes sense to sacrifice for a relative, but not why I should sacrifice for someone genetically unrelated. For societies to continue to develop beyond extended family units another mechanism is needed. In 1971, Robert Trivers provided a model for this with his theory of the evolution of reciprocal altruism. Simplistically, this is an "I'll rub your back, so that you will rub my back" strategy, or as Trivers put it, "Reciprocal altruism can also be viewed as a symbiosis, each partner helping the other while he helps himself" (39). Cooperative behavior pays off. My sacrificing some time and effort to help you now pays off when you help me later and so functions as an investment in my long-term fitness.

This model was extended and enriched by Richard Alexander, who developed the notion of indirect reciprocity (1987). My altruistic behavior toward you can benefit me even if you do not return the favor. Alexander points out that altruism can pay off indirectly in several ways: for one I may gain the reputation for being a cooperator and this may encourage others to be willing to cooperate with me; or an altruist may be rewarded by society, either economically or in terms of status-enhancement, for her contributions; or an altruist may improve his and/or his family's fitness by increasing the general fitness of the community (94).

Reciprocal altruism allows for the development of extended relationships of assistance and the development of more complex societies in a way consistent with the mechanics of Darwinian evolution. Given the challenges and the precariousness of human existence during our evolutionary history a strategy that promoted a system of mutual assistance would have had a great selective advantage.

Kin selection and reciprocal altruism (both direct and indirect) then are the twin pillars of evolutionary approaches to morality. Reciprocal altruism extends the influence of morality beyond the clan ethic supported by kin selection, and together they provide an effective means of explaining a wide range of moral phenomena.

Such models can only work, of course, if people do in fact reciprocate. There is, however, a great temptation not to reciprocate – to cheat or defect – since if you have already benefited from someone's cooperation any reciprocation on your part would just be an unnecessary cost. Given this, the ability to discriminate between cooperators and cheaters is crucial, as is the commitment to punish those who refuse to reciprocate. If cheating is costly, that is, "if cheating has later adverse affects. . .which outweigh the benefit of not reciprocating" (Trivers, 36) then there is a motivation to cooperate. However, if cheating is easy to get away with, then the cost, and personal risks, of cooperating with others is raised. If a society is to function at a level beyond the family it must develop a system of encouraging cooperation, and detecting and punishing cheating. From an evolutionary perspective this is just what a moral system does – it is a code of behavior that approves of and rewards certain behaviors necessary to cohesive social functioning, while condemning and punishing those behaviors contrary to cohesive social functioning. As Alexander puts it, "Moral systems are systems of indirect reciprocity" (142).

There is a large and growing body of literature supporting this theory of the evolution of morality, and suggesting, furthermore, that this has played a decisive role in the evolution of human cognition. It is important to bear in mind that when we speak of these evolutionary strategies we do not refer to necessarily conscious processes. We do not, at least not usually, consciously make decisions based on calculations of degree of kinship, or on probability of reciprocation. Evolution works by developing cognitive and emotional predispositions to favor kin, to respond favorably to reciprocal social interactions, and to respond with outrage toward those who cheat. It is these cognitive and emotional predispositions which form the ground upon which we build philosophical and religious moral systems.

Evolutionary Psychology and Moral Religions

One of the most common, ancient and powerful cultural institution for the promotion of group cohesion is, of course, religion. It is my thesis that religions can be explained as cultural institutions which regulate individual behaviors in a pro-social direction by triggering evolved cognitive/emotional mechanisms. Two of the more fundamental of these mechanisms are kin selection and reciprocal altruism and, as we shall see, a great deal of religious ethics can be understood as elaborations on these two evolutionary moral strategies. Religions also establish methods for promoting conformity to group behavior and heightening a sense of loyalty to the group. An evolutionary understanding of moral psychology will allow us insight into the power that religions wield, for good and woe, to shape human behavior, even in a post-modern, industrialized world very different from that inhabited by our earliest religious ancestors.

Religion is much more than morality, and in fact it is not always concerned with morality; the gods are not necessarily moral beings. However, under certain conditions the connection between morality and religion becomes significant. The larger and more complex a society becomes, the greater the temptation to defect from social cooperation, and the greater the chance of doing so successfully. This makes sacrificing for the social good more costly, and even for the socially conscientious, a less rational option. The danger of this spiraling out of control threatens the cohesion of such societies. Religion seems to provide a remedy to this situation. It is not the only solution, but it has been one of the more robust.

In order to see the suitability of religion to play this role it is necessary to consider how gods are conceptualized. Gods come, of course, in many variations. The most common, if not universal, feature attributed to such beings, however, is a mind (Boyer 144) and this has important implications. In our day-to-day interactions with other people, it is crucial to be able to distinguish potential cheaters from potential cooperators. To cooperate with another cooperative individual brings important rewards, but to cooperate with a cheater is to throw away my resources – so being able to distinguish the two is an essential social skill. To do this requires a wide range of information, not least of which concerns the mental states of these other individuals – what information do they have, what information do they lack, what are their intentions? Pascal Boyer points out that we conceive of other humans as "limited access strategic agents" (155). That is, we assume that others do not have access to complete or perfect information relevant to social interactions. Our information is limited, our ability to discern another's intention faulty, and this limitation is mutual.

Say, for example, I want to avoid a tedious assignment in order to enjoy a beautiful Spring day, but I also want to avoid being penalized for this choice. I decide to tell my boss that I must stay home to care for a sick child. I view this as a promising strategy because I view my boss as a limited-access-strategic-agent. That is, I assume she does not have access to my true intention, nor to the actual health status of my child.

People the world over, however, represent gods as "full access strategic agents" (158). That is, they view their gods or ancestors not necessarily as omniscient but as having access to all the information relevant to particular social interactions. The gods have access to all that is needed for making a sound judgment in any particular situation. They know my child is healthy and at school and that actually I intend to spend the day watching a baseball game. Now, not all gods may be represented as possessing this quality, but the ones that do have special significance. And beings that possess such a trait are in a particularly privileged position to assume a moral role.

Gods, as full access strategic agents, occupy a unique role that allows them to detect and punish cheaters, and to reward cooperators. In moral religions such gods are conceived of as "interested parties in moral choices" (Boyer 172). They are concerned with social interactions and fully cognizant of the behavior and motives of those involved. Communal belief in such beings, therefore, lowers the risk of cooperating and raises the cost of cheating by making detection more probable, and punishment more certain. Religion then becomes the vehicle for the moral code of a society required for that society to continue to function as a coherent unit as it grows in size and complexity.

Furthermore, religion not only supports evolved moral mechanisms by providing supernatural oversight, it also powerfully functions as a signal of willingness to cooperate. As noted it is imperative to be able to discriminate between potential cooperators and cheaters. Indeed, research is establishing the existence of a "cheater-detection module" as part of our cognitive repertoire for dealing with social interactions (e.g. Verplaetse, Vanneste, & Braeckman, 2007; Vanneste, Verplaetse, Van Hiel, & Braeckman, 2007). But as societies become larger and more anonymous, this task becomes increasingly difficult. I may know which of my colleagues I can trust, but what of those in the larger community? Belief in a moral god addresses this difficulty, but only if such belief is commonly shared. If you do not believe god is watching you, then you have less to fear from cheating and I have more to lose by cooperating. My belief that you will be punished someday for your lack of belief does little to protect me now. How can I trust you to reciprocate my cooperation? It has been pointed out that humans have developed a wide range of ways to signal a commitment to cooperate in order to encourage cooperative behavior from others, e.g. a smile and an open hand (see Frank, 1988; Nesse, 2001). However, humans also possess the ability to deceptively signal such an intention – e.g. a fake smile and a knife in a hidden hand – and so one must be wary of the sincerity of such signals. Given this, evaluating signals of commitment is an important element in determining who you can and cannot trust. A sound rule to guide this task is 'the harder it is to fake a signal to commit to cooperation the more trustworthy it is.' As William Irons notes, "For such signals of commitment to be successful they must be hard to fake. Other things being equal, the costlier the signal the less likely it is to be false" (2001, p. 298).

Religious rituals and rules can function as hard-to-fake signals, and indeed, Irons has characterized religion as a "hard-to-fake sign of commitment" (1996, 2001). Irons points out that religions are typically learned over a long span of time, their traditions are often sufficiently complex to be hard for an outsider to imitate, and their rituals provide opportunities for members to monitor each other for signs of sincerity. This is a costly and time-consuming process (2001, p. 298). Showing oneself to be a member of a religion by having mastered the rituals and practices of that religion signals that one has already made a significant contribution of time and energy to the group and a willingness to follow the code that governs the group. That is, it signals that one is a reliable partner in social interactions and can be trusted to reciprocate.

From an evolutionary perspective religious morality provides a vehicle for extending the evolutionary mechanisms for morality, i.e. kin selection and reciprocal altruism. Also by serving as a hard-to-fake-sign of commitment religions function to discriminate between in-group members (those who have invested in the religion and so can be trusted) and out-group members (those who have not invested in the religion and so cannot be trusted). If this is an accurate account of the evolution of religious morality then it should be possible to detect these evolutionary concerns embedded in religious moral traditions and so ground such ethical systems in an evolutionary matrix.

The Evolutionary Bases of Religious Ethics/Religious Violence

The goal of this discussion is to make the case that an evolutionary approach to understanding the bases of religious ethics is a worthwhile contribution to the study of both religion and ethics.While a more comprehensive analysis is called for by choosing representative moral principles that exemplify an evolutionary logic I hope to demonstrate the plausibility of this approach, and its potential for shedding new light on these issues. It is also crucial to recognize another limitation. Judaism and Christianity have incredibly complex, at times inconsistent, ever evolving moral traditions. The treatment here will be concerned with the bases of the earliest expressions of these moral traditions, specifically those found in the Torah and the Gospels. This is not to imply that an evolutionary account cannot be extended to later developments, only that this will not be attempted here.

Looking first at Judaism, we can see the stories of the Patriarchs as embodiments of the logic of kin selection. Jews are all children of Abraham. "Israel" is not merely the ancestral home of the Jewish people (bequeathed by God, himself) but was the father of the twelve tribes. All Jews through their tribal lineage are members of one, extended family. This extended family is the basis of Judaism, and Jewish morality. It is the basis, but of course not the whole.

The Law begins by establishing the pre-eminence of the Hebrew God over all other gods and connecting the prosperity of the community to obedience to God and His law. "I the lord your God am a jealous God, visiting the iniquity of the fathers upon the children to the third and fourth generation of those that hate me, but showing steadfast love to thousands of those who love me and keep my commandments" (Exodus 20: 5–6). This yoking of communal prosperity to obedience to God's Law is the defining characteristic of the relationship between God and his chosen people and serves the function of fortifying social bonds with divine sanction. A community cannot survive and prosper without a shared commitment to a moral system. The ancient Hebrews addressed the problem of extending the demands of morality by establishing the idea of a covenantal relationship, which both embodies and reinforces the logic of evolutionary morality.

Also significant is the description of God as "jealous." This human character flaw may seem a strange characteristic for God to possess, but it is the fact that Yahweh is capable of such emotional extremes that makes him such an effective enforcer of moral norms. In effect, God's jealousy is a signal-to-commit to enforcing the moral codes he oversees. Working from our human experience we know that some people are easier to get around than others; that some people, even in authority, can be a "soft touch" and not diligent in enforcing the rules; and that some people talk tough but are not willing/interested in following through. Each of these possibilities makes a person a less credible enforcer and so weakens the motivation to keep to our commitments/agreements. In other words, an enforcer of the law who is not credible because we are unsure of their commitment to enforce the law weakens the whole system. So how does Yahweh (or any other enforcer) signal his commitment to follow through and enforce the law?

Emotional responses work very effectively to signal willingness to punish behavior. From a purely rational calculation, it is sometimes not worthwhile to actually retaliate against a cheat. But the powerful emotions that often accompany a moral complaint – i.e. anger and jealousy – signal that a person is not responding from rational calculations and is prepared to retaliate even if it costs them more than ignoring the offense would. This command signals that Yahweh is not a purely rational agent – he is a jealous god. Jealous agents lash out irrationally, and often violently and impulsively. And not only does Yahweh promise to punish the wicked but he promises to visit "the iniquity of the fathers upon the children to the third and fourth generations." Note how effectively this command taps into our evolved psychology. Yahweh will take his anger out on my genetic legacy, making for naught all that I have sacrificed for my children and raising the specter of reproductive failure. Evolution has made sure that we are very sensitive to such threats.

As we read the specific rules set out in the Law what we find are not so much rules for spiritual purification as rules for social cooperation. We are commanded, "You shall not covet your neighbor's house; you shall not covet your neighbor's wife, or his manservant, or his maidservant, or his ox, or his ass, or anything that is your neighbor's" (Exodus 20: 17). We see prohibitions against murder, adultery, theft and perjury - all of which can be justified on purely practical grounds as minimal requirements for social living. And we also find very specific punishments for those who violate these prohibitions. These penalties are essential to promote and support reciprocity between members of a society.

This of course does not argue against a divine authorship. We can certainly imagine God legislates just those rules best designed to ensure peace among neighbors. However in certain laws we see a level of specificity more appropriate to a civil code than to a guidebook for moral perfection. This is, of course, what is to be expected from a religious ethics that ultimately functions as a social bond. For example we find this very particular law for dealing with an owner of an unruly ox:

> When an ox gores a man or a woman to death, the ox shall be stoned, and its flesh is not to be eaten; but the owner of the ox shall be clear. But if the ox has been accostumed to gore in the past, and its owner has been warned but has not kept it in...its owner also shall be put to death. If it gores a man's son or daughter, he shall be dealt with according to this same rule. If the ox gores a slave, male or female, the owner shall give to their master thirty shekels of silver, and the ox shall be stoned (Exodus 21: 28–32).

As a civil regulation this is sensible and fair. It tries to take into account both parties to a potential dispute and to administer justice proportionate to the situation. This type of regulation, unglamorous and mundane as it may be, is key to maintaining a sense of reciprocity within a community. As a precept of a higher moral law, however, one might wonder at the unquestioning acceptance of slavery, and the implicit commercialization of human flesh expressed in the compensation for a gored slave. But notice, a slave is not part of the community. A slave is not evaluated for potential to cooperate and reciprocate cooperation. A slave is a possession that is dealt with, however benignly, through coercion. The harm done to a slave threatens social cohesion by the harm done to its owner, and that is what must be addressed.

We also find much in the Law that attempts to keep the covenantal relationship central to Jewish life. Circumcision, Sabbath observations and the dietary and purification laws serve to bind the people to their God, but they also clearly define the boundaries of the Jewish community. In Genesis we read of the function of circumcision:

> And God said unto Abraham, "...This is my covenant, which you shall keep, between me and you and your descendants after you; Every man among you shall be circumcised. You shall be circumcised the flesh of your foreskins, and it shall be a sign of the covenant between me and you. Any uncircumcised male who is not circumcised in the flesh of his foreskin shall be cut off from his people..."(Genesis 17: 9–14).

These are costly and time-consuming rituals. The complexity of many of these practices requires a long-term immersion in the tradition in order to master. As such they functioned as hard-to-fake-signals of commitment to the group, identifying those who could be trusted to cooperate and those who may be lead into temptation to defect.

In considering Christianity we are faced with a more complex situation. This is not because of any qualitative difference between Judaic and Christian ethics, but is due to a more fluid and dynamic sense of what constitutes community. Christianity began as a sect within 1st century Judaism, but developed into a cosmopolitan, Hellenized religion. The moral teachings of Christianity reflect the tensions, contradictions and conflicts that characterized that historical process. For example, in Luke 10: 25–29, Jesus, after declaring "Love thy neighbor" a central moral requirement, is asked "and who is my neighbor?" This is a significant question not only because it seeks to establish the boundaries of the moral community but because it expresses confusion about those boundaries. This moral confusion is characteristic of large, complex, and increasingly anonymous societies and would be inconceivable in most tribal societies. Jesus replies with the story of the Good Samaritan in which the hero is a member of a reviled out-group who stops to aid a stranger in need, while two Jewish characters that one might have expected to be moral role models (a priest and a Levite) ignore the suffering of a fellow Jew (Luke 10: 29–37). The parable serves to indicate that a tribal-morality is no longer adequate.

An evolutionary analysis of early Christian ethics, therefore, requires a much more specific and detailed consideration of the developmental story than is possible here. Still, we can pull out examples from the canonical gospels that offer prima facie support for an evolutionary account. For the sake of simplifying the discussion I shall restrict my examples to the Gospel of Matthew.

As we turn to Christianity we find that Jesus summarizes the moral message of the Mosaic Law by declaring "So whatever you wish that men would do to you, do so to them; for this is the law and the prophets"(Matthew 7: 12).When later asked to identify the greatest of the commandments in the Law, Jesus replies:

> You shall love the Lord your God with all your heart, and with all your soul, and with all your mind. This is the great and first commandment. And a second is like it, You shall love your neighbor as yourself. On these two commandments depend all the law and the prophets. (Matthew 22: 37–40)

"Do unto others" is the classic expression of reciprocal altruism. This central principle of evolutionary morality is here declared by Jesus to be the basis of all the teachings of the Jewish Law and the basic moral rule for Christians. Significantly, it is subordinated to only one other commandment, a complete devotion to God – which is consistent with the evolutionary logic of religious ethics. God serves to uphold the laws that bind society together and enables reciprocal altruism to function in a large complex society. This supreme commandment signals a complete commitment to the being that oversees the good of the group and thus is a sign of commitment to that group.

Still, an objector can point out that this, along with Jesus' messages to "love your enemies" (Matt. 5: 44) and "if anyone strikes you on the right cheek, turn to him the other also" (Matt. 5: 39) seem at variance with the principles of reciprocal altruism. Indeed, it has been suggested that this moral stance represents a "protest against the principle of [natural] selection" (Theissen, 1984, p. 112, see also Hefner, 1996). How may an evolutionary account respond to this?

There are two responses and both are important to a fuller appreciation of an evolutionary study of ethics. One is to admit that Jesus is here moving beyond the constraints of evolutionary moral logic. This exposes an important limitation, though not refutation, of evolutionary morality. An evolutionary account of morality should not be understood to imply biological determinism. It does not deny the possibility of resisting these cognitive/emotional predispositions, nor does it reject moral innovation. The complexity of our social world and the flexibility of our cognitive abilities conduce to allow for an element of moral creativity. In Jesus' teaching we see an attempt to stretch our moral imagination. It may be argued that this is what characterizes the moral prophets in human history – their ability to articulate a moral vision which pushes against our ingrained predispositions.

A second response is to point out that as moral exhortation Jesus' teaching may move beyond evolutionary logic, but as a guide to behavior it is evolutionary logic that often holds sway. While this is an admittedly contentious point, I would claim that the history of Christianity is filled with examples (e.g. the crusades, the inquisition, the persecution of heretics and Jews, etc.) that speak to the power of the underlying evolutionary logic to overwhelm attempts to develop moral attitudes contrary to it (e.g. "turn the other cheek"). The response of Christians in history to enemies and to attacks has often been much more in line with the psychology of evolutionary morality than with these particular teachings of Jesus. This is not so much a condemnation of Christianity as it is a lesson on the difficulty of moving beyond these evolutionarily ingrained moral predispositions. Even Jesus himself, the Gospels show, fell prey to the pull of these predispositions as we can see when he condemns those who refuse to accept his teachings (Matt. 10: 15). So Jesus' use of the golden rule is consistent with an evolutionary analysis, even if that analysis does not allow a simple identification of the two principles (i.e. "do unto others" and reciprocal altruism.)

Christian morality is also filled with imagery to encourage kin selection. We are all children of God and so fellow members are brothers and sisters as well as neighbors (e.g."you are all brethren. . .for you have one Father who is in heaven." Matthew 23:8–9)

Here again, however, we can find evidence of a confusion over moral boundaries and christianity's attempt to clarify and extend the moral community. We read of an occasion when Jesus was teaching and was informed that his mother and brothers had come to speak with him. Jesus replied,

> "Who is my mother, and who are my brothers?" And stretching out his hand toward his disciples, he said, "Here are my mother and my brothers! For whoever does the will of my Father in heaven is my brother, and sister, and mother" (Matthew 12: 46–50).

This is a radical extension of the moral community, but it is formulated in a way that takes advantage of an evolutionarily ingrained moral predisposition toward kin. Throughout the Gospels, Jesus uses kin relationships as models for clarifying moral obligations.

Early Christians also developed a variety of ways to signal their identity with the group, and their willingness to reciprocate acts of altruism. Their rejection of circumcision and dietary laws distinguished Christians from their parent group and fellow monotheists, the Jews (e.g. on circumcision: "Now I, Paul say to you that if you receive circumcision, Christ will be of no advantage to you." Galatians 5: 2) while their sharing of the Eucharist and refusal to sacrifice to Roman gods set them apart from their pagan neighbors. This last act served as a very effective hard-to-fake-signal of commitment given the often drastic consequences that followed it.

The Evolutionary Logic of Religious Violence

Thus far we have been looking at morality as a means for establishing a sense of community and a code of in-group behavior. In serving this function morality also identifies an out-group, and implies an out-group ethic. The key consideration within a group is to promote pro-social behavior by ensuring reciprocation among group members. The flip side of this membership is of course exclusion. If you are not family or neighbor, then you are an outsider. Outsiders are not invested in the group and so have no motivation to cooperate or to reciprocate cooperation. Therefore they endanger the community. For all the constructive morality found in religion, we find an equally prominent place for warnings against outsiders (see also, Hartung, 1995).

In order to consider this flip side of morality let's consider the rule against killing. On Mt. Sinai God enshrines "You shall not kill" as a divine command (Exodus

20: 13) yet the first order Moses gives upon descending from the mountain is for the execution of those who had fallen into sin while he was gone.

> Then Moses stood in the gate of the camp, and said, "Who is on the Lord's side? Come to me." And all the sons of Levi gathered themselves together to him. And he said to them, "Thus says the Lord God of Israel, 'Put every man his sword on his side, and go to and fro from gate to gate throughout the camp and slay everyman his brother and every man his companion, and every man his neighbor.'" And the sons of Levi did according to the word of Moses; and there fell of the people that day about three thousand men. And Moses said, "Today you have ordained yourselves for the service of the Lord..." (Exodus 32:26–29)

Indeed, throughout the Mosaic Law we find numerous actions that are to be punished with death. Not only is a murderer to suffer the death penalty (Exodus 21: 12) but also those who commit adultery (Leviticus 20: 10), bestiality (Exodus 22: 19) blasphemy (Leviticus 24: 16), as well as those who profane the Sabbath (Exodus 31: 15) or curse their parents (Exodus 21: 17)–to name just a few.

After having received the law and communicated it to the people, Moses then leads the Hebrews on what can be described as a blood-soaked trek to the Promised Land. We are told, for example, that when God delivered the land of Heshbon to the Hebrews they "utterly destroyed every city, men, women, and children; we left none remaining" (Deuteronomy 2: 34). They then moved onto the land of Bashan where they "smote him until no survivor was left to him." The passage continues,

> And we took all his cities at that time-there was not a city which we did not take from them–sixty cities ... And we utterly destroyed them, as we did to Sihon the king of Heshbon, destroying every city, men, women, and children (Deuteronomy 3: 4–6).

In case we might be tempted to think that the extent of the killing was an excess brought on by the heat of battle, rather than a divinely sanctioned slaughter, we read in Numbers of a case in which Moses angrily chastises the army generals for not killing all the inhabitants of a city. In their defeat of the Midianites the Hebrews took as captives the woman and children, after slaying all the men. Moses, we are told,"was angry with the officers of the army" asking them "Have you let all the women live?"(Numbers 31: 14–15). He corrects their error by instructing them thus:

> Now therefore, kill every male among the little ones, and kill every woman who has known man by lying with him. But all the young girls who have not known man by lying with him, keep alive for yourselves (Numbers 31: 17–19).

What can we make of all of this? One is certainly tempted to charge the Mosaic Law with hypocrisy; aspects of which affront our sense of moral rightness. However, from the evolutionary perspective developed here there is less reason to be surprised.

Morality develops as a tool to promote within-group cohesiveness. This cohesiveness functions as an adaptive advantage in competition with other groups (Alexander, 1987, Wilson, 2002). Morality is a code of how to treat those in my group; it is not extended, at least not in the same way, to those outside the group. Since these others are not bound by the same code they must be treated as potential cheaters. Those outside the group are in fact a potential threat to my group's survival. The people the Hebrews encountered on their journey were obstacles that needed to be overcome in the interest of group survival. As such the moral injunction "you shall not kill" did not apply to them.

The evolutionary logic behind these actions is, perhaps, most apparent in Moses's instructions to spare the virgins among the Midianite prisoners. This was clearly not done from compassion for he had no compunction about ordering the death of older women and male children. In the brute terms of reproductive fitness the young girls were prime resources for the propagation of the community, while the older women and boys would have been a drain on the resources of a nomadic people. Moses's actions seem coldly calculating to modern readers and not what one would expect of a religious hero, but to the degree that morality serves evolutionary ends Moses ably fulfilled his role as moral leader of his community.

While this may explain the lethal behavior towards those in the out-group, we have seen that "you shall not kill" was often suspended within the group, also. However, the same logic supports the imposition of the death penalty toward in-group members. Morality sets the bounds of appropriate in-group behavior. It also serves as a signal of commitment to the group. Breaking this code poses two problems that need to be addressed. For one, say in the case of theft, it creates an imbalance that needs to be rectified. More importantly, perhaps, is that it signals a break from the group that may cast the perpetrator into the category of the out-group. As such, the former in-group member becomes a potential threat and is outside the bounds of moral treatment. Some such breaks can be rectified by a willingness to accept the punishment of the group, but some cannot.

We can see this logic at work by looking at two very different capital crimes: murder and profaning the Sabbath. In the case of murder the death penalty seeks to restore the balance disrupted by the crime. "An eye for an eye" is the flip side of "Do unto others" and so this punishment flows from the logic of reciprocal altruism. However, in profaning the Sabbath no member of the group is harmed. There is no imbalance to be corrected. To understand the punishment for this crime we need to remember that Sabbath observations serve as a signal of commitment to the group and mark an individual as one who can be trusted to reciprocate. By profaning the Sabbath one is signaling that he has opted out of this arrangement and is no longer a trustworthy member of the community. In the logic of evolutionary morality you are either in the group or out of the group, and if you are out, moral laws no longer apply to you.

Before leaving this discussion there is a question of translation that should be addressed. It has been noted that "you shall not kill," which is the traditional translation found in many Christian versions of the Bible, is really a mistranslation which should be read as "you shall not murder." Does this change any of the preceding argument? I think not, at least not in any significant sense. What distinction can be drawn between "kill" and "murder"? It seems the least controversial reading is that to kill is to take a life, while to murder is also to take a life, but with the added connotation that such an act is prohibited by the norms or laws of society. We can see this intuitively in that a soldier who takes the life of an enemy soldier in the course of a battle has killed, but would not be charged with murder. So too, a correctional officer who pushes the button that sends a lethal cocktail of drugs into the veins of a convict has taken a life but is not deemed a murderer. So "you shall not kill" is a more categorical prohibition against taking a life than is "you shall not murder." Now, how does this impact upon our discussion?

One important implication is that the charge of hypocrisy is unfounded. If the commandment is "you shall not murder" then imposing the death penalty upon blasphemers and such is not hypocritical because such an action is not murder. Killing those who violate certain commands of the law, killing those who betray the group or those outsiders who oppose the group, are divinely sanctioned instances of taking a life, and by definition are not instances of murder.

Still, other issues arise when we read the law as "you shall not murder." For one, it makes the law mundane, as it then merely tells us not to commit killings which have not been sanctioned. Such a prohibition is a nearly universal trait of organized societies, although there is a diversity of ways to determine which killings are to be sanctioned. It also exposes the tautologous nature of the law: you shall not murder = it is wrong to kill those who God/society deems it wrong to kill. "You shall not kill" would be a much more substantive and morally unique command (found elsewhere at the time, to my knowledge, only in Buddhism).

However, from the perspective of an evolutionary analysis things change not at all. The fact that "you shall not murder" is a nearly universal social prohibition is consistent with an evolutionary understanding of bio-cultural development. More pertinent is how the command "you shall not murder" is to be applied – which killings are sanctioned, and so not "murder," and which are prohibited. Here we can return to the previous discussion without any emendations for, as argued there, the distinction between sanctioned and prohibited killing falls along lines supported by the logic of evolutionary morality.

From a post-Enlightenment moral view, we might expect a principle against killing to apply categorically, and so prohibit both the death penalty (at least for non-violent crimes) and the slaughter of innocent children – hence the charge of hypocrisy against the ancient Hebrews. But, from an evolutionary view of morality there is no hypocrisy. "You shall not kill" is a moral rule, and as such applies to all members of the group. Those outside the group, whether members of a competing group or a lapsed member of the community, do not fall under the extension of this rule.

As we turn to Christianity we must again be sensitive to the social context in which Christianity developed. The early Christians were a minority group within a minority group, in a world dominated by a Roman-Hellenistic culture. As we have seen, this made the issue of boundary clarification a vital concern to the Christians. While the Christians set their group-boundary differently than their fellow monotheists, they nonetheless demonstrated the same in-group/out-group divide that we encountered in Judaism. One of the clearest expressions of this dichotomous approach is found in Christ's parable of the sheep and goats. Speaking of the final judgment, Christ tells us,

> Before him will be gathered all the nations, and he will separate them one from another as a shepherd separates the sheep from the goats, and he will place the sheep at his right hand, but the goats at his left. Then the King will say to those at his right hand, "Come,

> O blessed of my Father, inherit the kingdom prepared for you from the foundation of the world."...Then he will say to those at his left hand, "Depart from me, you cursed, into the eternal fire prepared for the devil and his angels" (Matthew 25: 32–41).

What is particularly notable about this passage is the severity of the treatment toward those in the out-group. In our examples from the Jewish Scriptures, those out-side the group merely suffered death, here they suffer eternal torment. Christianity raised the stakes for being on the wrong side of the divide. Throughout the Gospels the opponents of the Christians are categorized not merely as dangerous or evil, but as in league with the devil. Elaine Pagels (1995) points out that in "the ancient world, so far as I know, it is only Essenes and Christians who actually escalate conflict with their opponents to the level of cosmic war" (84). Understanding this escalation is a complicated task but an explanation rooted in evolutionary logic may be suggested. As radical, minority sects within 1st century Judaism both the early Christians and Essenes had little temporal power to exercise in defense of their group and so were less able to punish those who defected. If the cost of defection is low, the likelihood of defection increases. This raises the cost of cooperation. A group cannot survive under such circumstances. Divine retribution then assumes a more essential role. An individual could, theoretically, enjoy the benefits of membership in a Christian community, then defect before reciprocating and be protected from punishment by being absorbed back into the more powerful majority group. However, in doing so they were now aligning themselves with the enemy of God and could have no hope of escaping divine justice.

So we can understand this shift away from a physical punishment of opponents toward a spiritual punishment as an example of the same evolutionary moral logic found in our discussion of Judaism, applied to the specific environmental conditions of early Christianity, rather than as a repudiation of that logic. That this is so may be supported by the fact that as soon as Christianity acquired the role of the dominant group within Roman society it quickly resorted to the more familiar, mundane means of punishing defectors.

From these examples we can discern a certain logical structure to religious violence. The initial move is to discriminate between an in-group and an out-group; next is a differential in moral evaluation of the two sides of the divide. Thus far, this structure is not unique to religious violence. Religion's contribution is to align the in-group with god, and the consequent demonization of the out-group. This further distinguishes the moral evaluation of the two groups and leads to an escalation of the stakes. Conflict is no longer simply a competition between two groups seeking to promote their own interests, it is now a cosmic struggle with no middle ground available, and nothing short of victory acceptable.

We can find evidence of this logic at work in Islam, also. There is a clear divide between the faithful and the unbelievers, who by that fact are seen as being in league with the devil:

> "Allah is the guardian of those who believe. He brings them out of darkness into the light; and those who disbelieve, their guardians are Shaitans [Satan] who take them out of the light into the darkness (Surah 2 : 257).

We also find very different moral codes for those on either side of this divide. While a Muslim is prohibited from killing a Muslim no such prohibition applies to disbelievers:

> And whoever kills a believer intentionally, his punishment is hell, and Allah will send His wrath on him and curse him and prepare for him a painful chastisement (Surah 4 :93).

> When your Lord revealed to the angels: I am with you, therefore make firm those who believe. I will cast terror into the hearts of those who disbelieve. Therefore strike off their heads and strike off every fingertip of them (Surah 8: 12).

And as we would expect, the fortunes of those on either side of the divide continue to diverge into eternity:

> A parable of the garden which those guarding against evil are promised: Therein are rivers of water that does not alter, and rivers of milk the taste whereof does not change, and rivers of drink delicious to those who drink, and rivers of honey clarified; and for them therein are all fruits and protection from their Lord. Are these like those who abide in the fire [the disbelievers] and who are made to drink boiling water so it rends their bowels asunder (Surah 47: 15).

By heightening the different fates of in-group and out-group members Islam lowers the cost of investing in the good of the group. The grotesque nature of the price of defection is a particularly effective, and common, strategy for discouraging defection (see, e.g., the book of Revelation, and Dante's *Inferno*). The role of violence in religion is a vital issue. People puzzle over the apparently paradoxical morality found in Judaism, Christianity and Islam. Proponents characterize these religions as religions of peace, and then struggle to explain the abundant evidence to the contrary. But neither philosophy nor theology has been able to adequately reconcile these two aspects of religion. Typical answers to this paradox tend to dismiss one side or the other, i.e. deny that religion is ever responsible for the violence done in its name – it's the fault of aberrant individuals – or downplay the power of religion to generate true moral progress. Neuroscience is a latecomer to this issue and while I am eager for the insights it will shed on both moral and religious experiences, it is evolutionary psychology that provides a theory of religion that can set out the cognitive mechanisms underlying these two sides of religious behavior and address the apparent paradox.

Conclusion

I introduced evolutionary psychology as a methodology that may allow us to assess religion in a way that neuroscience does not. That is, neuroscience can identify the brain mechanisms involved in various religious experiences but cannot speak to the origin of such mechanisms, and so is of limited use in a normative evaluation of religion. As noted, neuroscientists have used the same data to come to widely different conclusions as to the nature of religious experience. How does evolutionary psychology fare in this task?

Evolutionary psychology can set out the connection between various religious beliefs and practices and the evolutionary goal of inclusive fitness. If the argument presented here is sound, i.e. religious moral traditions are cultural expressions of an underlying evolved psychology, the consequences for an assessment of religion are significant. We can then see religious moral traditions are means of establishing group coherence and identity, ultimately grounded in the need of individuals pursuing their inclusive fitness within a social environment. This stands in stark contrast to more traditional religious interpretations, at least as religion is understood in a modern, Western sense, as expressions of god's will. Furthermore, if we can also understand our notions of god and the nature of god as by-products of the brain's evolved mechanisms for processing, storing and retrieving information, then the belief in a divine being is undermined. Given this, the reductionistic interpretation of religion is supported in a more compelling manner than can be offered by neuroscience alone.

Before concluding we must note that this interpretation is resisted by some evolutionary thinkers. Most significant, I believe, is the alternative explanation of Justin Barrett. Barrett is a leading player in the burgeoning field of the evolutionary/cognitive study of religion. As such, he has a deep and sophisticated grasp of these issues. Yet when he addresses the question of the implications of this research for belief he adopts a conciliatory position, suggesting that it is reasonable to believe that a god who works through natural processes, and who desires that his creatures come to know him, might arrange for the evolutionary process to result in the cognitive mechanisms that evolutionary psychology has identified as the source of religious cognition (Barrett, 2004). This is consistent with Newberg's interpretation of his own research (although Barrett adopts a more traditional religious position than the mysticism of Newberg). If Barrett's interpretation is sound then evolutionary psychology is in the same boat as neuroscience, rather than moving the debate forward.

Needless to say, I do not find Barrett's position ultimately tenable. I say "ultimately" because on one level it is perfectly reasonable. God, as the supreme author of the laws of nature could certainly have designed those laws in such a way as to reveal himself to us; Or, in a more Spinozan vein, god understood as the laws of nature is discernable in those laws and their effects. Neither of these options is logically incompatible with an evolutionary account, but both would then lead to very different notions of the deity.

If god has designed the evolutionary process to result in the cognitive processes which we have uncovered as the basis of religious beliefs then it is those underlying processes that bear the stamp of divine favor, not the culturally constructed interpretations that are layered upon them. This "directed evolution" may be used to support some religious worldview but it undermines the claim of any specific religion to a privileged position. In this sense, the religious quest to discover the will of god would focus on the common structures that underlie varying religious traditions, and would not expect any one tradition to be the literal expression of the divinity.

So, it may in fact be possible to retain some sort of religious worldview, given an evolutionary analysis, but not, for example, the Christian worldview that Barrett seems to profess. The religious worldview we end up with is akin to the more spiritual/mystical religious view advocated by Newberg. However, Newberg was able to make that move because neuroscience gave him only the physical structures correlated with religious experience, leaving him free to speculate on a religious origin of those structures. By adding an evolutionary perspective we now have a naturalistic explanation for the origins of the physical structure of the brain; one that needs no supernatural intervention, or purpose. In this case positing that the evolutionary process was ultimately designed by god for a religious purpose, while not logically precluded is certainly logically gratuitous.

The other "religious" option is the Spinozan move, i.e. identify god and nature and see evolution as the unfolding of the divine reality. This is a form of pantheism that is appealing to some more empirically minded believers, and there is nothing in evolutionary science or neuroscience to preclude this interpretation. But note what follows: to understand god, now, is to understand the workings of the universe – science, not revealed religion, is the source of this understanding. This may add, for some, a spiritual feel to scientific work but it too undermines traditional religious systems.

Nothing said here is meant to minimize the continuing importance of the neuroscientific investigation of religion and morality. In fact, I believe the most promising path to a deeper understanding of both of these topics is through a partnership between these disciplines. Evolutionary psychology generates theories based on an understanding of the cognitive challenges that faced our ancestors. Empirical research is conducted to test whether these hypothesized cognitive tools match the way the brain is determined to function under experimental conditions. This provides the evidence to support or question evolutionary explanations. Neuroscience may be able to make a significant contribution to supporting, or challenging, such explanations. I believe this is particularly true for my thesis that religions function to support evolved moral mechanisms. As neuroscience comes to a greater understanding of the neurological correlates of moral cognition it creates new possibilities for testing the moral function of religion.

For example, much of the neuroscientific research into religion examines brain functioning during ostensible religious activities, such as meditation and prayer. My thesis is that, prayer and meditation notwithstanding, from an evolutionary perspective more significant religious activity includes attending to signals of commitment to a religion, as a component of cheater-detection strategies. If this is so then brain scans of individuals participating in cheater-detection experiments, such as Prisoner's Dilemma games, to determine if there is distinctive neural activity during such tasks, could provide a baseline for comparison of brain scans of individuals attending to religious images and activities. Does the brain activity of individuals detecting cheating in a Prisoner's Dilemma game match that of individuals viewing someone violating a religious taboo, or incorrectly performing a religious ritual? If so, this could provide supporting evidence for the costly-signaling component of an evolutionary/cognitive theory of religion.

Also, studies show a "strong correlation between non-cooperative behavior and fear-related emotions" (Vanneste et al., 272). I have argued that belief in a moral god supports cooperative behavior by making the cost of cheating higher. This then should mitigate the fear response to non-cooperative behavior. Using brain scans we might compare neural fear responses to non-cooperative behavior to detect different responses of religious and non-religious individuals. For religious individuals, is there a reduction in neural fear response if the non-cooperative individual is a member of the same religious community (and thus more susceptible to the fear of divine punishment)? The data from such experiments could provide important evidence, supporting or confuting, an evolutionary explanation.

On a more general level, but no less important, as neuroscience continues to identify the anatomical and chemical correlates of religious and moral cognition it solidifies the case for the biological nature of these phenomena. As biological phenomena, religion and morality are proper subjects of evolutionary analysis, which then can shed light on the psychological mechanisms underlying them. As we continue to struggle with the moral challenges of our day, challenges which often involve a religious worldview, we can no longer afford to rely on philosophical speculation uninformed by the best science available, or worse, on moral commitments grounded in ancient cultural traditions that present themselves as vehicles of revealed truth. Cognitive science gives us the prospect of a better method.

References

Alexander, R. D. (1987). *The biology of moral systems*. New York: Aldine De Gruyter.
Atran, S. (2002). *In gods we trust: The evolutionary landscape of religion*. Oxford: Oxford University Press.
Boyer, P. (2001). *Religion explained: The evolutionary origins of religious thought*. New York: Basic Books.
Dawkins, R. (1976). *The selfish gene*. Oxford: Oxford University Press.
Frank, R. H. (1988). *Passions within reason: The strategic role of the emotions*. New York: Norton.
Hamilton, W. D. (1964).The genetical evolution of social behavior, I, II. *Theoretical Biology. 7*, 1–52.
Hartung, J. (1995). Love thy neighbor: The evolution of in-group morality. *Skeptic, 3*, 86–89.
Hefner, P. (1996). Theological perspectives on morality and human evolution. *Religion and science: History, method, dialogue*. New York: Routledge.
Holy Bible. (1974) Revised Standard Version, Meridian Publishers.
Holy Qur'an. (1990) (M. H. Shakir, Trans.). Elmhurst, NY: Tahrike Tarsile Qur'an, Inc.
Irons, W. (1996). In our own self image: The evolution of morality, deception, and religion. *Skeptic, 4*, 50–61.
Irons, W. (2001). Religion as a hard-to-fake sign of commitment. In R. M. Nesse (Ed.), *Evolution and the capacity for commitment*. New York: Russell Sage Foundation.
James, W. (1902). *The varieties of religious experience*. New York: Modern Library.
Nesse, R. M. (Ed.). (2001). *Evolution and the capacity for commitment*. New York: Russell Sage Foundation.
Newberg, A., D'Aquili, E., & Rause, V. (2001). *Why god won't go away: Brain science and the biology of belief.* New York: Ballantine Books.
Pagels, E. (1995). *The origin of satan*. New York: Vintage Books.
Persinger, M. (1987). *The neuropsychological bases of god beliefs*. New York: Praeger Press.
Persinger, M. (1999). This is your brain on god. *Wired, 7*, November.
Ramachandran, V. (1997). *Society of neuroscience, 23*, Abstract No. 51.1, October.

Theissen, G. (1984). *Biblical faith: An evolutionary approach*. Philadelphia, PA: Fortress Press.
Trivers, R. L. (1971). The evolution of reciprocal altruism. *Quarterly Review of Biology, 46*, 35–57.
Vanneste, S., Verplaetse, J., Van Hiel, A., & Braeckman, J. (2007). Attention bias toward noncooperative people. A dot probe classification study in cheating detection. *Evolution and Human behavior, 28*, 272–276.
Verplaetse, J., Vanneste, S., & Braeckman, J. (2007). You can judge a book by its cover: The sequel. A kernel of truth in predictive cheating detection. *Evolution and Human Behavior, 28*, 260–271.
Wilson, D. S. (2002). *Darwin's cathedral: Evolution, religion, and the nature of society*. Chicago: Chicago University Press.

An Evolutionary and Cognitive Neuroscience Perspective on Moral Modularity

Jelle De Schrijver

Introduction

In the 19th century, the case of Phineas Gage suggested that our moral sense could be located in a particular area of the brain. Damage to a part of the prefrontal cortex seemed to have selectively bereft the railroad worker of his moral faculties, resulting in lawless and anti-social behaviour. Additional studies revealed that behaviour of Phineas Gage and patients with similar brain damage can be further characterized by the disturbance of social behaviour, a diminished response of social emotions such as compassion and failures in non-moral decision-making or planning (Anderson, Bechara, Damasio, Tranel, & Damasio, 1999). As the damage is not selective, the "moral" misbehaviour syndrome is apparently not confined to the moral sphere alone. This leads to the conclusion that there is no discrete "moral centre" or single morality module in the brain (Greene, 2005).

Yet, unravelling the relation between our brains and the capacity to engage in moral behaviour has been an ongoing challenge. First, we need to define morality. The philosophical record uncovers a vast amount of different approaches to morality. Some examples in a nutshell: Aristotle argued the moral life is related to living a balanced life, equilibrating on the golden mean between vices. For Kant a moral life is a life where duty supersedes inclination and where one follows principles which can be endorsed by all rational beings. For Mill the good is what gives the most happiness to the greatest number of people. Nevertheless, despite the variety of definitions, a common element in most moralities is the concern for and the harmony with others. The morally good or the morally bad are often defined in relation to the welfare of the group.

Actually, the word "moral" has a dual meaning. The first descriptive meaning of the word "moral" indicates a person's comprehension of morality and his capacity to put it into practice. The antonym of "moral" in this descriptive sense is "amoral", indicating an inability to distinguish between right and wrong. A rock,

J. De Schrijver (✉)
Department of Philosophy and Moral Sciences, Ghent University, Belgium
e-mail: jelle.deschrijver@ugent.be

J. Verplaetse et al. (eds.), *The Moral Brain*, DOI 10.1007/978-1-4020-6287-2_12,

a baby or an ant lack a moral sense simply because they do not have the necessary capacities to behave morally, therefore we can't expect them to respect certain rules or conventions regarding others' welfare and therefore they are regarded as amoral. The second normative meaning of the word "moral" refers to the active practice of values. In this sense, the antonym of "moral" is "immoral", referring to actions that violate principles related to others' welfare. The philosophical theories that have been presented in the last millennia are instances of this normative aspect of the word "moral". Each normative theory draws lines between moral and immoral behaviour. Whereas the rock, the ant or the baby may be regarded as amoral because they are unable to grasp moral rules, a murderer or cheater may be regarded as immoral when he or she (knowingly) transgresses rules regarding others' welfare. Yet in order to behave moral in the normative sense, one needs to be moral in the descriptive sense. Similarly, people without legs are not expected to walk in a straight line, since they cannot walk anyway.

When the relation between the brain and moral behaviour is investigated, one focuses foremost on morality in the descriptive sense. One zooms in on the architecture of the brain and mental capacities that allow us to behave morally. Studying certain structural deficits may demonstrate why amorality occurs, why for instance Phineas Gage loses his faculty to behave morally. This means that we wonder why Phineas Gage stops caring about certain social consequences of his acts.

Recent technological and theoretical developments enable at least two disciplines to focus on "the moral brain". On the one hand, technological progress allows brain scientists to bridge the gap between anatomy and behaviour. The use of fMRI allows linking "real-time" brain activity to decision making processes. Thus, cognitive neuroscience (CN) helps to associate particular brain areas with moral judgment processes. On the other hand, theoretical development in evolutionary biology demonstrated how altruistic behaviour can be regarded as the product of natural selection, the result of adaptations promoting our fitness (Hamilton, 1964; Trivers, 1971). Evolutionary psychology (EP) is interested in the way evolution shaped the human brain and mind (actually the concepts *brain* and *mind* have the same referent, yet it can either be described as a physical or a functional system, in the former case the structure is emphasized (brain), in the latter the information-processing capacity (mind)). One of EP's central assumptions is that the mind is built up of many independent mental organs, so called modules, each dedicated to a particular function. Applied to morality, this entails that a collection of modules are regarded to be the building blocks of our "moral sense". This thesis can be called moral modularity.

Cognitive neuroscientists try to answer the question *how* the physical structure of the brain is linked to function. In contrast, an evolutionary approach will focus first of all on the question *why* certain human capacities provided a selective advantage to our ancestors. Both in a cognitive neuroscience and an evolutionary approach, questions about the causation of behaviour arise. However, the level that is addressed differs. The level is either structural (zoomed in) or evolutionary (zoomed out). Tinbergen (1963) calls this the difference between respectively proximate and ultimate causation. Yet, despite this difference of perspective, both schools of thought advance hypotheses about brain organization. The central question in this chapter is whether this modular model of the moral brain in an EP perspective is validated from a CN perspective.

Evolutionary Psychology

Modularity Hypothesis

Organs perform different and well-defined functions. Some examples: the kidneys have evolved to filter the toxic products out of the body and the heart to pump blood through the arteries and veins. Evolutionary psychologists argue that not only "classic" organs, such as the heart and kidneys, have evolved to solve particular problems. The brain itself, they argue, can be regarded as a conglomerate of many independent mental organs, or modules, each adapted to solve a particular problem (Buss, 2004; Pinker, 1998; Tooby & Cosmides, 1995). In contrast to a purely empiricist approach, EP does not regard the mind as a central computer performing many different tasks with the same instrument. The ability to interpret and react to the outside world requires a large amount of pre-specified knowledge. Without it the world would be experienced as an undifferentiated mass of pixels that is impossible to make sense of, EP points out.

During his life an individual is confronted with many types of stimuli (visual, auditory, etc.) and he has a broad repertoire of responses to those stimuli (approach, avoid, seduce, eat, etc.). Yet to react appropriately to all the problems an individual encounters and connect to the right stimulus to the efficient response is very hard. Moreover as different problems need different content, it is argued the mind must consist of several computing mechanisms. These computing mechanisms - called modules, cognitive devices or mental organs – are said to be domain-specific (Buss, 2004; Kennair, 2002). This means that only a certain type of stimuli is interpreted by a specific module, just like eyes are only equipped to transmit visual stimuli and are insensitive to sound. But apart from this input-specificity, there is also a form of content-sensitivity. This means that not all types of content are treated equally: stimuli of handsome faces for instance are more attention grabbing than others, for they may have been more relevant to our ancestors as they signal possible mating success. After computation of the stimuli this process leads to a particular output that influences behaviour or is further converted by other modules (Buss, 2004).

In contrast to the view of Fodor (the founding father of modularity), in an EP perspective not only perception, but several higher cognitive functions are considered to be modularly organized. There are said to be specialized systems for face recognition, for recognizing emotions, eye direction or singling out cheaters (Buss, 2004; Pinker, 1998). In fact, EP argues "our cognitive architecture resembles a confederation of hundreds of thousands of functionally dedicated domain-specific computers (modules)" (Tooby & Cosmides, 1995). Because of the large amount of evolutionarily specified modules said to be active in the brain, Samuels dubs this thesis the massive modularity hypothesis (Samuels, 2000).

A module helps to filter useful information from the complex environment. Yet, to acquire knowledge and information is useless, unless the systems are coupled to motivational systems that generate adaptive choices and behaviours (Duchaine, Cosmides, & Tooby, 2001). These motivations are the output of some of the modules. For example: an attractive woman may be the input that leads to sexual desire in a man (output). The architecture of the system that assigns sexual value to the female body is shaped by the relative survival and reproduction of ancestral design variants, for only men that were attracted could have a large offspring. EP considers most of the human preferences to be underpinned by evolutionary adaptations. These preferences exist "because natural selection built neurocomputational circuitry into our minds to compute it as one of several kinds of representation necessary for regulation of our behaviour according to evolutionarily functional performance criteria" (Tooby, Cosmides, & Barrett, 2005). Crucial in the evolutionary psychological approach is that these modules are designed to solve adaptive problems endemic to our hunter-gatherer ancestors. Actually, a module is regarded as the collection of all the neurocomputational circuitry adapted to solve a particular problem (Tooby et al., 2005). The information processing in the modules is usually said to be quick and automatic, separate from the conscious decisions a person is taking.

Summarizing, one can discern three central tenets in EP: (1) much of the structure and function of the human brain is innate as a result of natural selection (a nativist-adaptationist claim); (2) Functions are embodied in domain-specific modules, focused on a particular type of information. Activity of the modules results in automatic, subconscious and quick flashes, preferences or motivations guiding our behaviour; (3) The brain is massively modular, many simple elements or modules underlie behavior. Massive modularity entails that *most* of human behaviour is guided by subconscious modular processes.

Moral Modularity

Humans are not only motivated to approach attractive mates, friends, or tasty food; furthermore they are concerned about other people suffering, or fairness, etc. Where does this motivation to care about the welfare of others come from? Evolutionary psychologists argue that we are motivated to do so because this provided evolutionary advantages to our ancestors. These motivations are regarded as the outcomes of modules specialized in socio-moral problems, the so called moral modules. By means of the three central tenets of EP, the perspective of EP on the moral modular organization of the brain will be unraveled.

(1) A collective action problem is a situation in which collective and individual interests conflict, and where individuals have an interest to free ride. Hardin's classical example concerns the common grass fields where cattle graze. It is in the advantage of each individual sheep farmer to let as many of his animals feed on the grounds, as this ground is free. Yet, when all farmers use these

grounds as intensively, the grounds will be exhausted very soon. Only cooperation and regulation of the individual temptation to make unrestricted use of the grounds, may save the common feeding grounds (Hardin, 1968). In fact, it is argued that moral systems provide adaptive responses to similar collective action problems encountered by our ancestors, allowing individuals to cooperate as opposed to competing with each other. Moral systems tend to regulate individual temptation with emotional responses designed to facilitate cooperation and incite aggression toward those who cheat (Hauser, Young, & Cushman, 2008; Wilson, 1975). The way these cooperation problems are solved, appears to be by moral emotions. In contrast to basic emotions endowed with personal relevance, a group of emotions are linked to the interest and welfare of other individuals and are therefore called "moral" emotions. Embarrassment, shame, pride, guilt, compassion, contempt, gratitude, awe and indignation are prototypical examples (Moll, Zahn, Oliveira-Souza, Krueger, & Grafman, 2005). In a way, these emotions can be regarded as some form of insurance. You have to pay a price, a premium: due to your emotion, you are concerned about and care for others, thus you lose resources that could have been invested in your offspring. But, in return, you are insured of help if something might happen to you. If you don't have these emotions, you are singled out as a "non-paying" and unreliable member of the society possibly resulting in punishment and making it unlikely you will receive help when you need it. Moral emotions may have evolved as an efficient mechanism to stabilize cooperation (Bowles & Gintis, 2004; Frank, 1988; Greene, 2008).

A strong argument for innately specified modules is the "poverty of stimulus" argument: if children learn more than they are taught, then the child must have some form of innate information already represented. Chomsky remarks that "Growing up in a particular society, a child acquires standards and principles of moral judgment. These are acquired on the basis of limited evidence, but they have broad and often precise applicability" (Bolender, 2003; Chomsky, 1988). Fitting in with this argument is the observation that there seems to be no equipotentiality in moral learning. Children raised in communes or kibbutzim had difficulty learning about certain moral standards. They had troubles overcoming the preference to share material goods only with their close kin, moreover they were averse to mate with people they were raised with. These observations make Haidt and Joseph conclude that "it seems likely that children enter the world with some initial settings in the social domain (a liking for fairness, a dislike of harm) which are then extended by cultural learning" (Haidt & Joseph, 2005). Thus they argue for some form of innateness of moral intuitions.

(2) Modules are processing systems that are activated specifically by certain input stimuli and result in an output (or reaction) influencing behaviour. The output of "moral" modules are the intuitions guiding human behaviour, creating flashes of (moral) approval or disapproval, or, if the stimulus is strong enough, full-fledged moral emotions. EP emphasizes that cognition and emotion are strongly intertwined. An intuition or a feeling may give rise to a judgment or

a conviction. Moral intuitions refer to fast, automatic, and usually affect-laden processes that result in an evaluative feeling of good or bad or like-dislike (about actions or about a person). For instance, aversion of incest may be resulting from these intuitive reactions. In hypothetical situations presented to research subjects resistance to incest remains, even when all the reasons for opposition to incest have been neutralized (both parties consent, the use of foolproof contraception). Haidt calls this moral dumbfounding: the fact that people have a moral conviction, yet can't explain why they have it (Haidt, 2001). Similarly, aversion to harm people in general, or loved-ones specifically, may prevent aggression. A person in distress may elicit an impulse in the observer to help this person. Each of these different tasks may be enabled by different modules, EP argues. In this regard a module is a brain-network specialized to solve a particular kind of problem. Accordingly moral modules imply the role of some brain regions especially involved in moral cognition.

(3) Most of the values and preferences humans pursue are in an EP perspective regarded as the product of evolution. The content of these automatic reactions must have been innately specified to work without the subjects noticing it. "How else can values be derived?", EP wonders (Tooby et al., 2005). Yet, morality is a broad domain covering many "independent" processes. It is argued that we possess several innate intuitions about right and wrong, fair and unfair, such as an aversion to incest and cheaters, the desire to punish free riders (Buss, 2004; Lieberman, Tooby, & Cosmides, 2003). Consequently, moral judgment is regarded as the result of a patchwork of moral modules, domain-specific and solving ancient adaptive problems. Many different modules contributing (almost exclusively) to moral judgment processes means the process is massively modular.

Cognitive Neuroscience

Whereas the main focus for evolutionary psychologists lies on behaviour as the result of a collection of evolutionary adaptations, cognitive scientists focus foremost on cognitive functions, behaviour and its relation to structure. Classical moral theories, such as Kohlberg's, state that moral behaviour is an exclusively rational type of affair, governed by deliberative and high cognitive processing. In contrast, current perspectives of cognitive and moral decision-making (Damasio, Grabowski, Frank, Galaburda, & Damasio, 1994; Greene & Haidt, 2002).

Cognitive neuroscientists' central focus of attention is the question how the brain enables the mind. In this chapter I regard CN as the study of moral decision taking by means of brain studies. Generally, in CN, the moral systems are more or less regarded as partially independent elements under the umbrella of morality. Greene, for instance, discerned either personal or impersonal moral dilemmas. Whereas a personal dilemma consists of a person being actively involved in injuring another person, in the impersonal case this only happens from a distance, inactively (Greene, Nystrom, Engell, Darley, & Cohen, 2004). Blair discerns multiple, partially separable, neuro-cognitive architectures that mediate specific aspects of morality: social convention, care-based morality, disgust-based morality and fairness or justice (Blair, Marsh, Finger, Blair, & Luo, 2006).

What stands out in the cognitive neuroscientists' approach is the role of (1) higher cognition and (2) learning mechanisms, and (3) apparently no modularity in the EP sense is observed.

The Role of Reasoning Processes in Some Forms of Moral Judgment

Despite the recent focus on affect in moral judgment, some fear that, in these early days of brain research, the pendulum may have swung too far. This may imply an underestimation of the role of higher cognitive processing and some form of rationality (Prehn & Heekeren in this volume). Several studies pointed at the role of brain regions involved in abstract cognition mediating moral judgment. These are said to refine the current emphasis on emotions in moral judgment.

Greene demonstrated personal moral dilemmas produced relatively greater activity in emotion-related areas (the posterior cingulate cortex and the medial prefrontal cortex) as well as in the superior temporal sulcus, a region associated with social cognition. Contemplation of impersonal moral dilemmas produced relatively greater neural activity in brain areas generally involved in "cognitive" functions associated with working memory function in the inferior parietal lobe and the dorsolateral prefrontal cortex, resulting in "utilitarian" moral judgment decisions (the sacrifice of one individual to save many others). This allows for abstract reasoning, including reasoning about moral matters (Greene et al., 2004). Thus, depending on the context or the type of moral situation, different brain regions are involved, either primarily related to emotions or abstract thinking (Greene, 2005). Greene concludes that, in general, deontological judgments (inspired by moral rules, rights or duties) tend to be driven by emotional responses. These moral rules are generally post hoc rationalizations. In contrast, utilitarian judgments (where the best overall consequences of an action are chosen) depend on the different psychological processes that are more "cognitive" and more likely to involve genuine moral reasoning (Greene, 2008). In fact Greene holds a dual-process view: both quick and fast emotional responses on the one hand and (slower) cognitive control and reasoning on the other hand add to the moral judgment processes. Reasoning processes come into play when the automatic processes run into difficulties: when prepotent emotions and utilitarian judgments conflict, brain areas related to higher cognition may be activated to override the emotions (Greene, 2007; Greene et al., 2004).

Similarly, Prehn and Heekeren argue that the relative involvement of emotional and cognitive processes, depends on the situational context in which a moral judgment is made. They emphasize particularly the role of the competence of the decision maker to integrate emotional responses and reasoning processes. For, as both cognitive (reasoning and factual knowledge about accepted standards of moral behaviour) and emotional processes play an important role in moral judgment, they infer and emphasize that decision makers "will need to have an ability or competence that integrates the emotional and cognitive components into moral judgments, decisions and behavior" (Prehn & Heekeren in this volume). Thus, they suggest the study of individual differences in moral judgment may further help us to understand the different types of integration between emotional and cognitive mechanisms involved in moral judgment.

In short, in CN it is argued that the role of higher cognition, or representations that have no motivational value in themselves, should not be overlooked.

Focus on Learning Processes

Cultures across the planet differ widely in the norms and values they endorse (Prinz, 2007). It may seem that a moral psychology emphasizing learning processes is better equipped to explain this variety. Blair discerns two types of moral learning: quick emotional learning and reinforcement learning.

Brain damage to limbic and paralimbic regions may lead to unprovoked rage, lack of empathy, and abnormal sexual behaviour (Moll et al., 2005). It causes exaggeration or attenuation of basic motivational and emotional states, thereby affecting moral behaviour. This points at the central role these regions play in learning social rules. The learning process enables rules to be produced that are stored in other parts of the brain. A lack of this equipment prevents from social rules to be incorporated.

In fact, the amygdala play a central role in emotional learning and especially in aversive conditioning. It allows conditioned stimuli (CSs; including representations of moral transgressions) to be associated with unconditioned stimuli (USs; including the victim's distress cues or fear). Blair argues that "an individual learns about both the 'goodness' and 'badness' of objects on the basis of moral socialization" (Blair, this volume). The amygdala is crucial for the formation of stimulus-reinforcement associations. It allows previously neutral objects to be valued as either good or bad, according to whether they are associated with punishment or reward (Everitt, Cardinal, Parkinson, & Robbins, 2003; LeDoux, 1998). Individuals that show reduced autonomic response to the distress of others, have problems with aversive conditioning (impaired amygdala functioning). This is observed in individuals with psychopathy (Blair, 2003).

The second type of learning mechanism is characterized by so-called somatic markers. Patients with a damaged ventromedial prefrontal cortex (such as Phineas Gage) have difficulty in regulating behavior as they find it very difficult to anticipate the future. In a famous study, patients with ventromedial prefrontal cortex damage (VMPFC) were asked to play a card game in which they could win or lose a reward. They had to figure out the best strategy as they went along. In order to win, the participants had to forego short-term benefits for long-term profit. Patients with VMPFC lesions showed no anticipatory stress response when making a risky choice, whereas the control group did. It was suggested that patients failed to associate a feeling of stress with a choice, thus inhibiting their ability to distinguish between good and bad outcomes in situations of uncertainty (Bechara, Tranel, Damasio, & Damasio, 1996).

This leads Damasio to deduce the "somatic marker hypothesis" which states that emotional responses that involve bodily changes and feelings, for instance a change in skin conductance or an increase in heart rate (so called somatic markers) are linked with reinforcing stimuli. Associating certain bodily changes as a response to certain outcomes of actions (for instance, a bad feeling when losing money), allows the brain areas that monitor these bodily changes to respond whenever a similar situation arises. Faced with this situation, the mere thought of a particular action becomes sufficient to trigger an "as if" response in the brain, elicitng the same bodily feelings that he would experience if he performed the particular action (Damasio, 1994). In fact, this type of learning that depends on reinforcement information given by the bodily states (either "good" or "bad" bodily sensations), allows decisions on approach or avoidance a certain object or enact or not a certain type of behavior.

The fear system is thought to be involved both in aversive conditioning and instrumental learning. Yet, though a lesion at the amygdala may switch the aversive conditioning system, it does not inhibit instrumental learning to occur and vice versa (Blair, 2005). Two partially separate neural systems are at play Blair argues. He unites these two learning systems in his Integrated Emotion Systems (IES), implying both somatic markers and the quick emotional learning systems help a person to develop his moral capacities. Cognitive neuroscientists suggest there is no strict stimulus dependency. This means that by way of aversive conditioning through the amygdala, or by use of somatic markers humans are able to learn a wide range of behaviour that are socially appreciated. In short, no apparent stimulus dependency is proposed in this model as one can learn to avoid a wide variety of situations, simply due to these associating mechanisms.

Neural Modularity

Thus far several brain regions are discerned playing key roles during moral judgment, ranging over the orbitofrontal cortex, the amygdala, the temporal poles, the superior temporal sulcus (Greene & Haidt, 2002; Moll et al., 2005). Most of these brain regions are involved in many different types of processes. As noted before, damage to VMPFC does not lead exclusively to a "moral" misbehaviour syndrome. It affects non-moral decision making processes as well. The diversity of brain regions involved in moral processing and the use of these brain regions in many different types of functions, seems to make it very hard to speak of neural modularity. That means there is no brain region exclusively focused on one type of moral reasoning.

Table 1 Differences in emphasis of the EP and CN approach towards the relation between our brain and the moral mind

Evolutionary psychology	*Cognitive neuroscience*
Nativist claim: innateness	Different learning processes
Domain-specific modules	The role of higher cognition
Automaticity, quick	slower, controlled
Massive modularity	Absence of neural modules

Massive Moral Modularity: Confronting EP and CN

Due to their dissimilar perspective, some significant overlaps and differences stand out between a CN and EP approach to the moral brain. Both disciplines agree that multiple, more or less independent systems underlie the moral sense. Furthermore, both approaches attribute an important role to moral emotions. On the one hand, the role of quick reactions guiding behaviour is regarded as the most effective evolutionary solution for guiding human behaviour. On the other hand, many neural imaging studies reveal the role of emotion related regions in the brain involved in moral judgment or moral cognition. "Morality would be reduced to a meaningless concept if it were stripped from its motivational and emotional aspects" (Moll et al., 2005). Yet some differences in emphasis are apparent. The emphasis on learning seems to undermine the nativist claim of EP. The emphasis on higher cognition appears to weaken the claim of automaticity and domain-specificity and the diversity of neural structures involved in moral judgment challenges the claim of modularity (Table 1).

Learning

Whereas EP emphasizes the innate capacity to develop certain moral intuitions, some cognitive neuroscientists emphasize the role of the social context that influences moral preferences. The cognitive neuroscientist Moll observes that "Westerners and East-Asians differ in categorization strategies when making causal attributions and predictions, and moral values and social preferences are shaped by cultural codification. The VMPFC has a central role in the internalization of moral values and norms through the integration of cultural and contextual information during development" (Moll et al., 2005). Moll's main focus lies on social rule learning. He focuses on the neurology of socializations and on the neurology of how society has an impact on individual's (moral) thoughts and behaviour. Yet, the difficulty with his social rule learning model is that it is hard to explain why children process moral transgressions differently from conventional transgressions (Blair in this volume). Whereas going to your job in pyjamas is regarded as a conventional transgression, a moral convention is characterized as a situation where some sort of physical harm occurs such as "pulling each others hair". Children from the age of four make a difference between both types of transgressions (Turiel, 1983). However, if morality relies on only one learning system, it is very hard to explain why this difference in judgment occurs. There needs to be some specification that can distinguish between different types of learning systems. And EP argues it is very probable this specification is innately determined.

Learning processes are generally characterized by the association of two types of stimuli: a conditioned and an unconditioned stimulus. An unconditioned stimulus is a sensation that automatically appeals or repels the subject. If you are hurt (unconditioned stimulus) every time you cross a line (conditioned stimulus), you will develop an intrinsic avoidance to cross that line. Similarly, sad, fearful or pained expressions of others are unconditioned aversive stimuli. To refrain from violence (conditioned stimulus) seems to result from the expression of suffering (unconditioned stimulus) observed by the actor (Blair, 2003). Thus, some form of negative feedback down-regulates the motivation to hurt. Yet, how could the observed suffering play the role of the unconditioned stimulus, unless an innate mechanism is present that values another's suffering as negative? Tooby and Cosmides observe that "if organisms have motivational systems and concepts that play an embedded role in them, then both motivational systems and the concepts they employ must be, at least in part, developmentally architecture derived" (Tooby et al., 2005).

Yet, the philosopher Jesse Prinz suggests that this difference can be explained quite easily by emphasizing the different types of punishment by parents when children either make a conventional or a moral transgression (Prinz, 2007). However, the observation that psychopathic individuals have difficulty to differentiate between both types of transgressions and the fact that psychopathy is at least partially genetically determined, seem to point out the involvement of an evolutionary process. This process prepared us to make a difference between both types of transgressions. Blair argues that the lack of morality observed in psychopaths can be attributed to disrupted learning mechanisms. An emotional impairment that interferes with socialization which causes that individuals do not learn to avoid antisocial behaviour (Blair, 2003). Specialized learning systems allow us to make distinctions between types of transgressions. Blair further emphasizes the role of independent moral systems (allowing us to attend to suffering and moral transgressions), a care-based, conventional and disgust-based morality (Blair et al., 2006). It seems the emphasis on "learning processes" does not lead to the exclusion of innate factors.

The focus on learning processes is not necessarily in contrast with an EP perspective as the different learning mechanisms in themselves can be innately specified. Yet, for EP, the environmental factors are treated as "triggers" that activate the development of a module in accordance with a "developmental" program that is coded in the genes. CN often holds a looser approach, whereby the effect of the input of socialization is bigger.

The key question is how much of these learning systems are present at birth. Based on observations across different cultures, Haidt formulates an intuitionist model and identifies five sets of concern, each linked to an adaptive challenge and to certain moral emotions, as the best candidates for the psychological foundations of the human moral sense. Haidt hypothesizes that just like our varied experiences of touch is the result of only three kind of receptors in the skin (for pressure, temperature, and pain) there might be a few different kinds of "moral receptors" that form the foundation of our highly elaborated and culturally diverse moral sense (Haidt & Joseph, 2004). The five foundations he identified are harm-care, fairness-reciprocity, ingroup-loyalty, authority-respect and purity-sanctity. These sets are regarded as modular and evolutionarily prepared. This means certain patterns of social appraisal lead to specific emotional and motivational output. In the intuitionist model, it is insisted that the moral mind is partially structured, in advance of experience. Thus, the five classes of social concerns are likely to become moralized during development. Haidt and others argue that social issues that cannot be related to one of the foundations are much harder to teach, or to inspire people to care about (Haidt & Joseph, 2005). These five innate modules are said to regulate moral behaviour, yet they are dependent on socialization and education to develop. Thus, they reconcile the nativist claim with an emphasis on learning processes.

Higher Cognition

The emphasis on higher cognition seems to undermine the claim of massive modularity, as the role of higher cognition entails some general abstract cognitive capacities that are said to be involved in moral judgment processes. Yet, whereas an evolutionary perspective emphasizes adaptive intuitions, this does not imply that *only* intuitions play a role (Tooby et al., 2005). Moral intuition may be quick and automatic, providing value and motivation to fulfil our "successful needs", just like a feeling of hungriness motivates us to eat nutritious food substances. Thus, it provides the motivation to act. But intuitive processes are not the only possible adaptive solution to help us choose what is "good" for us. General intelligence, as an abstract form of human cognition, may help us out. General intelligence is not necessarily contradictory to the modular organization of the mind. A modular approach does not exclude abstract cognitive capacities, but it emphasizes more capacities are modularly organized than is generally thought of (Tooby et al., 2005).

Greene also pointed out that we feel inclined to help an injured child when we see it, but that we are not inclined to help a similar child when it is on another continent. Only reason can make us realise we are in fact confronted with the same situation, albeit an evolutionary "new" situation. He suggests "that we are disposed to respond emotionally to the kinds of personal moral violations that featured prominently in the course of human evolution, as compared to moral violations that are peculiarly modern in ways that makes them more impersonal." (Greene, 2008). To solve the impersonal moral violations we can appeal to higher cognition.

Neural Modularity?

Bolhuis argues that we cannot speak of modularity unless we find a specific brain area devoted to one particular function (Bolhuis & Macphail, 2001). Similarly, it is hard to pinpoint one specific brain domain exclusively concerned with (some part) of moral motivation. Whereas damage to the VMPFC, as was the case with Phineas Gage, reduces moral functioning, it does not undermine moral faculties exclusively. A sceptic might conclude this undermines the modular approach of morality.

Marr's computational theory may shed light on this matter. He argued that several levels are involved in information handling mechanisms. The lowest level, the physical structure is the hardware. The second level is the algorithm that specifically takes care of the information processing. A third level is the computational theory, where the purpose of the action is specified. The first level of vision for instance refers to the individual neurons, the second concerns the processing mechanisms at a higher level and the third level refers to the purpose of vision: perception. Marr (1982) claimed that "[a]lthough algorithms and mechanisms are empirically more accessible, it is the top level, the level of computational theory, which is critically important from an information-processing point of view". And EP elaborate on this level: "Modern evolutionary biology constitutes, in effect, a foundational organism design theory, whose principles can be used to fit together research findings into coherent models of specific cognitive and neural mechanisms" (Tooby & Cosmides, 1995).

The way this third level is underpinned by the hardware may be quite variable. Similarly, the functional design of a computer program does not correspond to the structure of its underlying hardware. Flombaum observes that "we might argue that a particular program is modular if we limit the program to accessing certain databases in the computer, but not others (...) however, this claim tells us nothing about how we expect the computer hardware to be organized" (Flombaum 2002).

Furthermore EP's view on modularity does not imply every module or adaptation is underpinned by one structure alone. Tooby and Cosmides argue "that each adaptation is a collection of elements many of which are shared in different configurations among adaptations, some of them quite broadly. The specialization of an adaptation for a function does not lie in the specialization of all parts to its function. The specialization lies in the way the particular interrelationship of the parts is coordinated to solve the specialized adaptive problem with particular efficiency" (Tooby et al., 2005). They believe it is the specific combination of particular brain areas that determines the specific specialisation of one particular capacity.

Neurobiologist data cannot challenge the claim that the psychological mechanism underlying behaviour is modular, as modules do not need to be located in specific brain areas. In a way, EP and CN argue on different levels. The error arises when one assumes that the execution of a psychological module depends on a similarly "modular" piece of brain. In short, psychological modularity does not necessarily assume neural modularity. And a modular view does not entail that a brain region is only involved in one particular function. In fact, studying the neurobiological levels one focuses foremost on structure whereas the EP's approach focuses on function from an evolutionary perspective. Yet, of course, this does not entail that some sort of neurospecificity may be associated with psychological specificity.

Conclusion

Both CN and EP focus on the brain processes involved in moral judgment. In contrast to a purely cognitivist approach, these disciplines attribute an important role to moral emotions and the occurrence of multiple independent systems that contribute to moral judgment. From an EP perspective, moral psychology can be framed in the moral modularity hypothesis. This entails the occurrence of (1) innate, (2) domain-specific modules that are (3) organized in a massively modular way. Cognitive neuroscience seems to challenge each of these central tenets by focusing on (1) learning processes, (2) the absence of neural modules in the brain and (3) the input of higher cognition in moral judgment. Yet, these positions are not necessarily contradictory.

The difference between moral and conventional transgressions suggests innate specialization of these learning systems, reconciling both approaches. The role played by brain regions involved in higher cognition, does not explicitly contradict an EP approach as it can be cooperation between different systems. In turn, psychological modularity does not necessarily entail neural modularity.

The key difference between EP and CN is the amount of supposed innate specification of the moral modules that are involved in moral decision making processes. Whereas EP suggest that most of the preferences, motivations and values are determined by evolutionary forces, the CN view entails the occurrence of several more or less independent systems that "learn" a quite diverse set of motivations. Perhaps a part of the difference in opinion is due to the fact both disciplines approach the word "moral" from a different perspective. Whereas CN takes a descriptive approach and wonders what the equipment is to enable morality, EP wonders which specific rules are regarded as "moral" (e.g. disgust of incest). They take a more content-sensitive approach towards morality. In this approach to the word "moral" one wonders why certain moral rules are as they are.

It seems that CN and EP share a mutual project, though they approach their field of interest from a different perspective: either zoomed in, or zoomed out. From an EP perspective different hypotheses about moral mechanisms may be raised, these can be tested by the cognitive neurosciences. Moral judgment is understood as a cognitive-emotional process building on several contributing components. One of the challenges will be to understand how different brain regions interact to perform such complicated tasks.

Acknowledgments The writing of this chapter was supported by the Research Foundation – Flanders.

References

Anderson, S. W., Bechara, A., Damasio, H., Tranel, D., & Damasio, A. R. (1999). Impairment of social and moral behavior related to early damage in human prefrontal cortex. *Nature Neuroscience, 2*(11), 1032–1037.

Bechara, A., Tranel, D., Damasio, H., & Damasio, A. R. (1996). Failure to respond autonomically to anticipated future outcomes following damage to prefrontal cortex. *Cerebral Cortex, 6*(2), 215–225.

Blair, J., Marsh, A. A., Finger, E., Blair, K. S., & Luo, J. (2006). Neuro-cognitive systems involved in morality. *Philosophical Explorations, 9*(1), 13–27.
Blair, J. (2003). Neurobiological basis of psychopathy. *British Journal of Psychiatry, 182*, 5–7.
Blair, J. (2003). Facial expressions, their communicatory functions and neuro-cognitive substrates. *Philosophical Transactions of the Royal Society of London Series B-Biological Sciences, 358*(1431), 561–572.
Blair, J. (2005). Applying a cognitive neuroscience perspective to the disorder of psychopathy. *Development and Psychopathology, 17*(3), 865–891.
Bolender, J. (2003). The genealogy of the moral modules. *Minds and Machines, 13*(2), 233–255.
Bolhuis, J. J., & Macphail, E. M. (2001). A critique of the neuroecology of learning and memory. *Trends in Cognitive Sciences, 4*(10), 426–433.
Bowles, S., & Gintis, H. (2004). The evolution of strong reciprocity: Cooperation in heterogeneous populations. *Theoretical Population Biology, 65*(1), 17–28.
Buss, D. M. (2004). *Evolutionary psychology: The new science of the mind* (2nd ed.). Boston: Pearson.
Chomsky, N. (1988). *Language and problems of knowledge*. Cambridge, MA: MIT Press
Damasio, A. R. (1994). *Descartes' error: Emotion, reason and the human brain*: New York: G.P: Putnam.
Damasio, H., Grabowski, T., Frank, R., Galaburda, A. M., & Damasio, A. R. (1994). The return of gage, phineas – clues about the brain from the skull of a famous patient. *Science, 264*(5162), 1102–1105.
Duchaine, B., Cosmides, L., & Tooby, J. (2001). Evolutionary psychology and the brain. *Current Opinion in Neurobiology, 11*(2), 225–230.
Everitt, B. J., Cardinal, R. N., Parkinson, J. A., & Robbins, T. W. (2003). Appetitive behavior: impact of amygdala-dependent mechanisms of emotional learning. *Annual New York Academy of Sciences, 985*, 233–250.
Frank, R. H. (1988). *Passions within reason: The strategic role of the emotions*. New York: W.W. Norton.
Flombaum, J. I., Santos, L. R., & Hauser, M. D. (2002). Neuroecology and psychological modularity. *Trends in Cognitive Sciences, 6*(3), 106–108.
Greene, J. D. (2005). 19. Cognitive neuroscience and the structure of the moral mind. In P. Carruthers et al. (Eds.), *The Innate Mind* (Vol. 1, pp. 338–353). Oxford: Oxford University Press.
Greene, J. D. (2007). Why are VMPFC patients more utilitarian? A dual-process theory of moral judgment explains. *Trends in Cognitive Sciences, 11*(8), 322–323.
Greene, J. D. (2008). The secret joke of Kant's soul. In W. Sinnott-Armstrong (Ed.), *Moral psychology, vol. 3: The neuroscience of morality: emotion, disease, and development* (Vol. 3 pp. 35 79). Cambridge, MA: MIT Press.
Greene, J. D., & Haidt, J. (2002). How (and where) does moral judgment work? *Trends in Cognitive Sciences, 6*(12), 517–523.
Greene, J. D., Nystrom, L. E., Engell, A. D., Darley, J. M., & Cohen, J. D. (2004). The neural bases of cognitive conflict and control in moral judgment. *Neuron, 44*(2), 389–400.
Haidt, J. (2001). The emotional dog and its rational tail: A social intuitionist approach to moral judgment. *Psychological Review, 108*(4), 814–834.
Haidt, J., & Joseph, C. (2004). Intuitive ethics: How innately prepared intuitions generate culturally variable virtues. *Daedalus, 133*(4), 55–66.
Haidt, J., & Joseph, C. (2005). The moral mind: How 5 sets of innate moral intuitions guide the development of many culture-specific virtues, and perhaps even modules. In P. Carruthers, S. Laurence, & S. Stich (Eds.), *The Innate Mind* (Vol. 3, pp. 367–392). Oxford: Oxford University Press.
Hamilton, W. D. (1964). The genetical evolution of social behaviour. *Journal of Theoretical Biology 7*, 1–52.
Hardin, G. (1968). The tragedy of the commons. *Science, 162*, 1243–1248.

Hauser, M., Young, L., & Cushman, F. (2008). Reviving Rawls' linguistic analogy operative principles and the causal structure of moral actions. In W. Sinnott-Armstrong (Ed.), *The biology and psychology of morality* (pp. 107–143). New York: Oxford University Press.
Kennair, L. E. O. (2002). Evolutionary psychology: An emerging integrative meta-theory for psychological science and practice. *Human Nature Review, 2*, 17–61.
LeDoux, J. (1998). *The emotional brain*. New York: Weidenfeld & Nicolson.
Lieberman, D., Tooby, J., & Cosmides, L. (2003). Does morality have a biological basis? An empirical test of the factors governing moral sentiments relating to incest. *Proceedings of the Royal Society of London Series B-Biological Sciences, 270*(1517), 819–826.
Marr, D. (1982). *Vision: A computational investigation into the human representation and processing of visual information*. San Francisco: W.H. Freeman.
Moll, J., Zahn, R., Oliveira-Souza, R., Krueger, F., & Grafman, J. (2005). The neural basis of human moral cognition. *Nature Reviews Neuroscience, 6*(10), 799–809.
Pinker, S. (1998). *How the mind works*: New York: W.W. Norton.
Prinz, J. (2007). Is morality innate? In W. Sinnott-Armstrong (Ed.), *The evolution of morality* (Vol. 1, pp. 367–406). Cambridge: MIT Press.
Samuels, R. (2000). Massively modular minds: Evolutionary psychology and cognitive architecture. In P. Carruthers (Ed.), *Evolution and the human mind*. Cambridge: University Press.
Tinbergen, N. (1963). On aims and methods in ethology. *Zeitschrift für Tierpsychologie, 20*, 410–433.
Tooby, J., & Cosmides, L. (1995). Mapping the evolved functional organization of mind and brain. In M. S. Gazzaniga (Ed.), *The cognitive neurosciences* (pp. 1185–1197). Cambridge, MA: MIT Press.
Tooby, J., Cosmides, L., & Barrett, H. C. (2005). Resolving the debate on innate ideas: Learnability constraints and the evolved interpenetration of motivational and conceptual functions. In P. Carruthers, S. Laurence, & S. Stich (Eds.), *The innate mind: Structure and content* (pp. 305–337). New York: Oxford University Press.
Trivers, R. L. (1971). The evolution of reciprocal altruism. *Quarterly Review of Biology 46*, 35–57.
Turiel, E. (1983). *The development of social knowledge: Morality and convention*. Cambridge: Cambridge University Press.
Wilson, E. O. (1975). *Sociobiology: The new synthesis*. Cambridge, MA: Harvard University Press.

Index

A
A Clockwork Orange, 1, 37
Alexander, R. D., 216, 218–220, 237–238
Altruism, 5, 8, 13, 20, 22, 24–29, 31–32, 38, 46, 73, 77–80, 94, 113, 122–123, 160, 185–192, 194–197, 204, 211–226, 236–238, 240, 244–245, 247
Altruistic punishment, 29, 31–32, 185, 190
Antisocial behavior, *see* Immoral behavior
Antisocial personality disorder, *see* Psychopathy
Atran, S., 233
Attachment
  disorders, 71
  extended attachment, 35–36, 69–82, 207
  primitive attachment, 71–72, 74
Autism, 89, 94, 97–98, 166

B
Baldwin effect, 216, 220
Barrct, J., 233, 251
Benedikt, M., 3–4
Blair, J., 14, 18, 36, 89, 105–106, 261–263, 265
Boyer, P., 233, 239
Brain
  brain size, 32, 219
  cliff analogy, 30
  unique human features, 30, 33, 69
Brain imaging
  caveats, 7
  importance, 31–34
  limitations, 21
  technologies, 33
Brain lesions
  acquired psychopathy, 39, 162
  patient EVR, 75–76
Brain regions, *see* Figure
  basal ganglia, 30, 112, 118, 171
    globus pallidus, 30, 166
    nucleus accumbens, 11, 21, 30, 60, 112, 120, 159, 162, 165, 168–169, 171–172, 174
    nucleus caudatum, 21, 30
    putamen, 30, 95
    ventral striatum, 48, 60, 73–74, 159–160, 163, 165–166, 170
  frontal lobe, 4, 57, 117
    anterior insular cortex (AIC), 10–11, 14–15
    dorsolateral prefrontal cortex (DLPFC), 11, 13–16, 18, 30, 33, 56, 60, 91, 141, 145–146, 157–159, 162, 164, 166, 168, 170–172, 174–175, 261
    frontopolar cortex, 48, 162
    gyrus rectus, 49
    medial prefrontal cortex (MPFC)
    orbitofrontal cortex (OFC), 11, 48–49, 55, 58–59, 110, 120, 133–134, 138–139, 145, 157–159, 162, 165–166, 172–173, 263
    ventromedial prefrontal cortex (VMPFC), 13, 57, 60–61, 91, 93, 133–134, 138–139, 142, 145–146, 149, 157–158, 262
  limbic system, 30, 72, 111–112, 137, 146
    amygdala, 19, 30, 34, 48, 50, 54–56, 58–61, 74–76, 92–94, 99, 110, 112, 115–116, 118, 120, 138–140, 142, 157–158, 160–166, 168, 170–174, 262–263
    anterior cingulate cortex (ACC), 96, 110, 114–115, 141, 157, 159, 162, 165, 169–170

J. Verplaetse et al. (eds.), *The Moral Brain*, DOI 10.1007/978-1-4020-6287-2_BM2,

hippocampus, 56, 137, 157, 161–164, 173
hypothalamus, 74–76, 112, 165–166, 170–171
posterior cingulated cortex (PCC), 12–13, 55–56, 138–139, 142, 146, 157–159, 161, 163, 172, 261
precuneus, 146, 157, 159, 161, 166
thalamus, 158–159, 164–166, 234
occipital lobe, 3, 32
other regions
periaqueductal gray, 166
ventral tegmental area, 74, 159–160
parietal lobe, 234, 261
inferior parietal lobulus (IPL), 13, 16, 158, 261
temporal lobe, 56–57, 75–76, 165, 234
angular gyrus, 34, 48–49, 55–56, 61, 157–158, 161, 163
superior temporal sulcus (STS), 11, 49, 59, 115, 138–140, 142, 145–146, 158, 162, 261, 263
temporal pole, 50, 76, 138–140, 142, 145, 161–163, 263
Burgess, A., 1

C
Cheater detection, 59–60, 240, 252
Chimpanzees, 31–32, 72, 111–114, 192
Christianity, 176, 241, 243–244, 248–250
Cognitive neuroscience, 33–35, 39, 45, 69, 117, 123, 129, 207, 255–268
Conscience, 2–3, 5, 8, 19, 22, 93
Cooperation, 5, 26–27, 31, 34, 37–38, 40, 45–48, 51, 58, 61, 69–70, 73–75, 77–80, 94, 160, 186, 190, 195, 211, 216–221, 223–225, 238–240, 243, 245, 249, 259
Criminals, 1, 3–4, 8, 17, 19–20, 56, 156–157, 164, 176, 205

D
Darwin, C., 2, 22
Darwinian medicine, 205–206
Dawkins, R., 156, 192, 194, 211, 217, 236
Deep brain stimulation (DBS), 20
Deep history, 28–29, 31, 33
Desmodus rotundus (Vampire bat), 25, 218
De Waal, F., 72–73, 113
Dopamine, 73, 159–160, 169, 171
Dr. Jekyll and Mr. Hyde, 1
Dualistic worldview, 176

E
Emotion contagion, 113–114
Empathic concern, 3, 32, 36, 76, 109–111, 113, 116–119, 121–123, 224
Empathy
cold versus warm, 14
lack of empathy in psychopaths, 262
Evolutionary models, 185–187, 197, 236–238
Evolutionary psychology, 5, 33, 39, 90–91, 148, 207, 233, 236, 238–241, 250–252, 256–260
Experimental games
dictator game, 26
iowa gambling task, 17
prisoner's dilemma, 34, 45–48, 50–51, 58, 60, 121, 217–218, 252
trust game, 16, 74
ultimatum game, 15, 26, 60, 198

F
Fairness, 60, 72, 122, 132, 146, 191, 197, 204, 258–259, 262, 266
Fear, 18–19, 36, 51, 55, 57, 71, 80, 92, 99, 113, 118, 131, 147–148, 216, 240, 253, 262–263
Fehr, E., 15–16, 26–27, 146, 162, 172, 174, 186, 188, 190, 197–198, 211, 217, 220
Figure, 185, 198, 262
Frontotemporal dementia (FTD), 4, 57

G
Gage, Phineas, 3, 61, 133, 255–256, 262, 267
Genome, 29, 214
Gilligan, C., 131, 146
Gray matter, 18, 55–56, 162, 165
Greene, J., 139–140, 158
Group selection, 23–24, 27, 38, 191, 193–198, 204, 215, 220–221
Guthrie, S., 233

H
Haidt, J., 11, 36, 48, 71, 89–91, 94–96, 105–106, 132–133, 139–140, 149, 158–159, 190, 197, 207, 260, 263, 265–266
Hamilton, W., 22, 24–27, 78, 192, 194, 204, 211, 213, 216, 236–237, 256
Hardin, G., 258–259
Helping behavior, 31, 71, 113, 119, 121
Heterocephalus glaber, 25
Hierarchy, 87, 89, 91, 96–98, 223
Hume, D., 156, 202
Hymenoptera, 22, 24–25

I
Immoral behavior
  among animals, 31
  cheating, 46, 50–52, 59–60
  deception, 22, 45, 52, 58, 61, 204, 206, 214, 223, 225, 230
  evolution, 2, 50–52
  neuroimaging studies, 6, 7
  See also Criminals, Psychopathy
Indirect reciprocity, 26–28, 69–70, 77–81, 190–191, 193, 197, 237–238
Innate, 129, 132, 147, 159, 206, 258–260, 264–266, 268
Integrated emotion systems (IES) (Blair), 92, 263
Intelligent creator, 23
Irons, W., 233, 240
Islam, 249–250

J
James, W., 235–236
Judaism, 241, 243, 249–250
Justice, 8, 13, 16, 20, 22, 40, 61, 87, 94–95, 122–123, 130–131, 242, 249, 261

K
Kant, I., 156, 255
Kin selection, 26, 38, 78, 113, 189, 191–192, 194–197, 204, 211, 217, 224, 236–238, 240–241, 245
Kohlberg, L., 87, 105, 130–132, 144, 147, 149

L
Lack, D., 22–23
Lamina pyramidalis, see Vogt, O
Lateralization, 112, 166
Learning, 4, 19, 30, 36–37, 40, 72, 78, 88, 90, 92, 95, 98–99, 160, 168, 171, 203, 206, 216, 259, 261–266, 268
Limbic system, *see* Brain regions
Localization
  history, 3, 6, 33

M
MacLean, P., 111
Magnetic resonance imaging (MRI)
  functional MRI, 6, 7, 117, 137
Mammals, 16, 23, 25, 30–32, 36, 69–70, 73, 91, 111–113, 202, 215
Materialism, 2, 5, 21, 24–25, 138, 141, 175–176, 214, 259
Mealy, L., 34
Membrane temperature, 112
Miller, J., 186
Mirror neurons, 115, 157
Modularity, 39, 207, 255–268
Module, 11, 29, 39, 206–207, 240, 244, 255–260, 264–267
Moleschott, J., 21
Moore, G. E., 202
Moral brain
  breakthroughs, 9–19
  engineering, 1, 15–17
  evolution, 22–26
  history, 28–30, 33, 46, 90, 244
  limitations, 21, 244
  prospects, 16, 20–21
Moral center
  localization of, 9–19
Moral competence, 144
Moral dilemma
  personal versus impersonal, 48–49
  trolley dilemma, 137–138, 140
Moral emotions
  appeasement, 98–99
  embarrassment, 98–99
  guilt, 8, 14, 19–20, 35, 38, 46–48, 51, 53, 70, 81, 131–132, 134, 138–139, 224–225, 259
  in prisoner's dilemma, 35, 45, 46–47, 58–59, 252
  shame, 8, 14, 46, 51, 70, 131–132, 138–139, 149, 224, 259
  sociomoral disgust, 8, 11, 14, 32
Moral intuitions, 8, 12, 33, 38, 90, 95, 190–191, 259–260, 264, 266
Morality
  care-based morality, 90, 91–94, 97–99, 261
  disgust-based morality, 95, 99, 261, 265
  domain theory, 87, 89, 91–95
  individual difference, 4, 37, 58, 89, 121, 142–146, 148, 262
  as a natural kind, 39–40
  principled morality, 35, 81
  rule-based morality, 13, 49
  as a unitary system, 87, 88–89
Moral judgment, 13, 16, 20, 36–37, 48–50, 52–53, 55–57, 73, 89, 98, 105–106, 129–150, 156–158, 163, 256, 259–264, 266, 268
Moral Judgment Test (MJT), 20, 143–144
Moral modularity, 39, 207, 255–268
Moral sense, 2–4, 12, 17, 20, 22, 24, 33–40, 61, 129, 132, 256, 264–266
Multiple moralities approach, 36, 87, 105

N
Naturalistic fallacy, 202–203
Natural selection, 2, 22–24, 26, 38, 52, 111, 113, 147–148, 155, 191–194, 197, 200, 202–203, 205, 212–214, 221, 224–225, 244, 256, 258
Neocortex, 30, 78, 111
Neural modularity, 263–264, 266–268
Neuroethics, 45, 61, 176
Neuromodulation, 37, 157, 167, 173, 175–177
Neuropeptides, 73, 75, 111
Neurosurgical intervention, 156, 175–177
Newberg, A., 198, 235

O
Other-regarding utility functions, 187–191
Oxytocin, 16–17, 73–75, 111

P
Pain, 7–8, 11, 16, 19, 23, 35–36, 50–51, 54–55, 60, 77, 79, 109–123, 143, 156, 168, 175, 177, 206, 212, 224, 250, 265
Pedophilia
  neurosurgical intervention, 156, 175–177
Perception-action model, 115
Persinger, M., 234
Personal distress, 109–111, 116–122
Perspective-taking, 13, 109, 116–119
Phrenology, 4–5, 21
Positron emission tomography (PET), 6–7, 32, 118, 168, 170, 174, 234
Prichard, J., 18, 50
Prinz, J., 89, 262, 265
Prisoner's dilemma, *see* Experimental games
Prosocial behavior, 73–74, 80–81, 110, 113, 117
Proximate causation, 26
Psychopathy
  brain abnormalities, 18, 165
  core features, 52, 57
  developmental versus acquired, 18, 161–164
  fear insensitivity, 215–216
  neurosurgical intervention, 156, 175–177
Punishment, 4–5, 11, 17, 19, 27, 29, 31, 61, 75, 77, 79–80, 88, 92–93, 130–131, 133, 135, 139, 160–162, 185, 190–191, 206, 217–218, 240, 242, 247, 249–250, 259, 262
Purity, 87, 91, 95, 191, 266

R
Ramachandran, V., 234
Ramon y Cajal, S., 1
Rawls, J., 190
Reciprocal altruism, 26–27, 29, 31–32, 38, 46, 78, 94, 113, 237–238, 240, 244–245, 247
Reciprocity, 26–29, 35, 37, 69–70, 73, 77–81, 87, 89, 91, 94, 99, 130, 185, 190–191, 193, 197, 204, 207, 211, 217–218, 220, 224–225, 237–238, 242, 266
Religion, 38, 129, 197, 233–236, 238–241, 243, 249–252
Reward system, 37, 48, 74, 79, 154, 159–162, 165–166
Robida, A., 1
Robinson, J. H., 28
Runaway social selection, 38, 201, 211–226

S
Schadenfreude, 11
Science fiction, 1–3, 17, 20, 37
Self, 4, 6, 8, 13–14, 16, 23–25, 37, 48–49, 52–53, 56, 58, 70–71, 76–77, 81, 109–111, 113–119, 121, 130, 155, 157, 159–160, 163, 166, 172–174, 176, 188, 201, 204, 216–218, 224–226, 237
Self-awareness, 119, 155
Selfish genes, 26, 156, 189, 192, 236
Sexual selection, 38, 189, 191, 196–198, 212–215, 219, 224
Six-stage model (Kohlberg), 130
Sober, E., 193–194
Social neuroscience, 109–110, 123
Social response reversal (SRR) (Blair), 96–98
Sociomoral disgust, 8, 11, 14, 32
Somatic marker (Damasio), 4, 17–18, 133, 135, 263
Soul, 2, 5, 21, 244
Stevenson, R. L., 1
Stimulus-reinforcement learning, 36, 92, 98–99
Strong reciprocity, 26–29, 79–80, 220
Sympathy, 22, 70–71, 113–114, 132, 149, 155, 187, 193, 224–225

T
Theory of evolution
  cross-pollination with neurosciences, 28, 31, 33
Theory of mind (ToM), 32, 48–49, 53, 97–99, 139, 155, 157, 162, 224, 226
Third-party punishment, 29, 32
Tit-for-tat, 193, 217

Transcranial magnetic stimulation (TMS), 7, 15–17, 37, 115–116, 145, 156, 167–170, 173, 175
Transgressions
  conventional, 88–89, 95–98, 129, 264
  moral, 14, 48–49, 88, 92–93, 95, 129, 158, 262, 264–265
Trivers, R., 27, 46, 221
Trust game, 16, 74
Turiel, E., 87–89, 91, 95–96, 129, 264

U
Ultimate causation, 26, 256
Ultimatum game, *see* Experimental games
Utilitarianism, 12, 13, 36, 48–49, 134–135, 141, 149, 163, 261

V
Violence inhibition mechanism (Blair), 92
Vogt, O., 3
Von Economo neurons (VEN), 4, 32

W
West-Eberhard, M. J., 204, 212–213, 215–216, 219, 221
White matter, 58, 67, 165
Williams, G. C., 194, 204–205, 211
Wynne-Edwards, V. C., 23, 193, 215

GPSR Compliance
The European Union's (EU) General Product Safety Regulation (GPSR) is a set of rules that requires consumer products to be safe and our obligations to ensure this.

If you have any concerns about our products, you can contact us on

ProductSafety@springernature.com

In case Publisher is established outside the EU, the EU authorized representative is:

Springer Nature Customer Service Center GmbH
Europaplatz 3
69115 Heidelberg, Germany

www.ingramcontent.com/pod-product-compliance
Lightning Source LLC
Chambersburg PA
CBHW071351220926
51613CB00008B/89
* 9 7 8 1 4 0 2 0 6 2 8 6 5 *